2023 Seminar on Microelectronics, Dielectrics and Plasmas (MDP 2023)

Saint Petersburg, Russia
20 November 2023

IEEE Catalog Number: CFP23UU4-POD
ISBN: 979-8-3503-8358-4

**Copyright © 2023 by the Institute of Electrical and Electronics Engineers, Inc.
All Rights Reserved**

Copyright and Reprint Permissions: Abstracting is permitted with credit to the source.
Libraries are permitted to photocopy beyond the limit of U.S. copyright law for private
use of patrons those articles in this volume that carry a code at the bottom of the first
page, provided the per-copy fee indicated in the code is paid through Copyright
Clearance Center, 222 Rosewood Drive, Danvers, MA 01923.

For other copying, reprint or republication permission, write to IEEE Copyrights
Manager, IEEE Service Center, 445 Hoes Lane, Piscataway, NJ 08854. All rights
reserved.

****** This is a print representation of what appears in the IEEE Digital
Library. Some format issues inherent in the e-media version may also
appear in this print version.***

IEEE Catalog Number:	CFP23UU4-POD
ISBN (Print-On-Demand):	979-8-3503-8358-4
ISBN (Online):	979-8-3503-8357-7

Additional Copies of This Publication Are Available From:

Curran Associates, Inc
57 Morehouse Lane
Red Hook, NY 12571 USA
Phone: (845) 758-0400
Fax: (845) 758-2633
E-mail: curran@proceedings.com
Web: www.proceedings.com

Contents

Preface 2

Definition of Critical Currents in Superconducting Magnetic Energy Storage Systems 3
Bagan Gontrand Steve Sedjro Kalimov Alexander G., Vazhnov Sergey

XPS Studies of Adsorption Sites on the Surface of Gas Sensitive Zinc Oxide Based Nanostructures 7
Arkhipenko Viktoriya A., Guketlov Aslan M., Bezverkhniy Vladislav P., Shomakhov Zamir V., Nalimova Svetlana S., Moshnikov Vyacheslav A.

Modeling of Nanostructuring Processes of Lead Chalcogenide Layers under Sensitizing Oxidation 11
Bezverkhniy Vladislav P.

Investigation of Structural and Dielectric Properties of thin Films (Sr/Ba) Nb2O6 on Al2O3 Substrates in the Field
of Microwave Applications 14
Bogdan Alexey, Karamov Artem R. Sapego Evgeny N. Tumarkin Andrei V.

Research of Microwave Radiation Absorption of FDM Printing Materials 18
Chemerev Ilya

Design and Fabrication of Normally-on and Normally-off Ultrathin AlN-based Transistors 23
Chukanova Olga B., Egorkin Vladimir I., Zemlyakov Valery E., Zhuravlev Maxim N/

Investigation of Mechanical Characteristics of Pressuring Materials of Generators Stator Winding
in Thermal Aging Process 27
Fedotov Nikita A., Shikova Tatyana M., Belko Victor O., Kostelov Andrey M., Mannanov Emil R., Chernyshov Dmitriy A.

Causes of Resistance Decrease of Corona Protection Materials in Thermal Aging Process 31
Fedotov Nikita A., Shikova Tatyana M., Belko Victor O., Reznik Alexandr S., Kostelov Andrey M., Mannanov Emil R.

Optimization Analysis of the Main Insulation Structure of the Large Generator Stator Bar 35
Feng Shengxi, Belko Victor O.

Influence of Laser Radiation Power on Raman Spectra of Nanostructured Silicon Wires with Crystallographic
Orientation (111) 40
Gagarina Alena Yu., Kuznetsov Alexey, Spivak Julia M., Bolshakov Aleksey D., Moshnikov Vyacheslav A.

Electroplasma Technologies for Cleaning, Polishing and Welding of Metals 43
Gaysin Fivzat M., Bagautdinova Liliya N., Gaisin Almaz F., Gaisin Azat F., Mastyukov Karim Sh., Zakirov Dzhaudat U.

Modeling of the Magnetic Field and Current Density Distributions in HTS SMES Systems 47
Govor Vladislav M., Kalimov Alexander G., Kobzar Evgenii N.

The Effect of Liquid Dielectric Impregnation on the Self-Healing Characteristics 52
Ivanov Ivan O., Feklistov Efrem G.

The Phenomenon of Low Frequency Dielectric Losses Increasing in Metallized Film Capacitors 56
Ivanov Ivan O., Hojamov Ahmet A.

Investigation of Particles Obtained by Green Synthesis Using Plant Extract 60
Khalugarova Kamilya, Kondratev Valeriy M., Kuznetsov Alexey, Gagarina Alena Yu.

Different Approaches to Increasing the Continuous Operation Time of IPMC Actuators 63
Khmelnitskiy Ivan K., Testov Dmitriy O., Aivazyan Vagarshak M., Luchinin Victor V., Korlyakov Andrey V., Karelin Alexandr M.

Investigation of the Electro Physical Properties of the Components of Modern Paper-Impregnated Insulation 67
Kiesewetter Dmitry V., Zhuravleva Natalia M., Reznik Alexandr S., Litvinov Danila, Trubin Denis, Le Sun

Research of Low-Cost Technological Process for Manufacturing of One Side Silicon Interposer 71
Kochergin Mikhail D., Vertyanov Denis V., Belyakov Igor A., Zhumagali Raiymbek N.

LNA IC Development for LTE Bands Based on Bulk CMOS 180nm Digital Library 75
Kotlyarov Evgeny Yu., Putrya Mikhail, Mikhailov Viktor, Zubov Igor, Timoshin Sergey

Complex Dielectric Permittivity Measurement of 3D Printing Resin FTD Nano Clear in the 1-10 GHz Band 82
Migalin Mikhail M., Kovalev Andrey V., Gadzhiev Samir R., Kuzmin Vladislav S., Libin Lev N., Fleyteng Vladimir A.

Investigation of the Power Source Parameters Influence for the Plasma Jet of DC Plasma Torch 86
Murashov Iurii, Zhiligotov Ruslan, Obraztsov Nikita, Kurakina Natalia I.

Hardware Realization Version of a Neuron for a Self-routing Analog-to-digital Converter 91
Naborshikov Anton A., Posyagin Anton I., Artemyev Ilya A., Tsyganstsev Vyacheslav A.

Adaptive LUT - Decoder LUT for FPGAs 95
Oputin Nikita E., Tyurin Sergey F.

Charge Relaxation in Polylactide/Montmorillonite Composite Materials 99
*Pavlov Andrey A., Kamalov Almaz M., Borisova Margarita E., Kovalenko Mikhail A., Goldade Victor A.,
Zotov Sergey V.*

Experimental Study of Self-Healing Characteristics and Insulation Resistance of Metallized Polypropylene Film
under Various Inter-Layer Pressure 103
Pechnikov Alexey V., Hojamov Ahmet A.

Study of the Sensitive Element of a Resonant Pressure Sensor with Membranes of Various Shapes 109
Phyo Win Tun, Timoshenkov Sergey P., Simonov Boris M.

Investigation of the Aging of Electrical Insulating Paper in Polydimethylsiloxane Liquid 112
Reznik Alexandr S., Zhuravleva Natalia M., Kiesewetter Dmitry V., Poyarkov Pavel A., Smirnova Ekaterina G.

The Research of GaN HEMT Transistor Input Gate Capacitance Dependence as a Function
of Operating Mode in GHz Band 115
Rodionov Denis V., Khlybov Alexander I., Kotlyarov Evgeny Yu., Timoshenkov Pavel V., Guminov Nikolay V.

Concentric Topology for Sensitive Elements of Surface Acoustic Wave Devices 119
Sorvina Maria A.

Multi-function LUT for FPGAs 122
Sovetov Stanislav I., Tyurin Sergey F.

Arc Spot Movement and Its Effect on Flow in an AC Plasma Torch 127
Surov Alexander, Obraztsov Nikita, Bykov Nikolay, Nakonechny Ghennady

One-hot Programming LUT for FPGAs 132
Vasenin Ivan, Tyurin Sergey

Thermal and Thermoelectric Properties of Bi-Te-Sb and Bi-Te-Se Doped by Carbon Additive 136
*Voloschuk Irina A, Babich Alexey V., Rogachev Maxim S., Glebova Daria D., Bozhedomova Anastassiya S.,
Babich Tatiana A.*

Modeling and Analysis of the Functioning of the Sensitive Element of a Capacitive Microaccelerometer Based
on the Microstructure of Silicon Carbide(Sic) and Silicon(Si) 139
Ye Ko Ko Aung, Simonov Boris M., Timoshenkov Sergey P., Phyo Win Tun, Paing Soe Thu

Fractal Character of the Conductor Destruction in the Zone of Local Increase of the Pulsed Current Density 144
Zhukov Dmitry V., Krivosheev Sergey I., Magazinov Sergey G., Kiesewetter Dmitry V., Malyugin Victor I.

Autors Index

Proceedings
of the Seminar on Microelectronics, Dielectrics and Plasmas (theory and practical applications)

November 20, 2023

St. Petersburg
Russia
2023

Preface

The IEEE Russia North West Section and Saint Petersburg Electrotechnical University "LETI" (SPbETU "LETI") are pleased to present the Proceedings of the Seminar on Microelectronics, Dielectrics and Plasmas (theory and practical applications).

The Seminar will cover a broad area of novel issues related to the theory and new practical applications of microelectronics, dielectrics and plasmas.

Chair of the Seminar Organizing Committee:
Prof. Vladimir M. Kutuzov, President of St. Petersburg Electrotechnical University "LETI" (Russia)

Definition of Critical Currents in Superconducting Magnetic Energy Storage Systems

Gontrand Steve Sedjro Bagan
*Higher School of High Voltage
Engineering
Institute of Energy
Peter the Great St.Petersburg
Polytechnic University*
St. Petersburg, Russia
bagan307@yahoo.fr

Alexander G. Kalimov
*Higher School of High Voltage
Engineering
Institute of Energy
Peter the Great St.Petersburg
Polytechnic University*
St. Petersburg, Russia
alexanderkalimov@gmail.com

Sergey Vazhnov
*Higher School of High Voltage
Engineering
Institute of Energy
Peter the Great St.Petersburg
Polytechnic University*
St. Petersburg, Russia
seva_011249@mail.ru

Abstract— Superconducting magnetic energy storage (SMES) systems are expected to be very prospective and flexible energy storage elements of future electric grid interconnectors based on renewable power sources. Superconducting storage elements may be characterized by such important parameters as a very fast response on variation of the energy generation conditions and consumption as well as a very high efficiency reaching 95%. In practice the quality of the power generated in renewable power systems depends strongly on fast variations in such natural factors as solar radiation or the wind velocity and abrupt change in the load. The SMES device is one of the best choices to overcome the corresponding fluctuation of a generated power. The basis of future SMES systems is assumed to be a coil set wound with high temperature superconducting tape. One of the problems to be solved in such a case is definition of the coil critical current. It is well known that the critical current density de-pends on the induced magnetic field intensity. The main difficulty in theoretical prediction of these characteristics is essential non-uniform current density distribution in the superconducting tapes. In this paper we propose a new method of the SMES coils description using a solution of an integral equation based on the critical state theory of 2-nd type superconductors and magnetic field laws.

Keywords— *Magnetic field, superconductivity, critical current, renewable energy sources, energy storage, integral equations.*

I. INTRODUCTION

Wind and solar energy sources are widely used nowadays in electrical energy production all over the world. The main reasons for their growth are their renewable nature and pollution-free characteristics. Main trends in modern economics imply a fast development of corresponding technologies in nearest future. Basic components of this branch of power engineering are the solar and the wind energy utilization. An important drawback of these energy resources is their instability especially in the case when they are used in local electrical grids. Sometimes the generated power may exceed capabilities of energy transfer system connecting power plant with the energy consumer while in other time periods, such as night time in solar power plants, the energy production degrades strongly. To ensure a stable energy production they require energy storage systems capable to compensate lack of energy in the periods when the energy production is affected by objective reasons. Different energy storage systems are used in such cases [1-2]. There are two contradictive requirements to such systems. One of them is capability to store maximum possible energy per unit volume of the storage system. The second one is a possibility to generate as high power as possible. Usually, chemical batteries are used in such storage system. However, their slow reaction on variable conditions of power generation and consumption

requires installing additional elements capable to overcome this drawback. Superconducting magnetic energy storage (SMES) coils having fast response, high charge and discharge efficiency may be regarded as a good solution in such a situation [3-8].

Quick development of superconducting technologies nowadays brought a new solution for building SMES systems. Recently several research groups report about new achievements in creating and investigating SMES coils made of 2-nd generation high temperature superconductors (2G HTS) [9-11]. Such solution ensures relatively cheap and simple production and exploitation of the SMES device. The main specific feature of the 2G HTS conductors is their shape. Industrially produced 2G superconductors are thin tapes with the width of 4 or 12 mm, the thickness of about 0.1 mm and the superconducting layer of about 1 micrometer thickness. In the real situation the current is distributed non-uniformly over the tape width and its description requires detailed analysis of the magnetic field distribution together with specific properties of the 2-nd type superconductors with strong pinning effect.

When the SMES based on 2-nd generation HTS coil system is developed, the maximum possible transport current is estimated by the parameters of superconducting band provided by the superconductor manufacturer. This information usually takes into account properties of a superconducting sample immersed in external magnetic field. But in practice, when the conductor is used in multi-turn windings, the magnetic field distributions and consequently the critical currents may differ strongly from the preliminary defined experimental values. A purpose of this work is a considering a new method of the current density distribution modelling in HTS 2G coils.

II. MATHEMATICAL MODEL OF THE 2G HTS COIL

A. Critical model of a superconducting material

The 2-nd type superconductors with a strong pinning can carry transport current with the engineering (average) density exceeding 100 A/mm^2. Such properties together with stability in strong magnetic fields allow to use these materials for building coils with high energy density per unit volume necessary for creating compact and powerful storage units. The relation between the magnetic flux density and the current density in superconductors is described by Kim's critical state theory [12]. This theory climes that the current density J in 2-nd type superconductors may be equal to zero or to the critical value only. Exact dependence of the critical current density on the magnetic field intensity is defined by a critical state model. The literature reports several models of this type, such as the London's - Bean's model, and different versions of the Kim's

979-8-3503-8358-4/23 $31.00 © 2023 IEEE

one. The most often used model for describing properties of 2G HTS tapes is an anisotropic Kim's model [13-16] with the main relation between the magnetic flux density and the critical current density of:

$$J_C = \frac{J_{C0}}{\left(1 + \frac{\sqrt{k^2 B_=^2 + B_\perp^2}}{B_0}\right)^\alpha} \qquad (1)$$

B_0, J_0, k, α being the material constants. $B_=$ and $B_\perp$ are respectively the parallel and perpendicular components of the magnetic field relative to the tape surface. Main parameters of this dependence are usually defined experimentally. The coefficient k is usually much less than a unit. So, the superconductivity in 2G HTS tapes is destroyed by mainly the normal component of the magnetic flux density.

B. An integral method of the current density modelling in 2G HTS coils

The important problem to be solved by the designer of a 2G HTS superconducting system, is the determination of distributions of the current density and magnetic field in superconducting tape windings. Different modelling methods are formulated according to the selected variable basis. One of them is the T-Ω formulation [17]. Within this approach the vector current potential T is combined with the scalar magnetic potential. The method is advantageous because of relatively small number of unknowns. Alternative approaches to the magnetic field modelling are: the A-V formulation [18], the T-A formulation [19] and the H formulation [20-21]. All these approaches use the finite element method to solve time dependent differential equations of electromagnetic field theory together with the critical state relation (1). Typically, a high discretization of the problem domain is required to achieve a reasonably good accuracy.

In this paper we consider alternative approach to modelling critical current and magnetic field distribution in 2G HTS coils. The magnetic field in the arbitrary point of the space may be expressed by the Biot-Savart Law:

$$\vec{B}(\vec{r}) = \frac{\mu_0}{4\pi} \int_V \frac{\vec{J}(\vec{r}') \times (\vec{r} - \vec{r}')}{|\vec{r} - \vec{r}'|} \, dV \qquad (2)$$

Together with the equation (1) the last integral relation forms a non-linear integral equation for the current density J in a system of several superconducting tapes connected in parallel. The currents in thin superconducting layers are averaged over the whole volume of a problem domain. Such transformation is approved by specially undertaken simulations [21]. We used this mathematical formulation to find a dependence of the critical currents in superconducting structures with a set of straight current carrying tapes.

III. RESULTS OF THE CRITICAL CURRENT MODELLING

We have applied a developed method for analysing the current density distributions in a system of the

superconducting tapes industrially produced by the company Superpower [3]. The width of the chosen tape is 12 mm, the thickness is 0.15 mm (without insulation), the critical current is 240 A. The parameters of the Kim's model defined experimentally [3] are: $k = 0$; $\alpha = 1$, $J_0 = 1.11 \cdot 10^8$ A/m²; and $B_0 = 0.12$ T.

The investigated system consists of several straight tapes connected in parallel. To validate the developed method, we applied it to the system with infinitely big number of such tapes. In this case the problem may be reduced to a solution of one-dimensional integral equation. This problem has also analytical solution described in [12]. Comparison between numerical and analytical results shown in Fig.1 demonstrates a good agreement. The distribution of the magnetic field corresponds to the case of the critical transport current in the conductor.

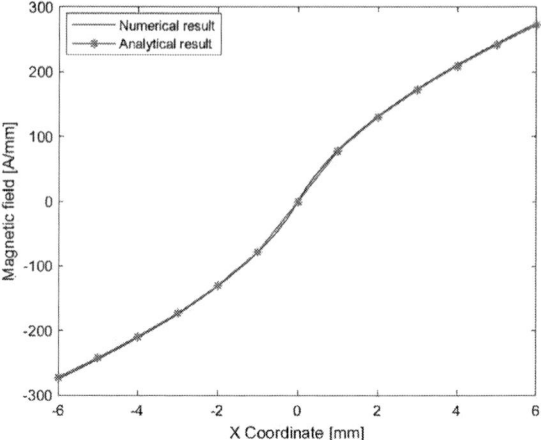

Fig. 1. Comparison of numerically derived distribution of the magnetic field intensity in the infinitely high superconducting plate with analytical results.

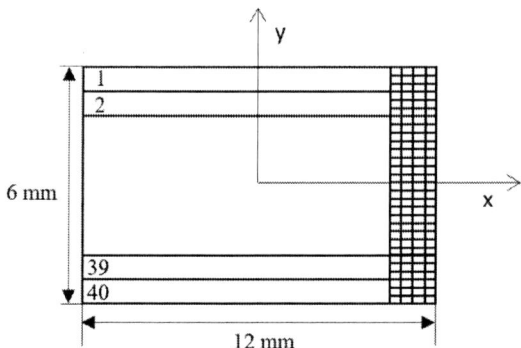

Fig. 2. Cross section of the superconducting system consisting of 40 parallel superconducting tapes of 12 mm width.

979-8-3503-8358-4/23 $31.00 © 2023 IEEE

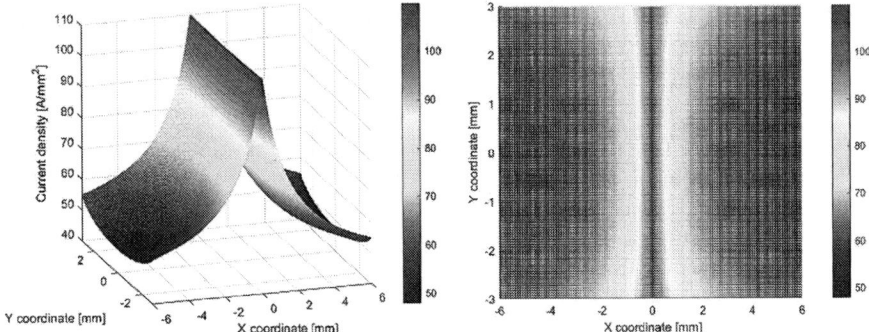

Fig. 3. Distribution of the current density in the superconducting system consisting of 40 parallel tapes.

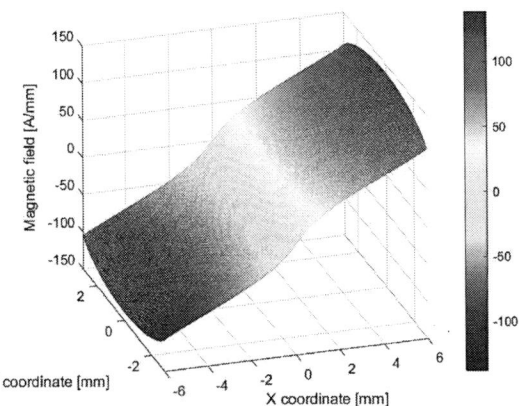

Fig. 4. Distribution magnetic field intensity in the superconducting system consisting of 40 parallel tapes.

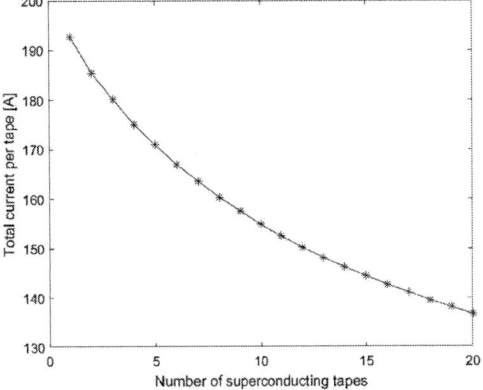

Fig. 5. Dependence of the effective critical current on the number of 2G HTS tapes connected in parallel.

A similar problem is solved for a set of infinitely long superconducting tapes connected in parallel. To approximate the current density distribution in such a system, its cross section is split into a set of small rectangular elements as it is shown in Fig. 2. The current density inside each element was considered to be constant. The integral equation (1) – (2) was approximated by a set of non-linear algebraic equations by applying the collocation method. The magnetic field intensity induced by simple elements was calculated analytically.

To solve the nonlinear system of algebraic equations approximating the integral one we used a simple iterative procedure.

The results of calculations corresponding to the case of 40 parallel tapes are shown in Fig. 3 – Fig.4.

The most important consequence of non-uniform current density distribution in superconducting tapes is degrading of their critical currents. The dependence of the current reduction on the number of parallel tapes is shown in Fig. 5. We see that the critical current degradation may reach a value of about 20-30%. This effect should be taken into account when optimizing a SMES coil structure.

The important advantage of the proposed integral approach with respect to typically used differential methods is a necessity to solve a static problem instead of time dependent ones [15-21].

IV. CONCLUSION

We have demonstrated that the magnetic field intensity and the current density distributions in the 2G HTS coils may be described by solution of a non-linear integral equation based on the Biot-Savart Law and the anisotropic Kim's model of the critical state of superconductors. This approach is successfully applied for calculating critical currents in the pack of tapes and later is supposed to be used for optimization of the SMES coil shape. The derived results are verified at the case of infinitely long superconducting plate with the current for which the analytical solution is available. At this stage of the research work we considered a superconducting current lead with a certain number of tapes connected in parallel. We demonstrated that critical currents in a system of superconducting 2G HTS conductors degrades with increasing a number of tapes connected in parallel. In future we plan to apply the developed numerical technology for description of the multi-turn superconducting structures.

REFERENCES

[1] B. Sorensen, Renewable energy: physics, engineering, environmental impacts, economics and planning, 5th ed., Academic Press, Cambridge, 2017.

[2] J. Twidell, and T. Weir, Renewable Energy Resources, 3rd ed. Routledge, Oxford-shire, 2015.

[3] W. Yuan, "Second-generation high-temperature superconducting coils and their applications for energy storage", University of Cambridge, 2011.

[4] L Chen., H. Chen, Y. Li, G. Li, J. Yang, X. Liu, Y. Xu, L. Ren, and Y. Tang, "SMES-battery energy storage system for stabilization of a photovoltaic-based microgrid", IEEE Trans. Appl. Supercond., vol. 28 n° 4, 2018, pp. 1–7. DOI: 10.1109/TASC.2018.2799544.

[5] I. Ngamroo, "An optimization of superconducting coil installed in an hvdc-wind farm for alleviating power fluctuation and limiting fault current", IEEE Trans. Appl. Supercond., vol. 29 n° 2, 2019, pp. 1–5. DOI: 10.1109/TASC.2018.2881993.

[6] H. Jiang, and C. Zhang, "A method of boosting transient stability of wind farm connected power system using superconducting magnetic energy storage unit", IEEE Trans. Appl. Supercond., vol. 29, n° 2, 2019, pp. 2–6. DOI: 10.1109/TASC.2019.2892291.

[7] A. M. S. Yunus, M. A. S. Masoum, and A. Abu-Siada, "Application of SMES to enhance the dynamic performance of DFIG during voltage sag and swell", IEEE Trans. Appl. Supercond., vol. 22, n° 4, 2012. DOI: 10.1109/TASC.2012.2191769.

[8] L. Chubraeva, and T. Sergey, "Project of autonomous power plant with high-temperature superconductive devices", 2018 Int. Multi-Conference Ind. Eng. Mod. Technol., Vladivostok, 2018, pp. 1–5. DOI: 10.1109/FarEastCon.2018.8602671.

[9] A. W. Zimmermann, and S. M. Sharkh, "Design of a 1 MJ/100 kW high temperature superconducting magnet for energy storage", Energy Reports, vol. 6, n° 5, 2020, pp. 180–188. DOI: 10.1016/j.egyr.2020.03.023.

[10] M. A. Al Zaman, M. R. Islam, and H. M. A. R. Maruf, "Study on conceptual designs of superconducting coil for energy storage in SMES", East Eur. J. Phys., vol. 1, 2020, pp. 111–120. DOI: 10.26565/2312-4334-2020-1-10.

[11] M. H. Ali, B. Wu, and R. A. Dougal, "An overview of SMES applications in power and energy systems", IEEE Trans. Sustain. Energy, vol. 1, n° 1, 2010, pp. 38–47. DOI: 10.1109/TSTE.2010.2044901.

[12] M. N. Wilson, Superconducting Magnets. Clarendon Press, Walton Street, 1987.

[13] W. Yuan, A. M. Campbell, and T. A. Coombs, "A model for calculating the AC losses of second-generation high temperature superconductor pancake coils", Supercond. Sci. Technol., vol. 22, n° 7, 2009, pp. 12-21. DOI: 10.1088/0953-2048/22/7/075028.

[14] V. M. Pan, A. L. Kasatkin, V. L. Svetchnikov, V. A. Komashko, A. G. Popov, A. Y. Galkin, H. C. Freyhardt, and H. W. Zandbergen, "Critical current density in highly biaxially-oriented YBCO films: Can we control JC(77K) and optimize up to more than 106 Amp/cm2?", IEEE Trans. Appl. Supercond., vol. 9, n° 2, 1999, pp. 1535–1538. DOI: 10.1109/77.784686.

[15] D. Yu, H. Liu, X. Zhang, and T. Gong, "Critical current simulation and measurement of second generation, high-temperature superconducting coil under external magnetic field", Materials (Basel), vol. 11, n° 3, 2018, pp. 339-349. DOI: 10.3390/ma11030339.

[16] Z. Jiang, K. P. Thakur, M. Staines, R. A. Badcock, N. J. Long, R. G. Buckley, A. D. Caplin, and N. Amemiya, "The dependence of AC loss characteristics on the spacing between strands in YBCO Roebel cables", Supercond. Sci. Technol., vol. 24, n° 6, 2011. DOI: 10.1088/0953-2048/24/6/065005.

[17] N. Amemiya, S. I. Murasawa, N. Banno, and K. Miyamoto, "Numerical modelings of superconducting wires for AC loss calculations", Phys. C Supercond. its Appl., vol. 310, 1998, pp. 16–29. DOI: 10.1016/S0921-4534(98)00427-4.

[18] N. Nibbio, S. Stavrev, and B. Dutoit, "Finite element method simulation of AC loss in HTS tapes with B-dependent E-J power law", IEEE Transactions on Applied Superconductivity, vol. 11, n° 1, 2001, pp. 2631–2634. DOI: 10.1109/77.920408.

[19] E. Berrospe-Juarez, V. M. R. Zermeño, F. Trillaud, and F. Grilli, "Real-time simulation of large-scale HTS systems: multi-scale and homogeneous models using T-A formulation", Supercond. Sci. Technol., vol. 32, n° 6, 2019. DOI: 10.1088/1361-6668/ab0d66.

[20] G. G. Sotelo, M. Carrera, J. Lopez-Lopez, and X Granados., "H-formulation fem modeling of the current distribution in 2G HTS tapes and its experimental validation using hall probe mapping", IEEE Transactions on Applied Superconductivity, vol. 26, n° 8, 2016, pp. 1-10. DOI: 10.1109/TASC.2016.2591825.

[21] V. M. R. Zermeno, A. B. Abrahamsen, N. Mijatovic, B. B. Jensen, M. P. Sørensen, "Calculation of alternating current losses in stacks and coils made of second generation high temperature superconducting tapes for large scale applications", J. Appl. Phys., vol. 114, n° 17, 2013. DOI: 10.1063/1.4827375.

979-8-3503-8358-4/23 $31.00 © 2023 IEEE

XPS Studies of Adsorption Sites on the Surface of Gas Sensitive Zinc Oxide Based Nanostructures

Viktoriya A. Arkhipenko
Department of Micro- and nanoelectronics
St. Petersburg Electrotechnical University "LETI"
St. Petersburg, Russia
va_arkhipenko@mail.ru

Aslan M. Guketlov
Department of Mechatronics and Robotics
Kabardino-Balkarian State University
Nalchik, Russia
guketlovaslan3@gmail.com

Vladislav P. Bezverkhniy
Department of Micro- and nanoelectronics
St. Petersburg Electrotechnical University "LETI"
St. Petersburg, Russia
vlad150897@yandex.ru

Zamir V. Shomakhov
Department of Electronics and Information Technologies
Kabardino-Balkarian State University
Nalchik, Russia
shozamir@yandex.ru

Svetlana S. Nalimova
Department of Micro- and nanoelectronics
St. Petersburg Electrotechnical University "LETI"
St. Petersburg, Russia
sskarpova@list.ru

Vyacheslav A. Moshnikov
Department of Micro- and nanoelectronics
St. Petersburg Electrotechnical University "LETI"
St. Petersburg, Russia
vamoshnikov@mail.ru

Abstract— **In recent years, approaches to study and control the adsorption sites on the surface of gas-sensitive materials have been actively developing. The specificity of chemical reactions and, consequently, sensor properties largely depend on the concentration and localization of active sites. The X-ray photoelectron spectroscopy technique affords the possibility to analyze the energy characteristics of a surface. Using this method, the effect of modification of sensor layers consisting of zinc oxide nanowires upon post-treatment in solutions of other metal (tin and iron) compounds on the chemical composition of their surface was analyzed. It was shown that under optimal technological conditions, the formation of ternary oxide compounds and a significant change in the ratio of oxygen in various bound states occur. It was demonstrated that such a change in the composition of the surface leads to a significant increase in the sensor response of the functional layers to isopropyl alcohol vapors.**

Keywords— *X-ray photoelectron spectroscopy, gas sensor, zinc oxide, hydrothermal synthesis, nanostructures, surface states, defects*

I. INTRODUCTION

Currently, chemisorption resistive gas sensors are of great interest in gas sensors, because their production methods are usually simple and do not require a lot of financial investments [1]. Sensors based on metal oxide semiconductors are characterized by fast response and low power consumption [2, 3]. One of the metal oxides widely used to create active areas of such sensors is zinc oxide [4]. Now both ZnO and the possible processes of its synthesis are well studied, but gas-sensitive structures based on it do not meet the modern requirements of gas sensors, so the question arises about the possible modification of such structures to improve sensor properties [5-7].

Since the response of the adsorption semiconductor sensor directly depends on the specific surface active area, its increase can lead to a significant improvement in the sensor properties of gas sensors. This can be achieved by forming composite systems based on ZnO nanorods by modifying them to multicomponent metal oxides, for example, Zn-Fe-O.

Zinc-tin oxide structures ($ZnSnO_3$ and Zn_2SnO_4) have advantages in detecting gases compared to binary zinc and tin oxides [8] because of their high electron mobility, low resistance, better stability and enhanced optical characteristics [9]. For example, it was shown that Zn_2SnO_4 nanorods have increased sensitivity and selectivity to NO_2 compared to zinc oxide nanorods[10].

The increase in the sensor signal of multicomponent metal oxides is as a result of different energy characteristics of various metal atoms on the surface [11, 12]. Another reason is the formation of additional structural defects, for instance, oxygen vacancies. Oxygen vacancies are known as adsorption sites for O_2 in the charged state. The reactions of chemisorbed oxygen with gas molecules lead to the appearance of a sensor response due to the formation of free electrons and their transition to the conduction band of the material.

The purpose of this work is to analyze the adsorption sites on the surface of zinc/iron and zinc/tin multicomponent oxide nanostructures. The surroundings of atoms on the surface of materials can be accurately studied by X-ray photoelectron spectroscopy.

II. EXPERIMENT

Ceramic chip with gold contacts was used as a substrate. The composite nanostructure was produced in two stages. The first stage was the formation of ZnO nanorods by spin-coating (seed layer) of nanoparticles with subsequent hydrothermal growth of 1D nanostructures [13, 14]. The second stage was the fabrication of Zn-Fe-O and Zn-Sn-O composite systems.

A solution of zinc acetate with distilled water was made for deposition of ZnO seed layer, on which zinc oxide nanorods are subsequently grown. Concentration of $Zn(CH_3COO)_2 \cdot 2H_2O$ was 5 mM, the solvent volume was 50 ml. An ultrasonic bath was used to accelerate the dissolution process. As a result, a solution with a dispersed phase of zinc hydroxide nanoparticles was formed.

Spin-coating (3000 rpm, 15 s) was used for deposition of $Zn(OH)_2$ nanoparticles on the surface of the substrate. The process was carried out 3 times to achieve uniform coverage. Zinc oxide seed layer was formed during annealing in a muffle furnace at 500 ° C for 15 min. ZnO and H_2O are produced from zinc hydroxide, followed by evaporation of water.

A solution of zinc nitrate and a solution of hexamethylenetetramine in distilled water were made for the growth of zinc oxide nanorods. The concentrations were 25 mM, and the volume of the solvent was 50 ml. The solutions were mixed using a magnetic stirrer. The resulting solution and samples with applied seed layers were placed in an autoclave. Synthesis conditions were as follows: time – 1 hour, temperature – 80 °C. Then the samples were dried, followed by annealing in a muffle furnace at 500 °C for 15 minutes.

The synthesis of Zn-Fe-O composite structure was carried out on two identical samples with grown zinc oxide nanorods by precipitation from a solution under normal conditions. Two solutions of iron sulfate heptahydrate in distilled water with different concentrations were produced: 0.025 M (ZnO-Fe-1) and 0.05 M (ZnO-Fe-2). The samples were immersed in solutions. The formation of the composite structure was carried out at room temperature (25 ° C) for 30 min. After that, the samples were dried, followed by annealing in a muffle furnace at a temperature of 500 ° C for 15 minutes.

A solution containing 40% isopropyl alcohol and 60% water was used to form Zn-Sn-O nanostructures. Potassium stannate trihydrate and urea were added to 20 ml of this solution. For the ZnO-Sn-1 sample, the masses of potassium stannate and urea were 0.0287 g and 0.187 g, respectively. To synthesize the ZnO-Sn-2 sample, 0.01433 g of potassium stannate and 0.0936 g of urea were added to the solution. Hydrothermal synthesis was carried out at 170 ° C for 30 minutes.

The chemical composition of zinc oxide nanorods and multicomponent oxide nanostructures was studied by X-ray photoelectron spectroscopy (K-Alpha, Thermo Scientific (USA)) using AlKa radiation with a photon energy hv = 1350 eV. The binding energy values of the core levels are determined relative to the C1s level with an energy of 284.6 eV.

A specially designed laboratory set-up was used to study the gas-sensitive properties of the samples [15]. The response was calculated as the ratio of the sample resistances in the air atmosphere and in the presence of isopropyl alcohol vapors.

III. RESULTS AND DISCUSSIONS

Figure 1 shows the spectra of samples modified in iron sulfate solutions with concentrations of 0.025 M and 0.05 M.

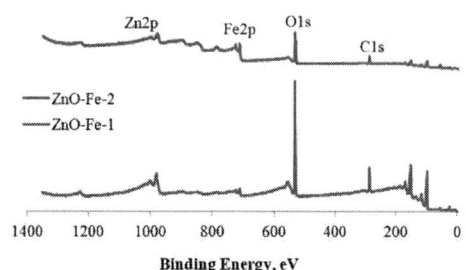

Fig. 1. Survey X-ray photoelectron spectra of Zn-Fe-O samples.

For a sample synthesized from a solution of iron sulfate of a higher concentration (0.05 M), there are no peaks corresponding to the binding energies of zinc atoms. Possibly, this is a result of the deposition of iron oxide particles on the surface in parallel with the incorporation of iron atoms into the structure of zinc oxide. On the spectrum of a sample synthesized from a solution of iron sulfate of a

lower concentration (0.025 M) peaks of oxygen, iron and zinc are observed. This allows us to conclude that the composite structure was formed during synthesis.

The peaks of oxygen core levels of Zn-Fe-O multicomponent oxide nanostructures are shown in Fig. 2. Two peaks are observed on the spectra, which correspond to the oxygen atoms of the crystal lattice [16] and vacancies in the oxygen sublattice [17]. For ZnO-Fe-1 sample, the corresponding binding energies are 529.9 eV and 532.1 eV. For ZnO-Fe-2 sample, the oxygen peak of the crystal lattice is observed at 530 eV, and the peak of the adsorbed oxygen is at 532.4 eV. At the same time, for zinc oxide sample, the binding energy of the crystal lattice oxygen is 530.8 eV, and the adsorbed oxygen is 532.3 eV. Thus, it can be seen that as a result of interaction with iron atoms, a significant change in the crystal lattice occurs, reflected in a significant shift in the oxygen binding energy of the crystal lattice. At the same time, the change in the binding energy corresponding to oxygen vacancies is insignificant. When the chemical composition of zinc oxide nanorods changes and Zn-Fe-O composite structure is formed, the content of oxygen vacancies on the surface increases significantly (48% for ZnO sample and 68% for the ZnO-Fe-1 sample). In the ZnO-Fe-2 sample, mainly adsorbed oxygen is observed on the surface, its content is 92%.

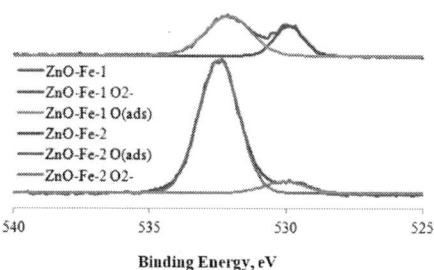

Fig. 2. Spectra of oxygen core levels in ZnO-Fe-1 and ZnO-Fe-2 samples.

The survey spectra of the ZnO-Sn-1 and Zn-Sn-2 samples are shown in Fig. 3.

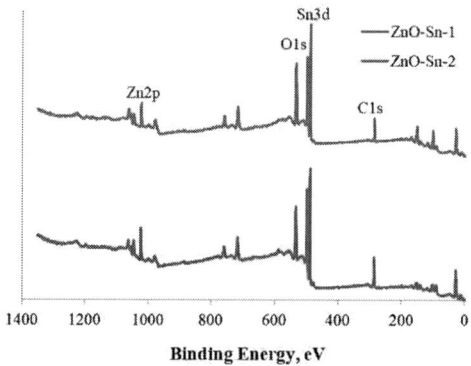

Fig. 3. Survey X-ray photoelectron spectra of Zn-Sn-O samples.

The charged states of the elements of the ZnO-Sn-1 sample are analyzed. For tin, peaks are observed at 486.6 eV and 494.9 eV. This position of the peaks indicates Sn^{4+} state [18]. For zinc, relative to the ZnO structure, the peaks shifted towards a higher binding energies (1021.7 eV and 1045 eV). Despite the slight shift, the charged state remains Zn^{2+}. Oxygen is also observed in the crystal lattice with peak at a binding energy of 530.3 eV and in the form of adsorbed oxygen-containing particles or oxygen vacancies at 531.9 eV. Compared to zinc oxide, oxygen peaks are also shifted towards high binding energies.

In ZnO-Sn-2 sample, tin peaks are observed at 486.4 eV and 494.8 eV. The peaks shifted towards lower binding energies compared to ZnO-Sn-1, the state remains the same – Sn4+. For zinc, the position of the peaks did not change.

The components of the O1s spectrum for Zn-Sn-O samples were analyzed (Fig. 4). Surface oxygen also exists in two states. The oxygen of the crystal lattice has a binding energy of 530.3 eV, and the adsorbed oxygen has a binding energy of 531.9 eV. Regarding zinc oxide, oxygen peaks also shifted towards high binding energies. At the same time, the content of adsorbed oxygen in the ZnO-Sn-1 sample is 65%, and in the ZnO-Sn-2 sample it is 56%.

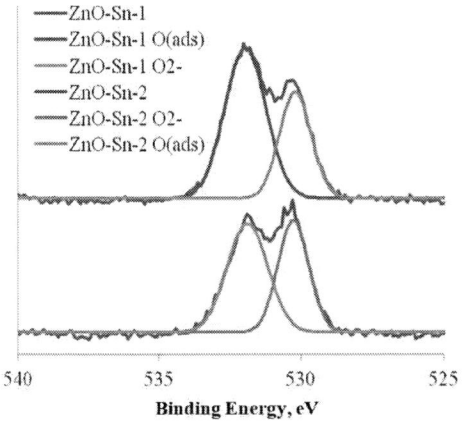

Fig. 4. Spectra of oxygen core levels in ZnO-Sn-1 and ZnO-Sn-2 samples.

Figure 5 shows the dependence of the response of the ZnO-Fe-1 sample on the concentration of isopropyl alcohol vapors. The value of the sensor response increases almost linearly with an increase in the concentration of isopropyl alcohol in the range of 200 – 1000 ppm (linearity coefficient R2 is 0.946).

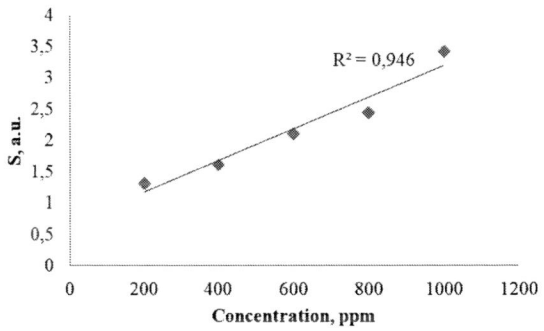

Fig. 5. The dependence of the sensor signal on the concentration of isopropyl alcohol.

It should be noted that the value of the sensor response of the ZnO-Fe-1 sample to 1000 ppm of isopropyl alcohol at 250 °C is 3.42, while for the sample of the initial zinc oxide nanorods this value is about 2.5. The responses of ZnO-Sn-1 and ZnO-Sn-2 samples to 1000 ppm of isopropyl alcohol at 310 °C were 5.26 and 8.24 respectively.

Thus, for all the studied multicomponent oxide nanostructures, an increase in the sensor signal is observed compared to ZnO nanorods.

A sample modified in a solution with a lower concentration of potassium stannate shows a higher sensor response. Perhaps this is due to the optimal ratio of adsorption sites which are the cations of various metals.

The improvement of sensor properties may be due to the influence of vacancies in O lattice sites on the processes of interaction of the oxide surface with gas molecules. The presence of vacancies in the oxygen sublattice leads to the modulation of the properties of oxides, which enhances the gas-sensitive characteristics of chemoresistor. Having a large number of localized electrons, they usually act as electron donors and determine the density of charge carriers of semiconductor oxide materials. At the phase interface, the presence of vacancies in O lattice sites promotes the adsorption of O2 particles and increase the concentration of charged oxygen on the surface. In addition, gas molecules can be adsorbed on such vacancies, which are the active centers of this process [19].

IV. Conclusions

Using the examples of zinc/iron and zinc/tin mixed oxides it is shown that an increase in the sensitivity of multicomponent materials can be associated with the formation of additional oxygen vacancies as a result of the synthesis of composite structures, as well as with the achievement of an optimal content of adsorption sites which are the various metal cations on their surface.

References

[1] Moshnikov V.A., Gracheva I.E., Kuznezov V.V. Maximov A.I., Karpova S.S., Ponomareva A.A. Hierarchical nanostructured semiconductor porous materials for gas sensors. Journal of Non-Crystalline Solids, 2010, V. 356, pp. 2020-2025.

[2] Mahajan S., Jagtap S. Metal-oxide semiconductors for carbon monoxide (CO) gas sensing: A review. Applied Materials Today, 2020,V. 18, art. no 100483.

[3] Mirzaei A., Kim S.S., Kim H.W. Resistance-based H2S gas sensors using metal oxide nanostructures: A review of recent advances. Journal of Hazardous Materials, 2018, V. 357, pp 314-331.

[4] Bobkov A., Moshnikov V., Varezhnikov A., Plugin I., Fedorov F.S., Goffman V., Sysoev V., Trouillet V., Geckle U., Sommer M. The multisensor array based on grown-on-chip zinc oxide nanorod network for selective discrimination of alcohol vapors at sub-ppm range. Sensors, 2019, V. 19, art. no 4265.

[5] Ryabko A.A., Nalimova S.S., Mazing D.S., Korepanov O.A., Guketlov A.M., Aleksandrova O.A., Maksimov A.I., Moshnikov V.A., Shomakhov Z.V., Aleshin A.N. Sensitization of ZnO nanorods by AgInS2 colloidal quantum dots for adsorption gas sensors with light activation. Technical Physics, 2022, V. 92, pp. 845-851.

[6] Yang Z., Jiang L., Wang J., Liu F., He J., Liu A., Lv S., You R., Yan X., Sun P., Wang C., Duan Y., Lu G. Flexible resistive NO2 gas sensor of three-dimensional crumpled MXene Ti3C2Tx/ZnO spheres for room temperature application. Sensors and Actuators B, 2021, V. 326, art. no 128828.

[7] Bobkov A., Luchinin V., Moshnikov V., Nalimova S., Spivak Y. Impedance Spectroscopy of Hierarchical Porous Nanomaterials Based on por-Si, por-Si Incorporated by Ni and Metal Oxides for Gas Sensors. Sensors, 2022, V. 22, art. no 1530.

[8] Yan Y., Liu J., Zhang H., Song D., Li J., Yang P., Zhang M., Wang J. One-pot synthesis of cubic ZnSnO3/ZnO heterostructure composite and enhanced gas-sensing performance. Journal of Alloys and Compounds, 2019, V. 780, pp. 193-201.

[9] Nalimova S.S., Maksimov A.I., Matyushkin L.B., Moshnikov V.A. Current state of studies on synthesis and application of zinc stannate (review). Glass Physics and Chemistry, 2019, V. 45, pp. 251-260.

[10] Hung C.M., Phuong H.V., Duy N.V., Hoa N.D., Hieu N.V. Comparative effects of synthesis parameters on the NO2 gas-sensing performance of on-chip grown ZnO and Zn2SnO4 nanowire sensors. Journal of Alloys and Compounds, 2018, V. 765, pp. 1237-1242.

[11] Karpova S.S., Moshnikov V.A., Mjakin S.V., Kolovangina E.S. Surface functional composition and sensor properties of ZnO, Fe_2O_3, and $ZnFe_2O_4$. Semiconductors, 2013, V. 47, pp. 392-395.

[12] Karpova S.S., Moshnikov V.A., Maksimov A.I., Mjakin S.V., Kazantseva N.E. Study of the effect of the acid-base surface properties of ZnO, Fe_2O_3 and $ZnFe_2O_4$ oxides on their gas sensitivity to ethanol vapor, Semiconductors, 2013, V. 47, pp. 1026-1030.

[13] Ryabko A.A., Bobkov A.A., Nalimova S.S. et al. Gas sensitivity of nanostructured coatings based on zinc oxide nanorods under combined activation, *Technical Physics*, 2022, vol. 92, pp. 758-764. DOI: 10.21883/JTF.2022.05.52382.314-21.

[14] Ryabko A.A., Mazing D.S., Bobkov A.A., Maksimov A.I., Levitskiy V.S., Lazneva E.F., Komolov A.S., Moshnikov V.A., Terukov E.I. The effect of interface alloying of the zinc oxide nanorods system. Physics of Solid State, 2022, V. 64, pp. 1681-1689.

[15] Anikina M.A., Ryabko A.A., Nalimova S.S., Maximov A.I. Synthesis and study of zinc oxide nanorods for semiconductor adsorption gas sensors. Journal of Physics: Conference Series, 2021, V. 1851, art. no 012010.

[16] Jing L., Xu Z., Sun X., Shang J., Cai W. The surface properties and photocatalytic activities of ZnO ultrafine particles. Applied Surface Science, 2001, V. 180, pp. 308-314.

[17] Xie F., Yang M., Song Z.-Y., DuanW.-C., Huang X.-J., Chen S.-H., Li P.-H., Xiao X.-Y., Liu W.-Q., Xie P.-H. Highly sensitive electrochemical detection of Hg(II) promoted by oxygen vacancies of plasma-treated ZnO: XPS and DFT calculation analysis. Electrochimica Acta, 2022, V. 426, art. no 140757.

[18] Shomakhov Z.V., Nalimova S.S., Kalmykov R.M., Aubekerov K., Moshnikov V.A. Study of the microstructure and composition of tin dioxide layers modified by silver nanoparticles, Physical and chemical aspects of the study of clusters, nanostructures and nanomaterials, 2021, no. 13, pp. 447-456.

[19] Zhang B., Wang J., Wei Q., Yu P., Zhang S., Xu Y., Dong Y., Ni Y., Ao J., Xia Y. Visible Light-Induced Room-Temperature Formaldehyde Gas Sensor Based on Porous Three-Dimensional ZnO Nanorod Clusters with Rich Oxygen Vacancies. ACS Omega, 2022, V. 7, pp. 22861−22871.

Modeling of Nanostructuring Processes of Lead Chalcogenide Layers under Sensitizing Oxidation

Vladislav P. Bezverkhniy
Department of Micro- and Nanoelectronics
Saint Petersburg Electrotechnical University "LETI"
Saint Petersburg, Russia
vlad150897@yandex.ru

Abstract—**IR emitters built on nanostructured layers of lead chalcogenides are the subject of this study. The goal of the work was to develop a model of the mechanisms involved in the nanostructuring and sensitizing oxidation of lead chalcogenides.**

Keywords —*lead chalcogenides, nanostructured materials, the Kirkendall-Frenkel effect*

I. INTRODUCTION

Atomic-molecular design and nanoarchitectonics are new directions in science and technology. The task of nanoarchitectonics is to control the structure of nanomaterials based on nanoscale units with certain characteristics and functions. Currently, the materials obtained with the help of concepts that are used in these areas are already widely used in various fields: biomedicine, energy, photonics and many others.

Such structures obtained using the concept of nanoarchitectonics are nanostructured layers of lead chalcogenides subjected to sensitizing oxidation to form shell structures. Currently, they are used in IR technology as receivers and emitters. The aim of the work is to model the processes occurring during nanostructuring and sensitization of lead chalcogenide layers. This is an urgent task, because modeling can allow you to control the properties of the materials obtained and improve their quality.

II. NANOSTRUCTURED LAYERS OF LEAD CHALCOGENIDES

$Pb_{1-x}Cd_xSe<I>$ layers are the main material based on the nanostructuring processes of which modeling was carried out [1]. Figure 1 shows an image of the grain of this material.

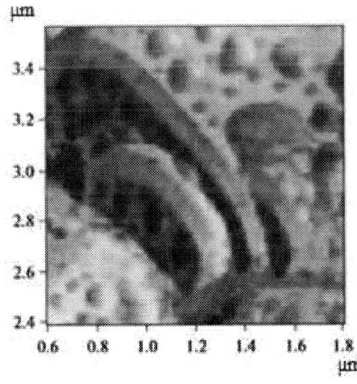

Fig. 1. The image of the cut grain Pb1-xCdxSe<I> obtained by AFM by lateral force microscopy [1].

The description of the process of obtaining $Pb1_{-x}Cd_xSe<I>$ layers is described in detail in [1, 2]. Based on these works, a method for producing lead chalcogenide was considered and model representations of this process were studied. These data are used to develop model ideas about these structures. The PbCdSe layers considered have grains with pronounced radial symmetry on the surface and can be modeled as spheres.

The main stages of nanostructuring and sensitization of lead chalcogenide layers:

1) When treated in an oxygen-containing atmosphere, an outer shell of oxide phases is formed on the layers of lead chalcogenides.

2) Due to the effects of Kirkendall and Frenkel, as well as the chemical reactions of lead and cadmium with iodine, pores are formed in the atmosphere, and stoichiometry changes in the direction of deeper layers.

3) Self-organization of the porous structure as a result of recrystallization in the presence of a liquid phase with iodine added to the initial charge. This reduces the concentration of surface defects.

In [2] it was also shown that the layer under the oxide coating has a p-type of conductivity. Closer to the center of the grain, the sign of conductivity changes, and the central region itself has an n-type of conductivity. Thus, this structure has an internal p-n junction.

III. MODELING OF NANOSTRUCTURING PROCESSES

The main processes occurring during the processes of nanostructuring and sensitization:

1) Diffusion of lead and cadmium.

2) Formation of vacancies.

3) Diffusion of vacancies.

4) Self-organization of pores from vacancies.

5) Growth of the oxide shell.

The following factors were taken into account during these processes:

1) Kirkendall and Frenkel effect. As a result of rapid diffusion and chemical reactions of cadmium and lead with iodine, vacancies are formed.

2) Due to the difference in the diffusion rate of cadmium and lead, the stoichiometry changes in the direction deep into the grain, a varizonal structure is formed.

3) The formation of vacancies is most likely to occur at the layer-atmosphere phase boundary. And only then does the diffusion of vacancies occur deep into the grain. Self-organization of pores from vacancies.

4) The growth of the oxide shell slows down the diffusion processes of all components, the processes of vacancy formation. At the same time, the factors leading to the process of self-organization of pores gradually become stronger than diffusion ones.

5) The growth of the oxide shell occurs heterogeneously, "islands" appear on the surface of the oxide shell.

The growth rate of the oxide shell is determined by the law:

$$h^a = b \cdot t$$

where h – thickness of the shell, t – time, a – coefficient from 1.5 to 2 and b – empirical coefficient.

The PbCdSe crystal lattice has a cubic structure, which means that the diffusion coefficients are isotropic. This allows the use of a one-dimensional diffusion approximation. As a result , the following diffusion equations through the core are obtained:

$$\frac{\partial C_{Pb}}{\partial t} = -D_{11}\frac{\partial^2 C_{Pb}}{\partial r^2} - D_{12}\frac{\partial^2 C_{Cd}}{\partial r^2} - D_{13}\frac{\partial^2 C_{Se}}{\partial r^2} - D_{14}\frac{\partial^2 C_O}{\partial r^2}$$

$$\frac{\partial C_{Cd}}{\partial t} = -D_{21}\frac{\partial^2 C_{Pb}}{\partial r^2} - D_{22}\frac{\partial^2 C_{Cd}}{\partial r^2} - D_{23}\frac{\partial^2 C_{Se}}{\partial r^2} - D_{24}\frac{\partial^2 C_O}{\partial r^2}$$

$$\frac{\partial C_{Se}}{\partial t} = -D_{31}\frac{\partial^2 C_{Pb}}{\partial r^2} - D_{32}\frac{\partial^2 C_{Cd}}{\partial r^2} - D_{33}\frac{\partial^2 C_{Se}}{\partial r^2} - D_{34}\frac{\partial^2 C_O}{\partial r^2}$$

$$\frac{\partial C_O}{\partial t} = D_{41}\frac{\partial^2 C_{Pb}}{\partial r^2} + D_{42}\frac{\partial^2 C_{Cd}}{\partial r^2} + D_{43}\frac{\partial^2 C_{Se}}{\partial r^2} + D_{44}\frac{\partial^2 C_O}{\partial r^2}$$

$$\frac{\partial C_v}{\partial t} = -\frac{\partial C_{Pb}}{\partial t} - \frac{\partial C_{Cd}}{\partial t} - \frac{\partial C_{Se}}{\partial t} - \frac{\partial C_O}{\partial t}$$

The exact values of the diffusion coefficients can only be found empirically. Moreover, they will depend on the thickness of the oxide shell. As the diffusion intensity decreases as a result of the growth of the oxide shell, the diffusion coefficients will tend to zero at some critical thickness. They will also depend on the concentration of vacancies that have not formed large pores, strengthening the vacancy mechanism of diffusion. The model uses in the following equation:

$$D(h) = D_o\left(1 - \frac{h}{h_{cr.}}\right)(C_v - C_{v\,cor})$$

where $h_{cr.}$ – critical thickness of the oxide, $C_{v\,cor}$ – concentration of vacancies that have formed pores.

The Baker- Lonsdale model is used to estimate the amount of substance that has "left" as a result of chemical reactions:

$$\frac{3}{2}\left(1 - \left(\frac{Q_t}{Q_\infty}\right)^{\frac{2}{3}}\right) - \frac{Q_t}{Q_\infty} = K \cdot t$$

Q_t – the amount of substance that has left the volume during the time t, Q_∞ – the total amount of substance that has left the volume, K – empirical coefficient.

The atomic-molecular design of the porous space is similar to the well-known Witten-Sender model for the formation of fractal structures by diffusion-limited aggregation (at the initial stage of growth) and cluster-cluster aggregation at the final stages of nanostructuring.

IV. SIMULATION RESULTS

This section presents the simulation results obtained by software methods. Figure 2 shows a model of pore formation in the middle of the process. Gray spheres are pores, red dots are vacancies.

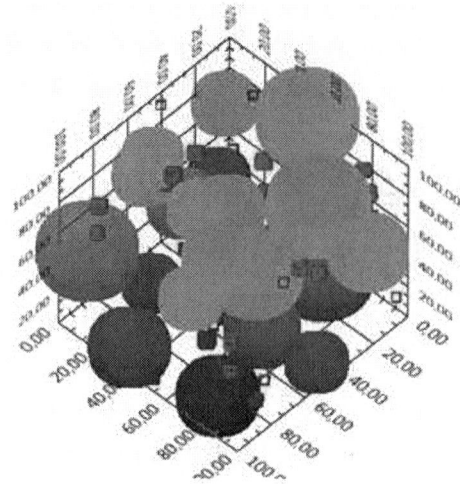

Fig. 2. Image of modeling the formation of pores and vacancies in the model.

Figure 3 shows the intermediate stage of the simulated nanostructuring process, which uses simple structures for better computer operation speed. In Fig. 3. The final stage of modeling is depicted with the replacement of simple structures with visually more informative structures.

Fig. 3. Image of nanostructuring modeling of lead chalcogenide layers with simple structures.

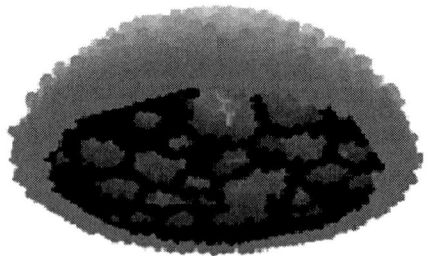

Fig. 4. The final stage of modeling with the replacement of simple structures with visually more informative structures.

V. CONCLUSION

A model of the process of nanostructuring and sensitizing oxidation of lead chalcogenides is presented. The images obtained by software methods using this model are presented at different stages of the process. The simulation results explain the observed experimental data, namely: the formation of an oxide shell, the appearance of a developed porous system, the appearance of "islands" on the surface of the oxide shell, the formation of a varizonal structure.

The processes of model experiments provide an opportunity to evaluate the optimization of IR emitters, especially the formation of p-n transitions and a decrease in the probability of non-radiative transitions (and, accordingly, an increase in the efficiency of photoluminescence).

ACKNOWLEDGMENT

The author expresses gratitude to the scientific supervisor Prof. Moshnikov V.A. and assoc. Spivak Y.M. for constant attention to work and fruitful discussion.

REFERENCES

[1] Spivak Y.M., Moshnikov V.A. Features of photosensitive polycrystalline PbSe layers with a mesh structure. Poverhnost'. Rentgenovskie, sinchrotronnie I neitronnie issledovania. [Surface. X-ray, synchrotron and neutron studies].2010, № 4, pp. 71–76.

[2] Spivak Y., Kononova I., Kononov P., Moshnikov V., Ignat'ev S. The Architectonics Features of Heterostructures for IR Range Detectors Based on Polycrystalline Layers of Lead Chalcogenides. *Crystals* 2021, no. 11 (9), 1143. https://doi.org/10.3390/cryst11091143

[3] Smerdov R.; Mustafaev A.; Spivak Y.; Moshnikov V. Functionalized nanostructured materials for novel plasma energy systems. *In Topical Issues of Rational Use of Natural Resources 2019*; CRC Press: London, UK, 2019; pp. 434–441.

[4] Bezverkhniy V.P., IO Spivak Y.M., Moshnikov V.A. Analysis of ways to improve the efficiency of photoluminescence in nanostructured layers of lead chalcogenides due to varizonicity. «Nanophisica i nanomateriali – 2021» [Nanophysics and nanomaterials – 2021], 2021, pp. 33-37.

[5] Maraeva E.V., Moshnikov V.A, Tairov Y.M. Models of formation of oxide layers in nanostructured materials based on lead chalcogenides during processing in oxygen and iodine vapors. Phisica i technologia poluprovodnicov [Physics and technology of semiconductors]. – 2013, v. 47, № 10, pp. 1431-1434.

[6] Maraeva E.V., Moshnikov V.A, Tairov Y.M , Petrov A.A. On the model of oxidation of polycrystalline layers of lead chalcogenides in a water-containing medium. Phisica i technologia poluprovodnicov [Physics and technology of semiconductors], 2016, v. 50, № 6, pp. 791-793.

[7] Alecsandrova O.A., Maximov A.I., Moshnikov V.A.m Chesnokova D.B. Chalcogenidi I ocsidi elementov IV gruppi. Poluchenie, issledovanie, primenenie. [Chalcogenides and oxides of group IV elements. Obtaining, research, application] Санкт-Петербург: ООО "Технолит", 2008. – 240 с. – ISBN 5-7629-0908-5.

Investigation of Structural and Dielectric Properties of thin Films (Sr/Ba)Nb$_2$O$_6$ on Al$_2$O$_3$ Substrates in the Field of Microwave Applications

Alexey Bogdan
Department of Physical Electronics and Technology
Saint Petersburg Electrotechnical University "LETI"
Saint Petersburg, Russia
alexey.bogdan98@gmail.com

Artem R. Karamov
Department of Physical Electronics and Technology
Saint Petersburg Electrotechnical University "LETI"
Saint Petersburg, Russia
temkakaramov@gmail.com

Evgeny N. Sapego
Department of Physical Electronics and Technology
Saint Petersburg Electrotechnical University "LETI"
Saint Petersburg, Russia
eugenysapego@yandex.ru

Andrei V. Tumarkin
Department of Physical Electronics and Technology
Saint Petersburg Electrotechnical University "LETI"
Saint Petersburg, Russia
avtumarkin@yandex.ru

Abstract— **Thin Sr$_{0.75}$Ba$_{0.25}$Nb$_2$O$_6$ ferroelectric films on monocrystalline Al$_2$O$_3$ substrates at different temperatures were obtained by magnetron sputtering of a ceramic target of stoichiometric composition. The main structural properties of the resulting ferroelectric films were determined. Their significant dependence on the choice of substrate and temperature during deposition was revealed. Based on the obtained ferroelectric films, planar capacitors for microwave applications were created. The electrical properties and dielectric characteristics of the elements based on the obtained films in the ultrahigh frequency range were determined.**

Keywords— *thin films, microwave application, properties ferroelectric films*

I. INTRODUCTION

The material Sr$_x$Ba$_{1-x}$Nb$_2$O$_6$ (SBN), where (0.2<x<0.8) is a representative of the family of tungsten bronzes with a tetragonal phase (4mm) at room temperature [2-4]. SBN thin films are promising in many fields of application [2,6], in particular non-volatile memory, microelectromechanical systems, phase shifters, etc. This material is of special interest in a thin-film version in the field of microwave applications [3].

Currently, SBN thin films are mainly grown on MgO substrates [1,4-6], with the exception of a few works [2,3,7] where Si and Al$_2$O$_3$ substrates with an electrode layer of Pt were used. Data on sufficiently low dielectric losses of SBN films grown on an MgO substrate at frequencies of 0.5-20 GHz have been published [4].

In this paper, the possibility of obtaining SBN films on single-crystal Al$_2$O$_3$ substrates for further use at microwaves as part of capacitive structures is investigated, the structural characteristics of thin films obtained at different substrate temperatures are compared. The effect of annealing on the structural properties of final samples is studied. The desired electrical characteristics of the thin-film structure on Al$_2$O$_3$ substrates are expected to be achieved in SBN thin films at x=0.75 through the a sufficiently low phase transition temperature [3], that will avoid high losses in the microwave range.

II. EXPERIMENT

The SBN thin films studied in this work were obtained by high-frequency magnetron sputtering of a ceramic target Sr$_{0.75}$Ba$_{0.25}$Nb$_2$O$_6$. Deposition was carried out on a substrate of monocrystalline sapphire (r-cut). The temperature of the substrate during deposition varied between 650-880°C. Temperature control was carried out using a thermocouple located under the substrate holder. Oxygen was used as the working gas. Gas discharge parameters: discharge voltage U=1kV , discharge current $I \approx$180 mA , working gas pressure P=2 Pa . After deposition, the films were cooled in an oxygen atmosphere at a rate of 2-3°C/min. Further, the films were annealed at temperatures of 1000-1100°C.

The crystal structure and phase composition of the obtained films before and after annealing were studied by X-ray diffraction (XRD) using a DRON-6 diffractometer (CuKα radiation).

III. RESULTS

Comparative diffractograms of the obtained SBN thin films deposited at temperatures of 650-880 °C on sapphire substrate are shown in Figure 1.

It can be seen from Figure 1 that the angular positions of the reflexes of the studied films do not actually shift when the temperature of the substrate changes. It indicates that the composition of the solid solution remains unchanged. At temperatures of 750-880°C, SBN increases in the direction of [410] [2 – 5, 9]. The obtained diffractograms indicate an increase in the intensity of the reflexes of the deposited films with an increase in the temperature of the substrate.

When studying the crystal structure of the film obtained by sputtering the target onto a substrate heated to a temperature of 650°C, no obvious reflexes from the material under study were observed. This fact correlates with the data on the growth of SBN films in the study [2]. Diffractograms of samples obtained at deposition temperatures of 750°C and 850°C are marked by the presence of a peak (410) at 29° of similar intensity. In addition, reflexes are observed in the angular interval of 31-33°, characteristic of the secondary phases of barium-strontium polyniobates.

The best, from the point of view of structural quality, is a sample deposited at a temperature of 880°C. The intensity of the reflex (410) is noticeably higher compared to low-temperature samples, despite the high background from the substrate. An additional reflection from the plane (311) also appears, which is also noted in [3,4] for high-temperature deposition of the studied films.

979-8-3503-8358-4/23 $31.00 © 2023 IEEE

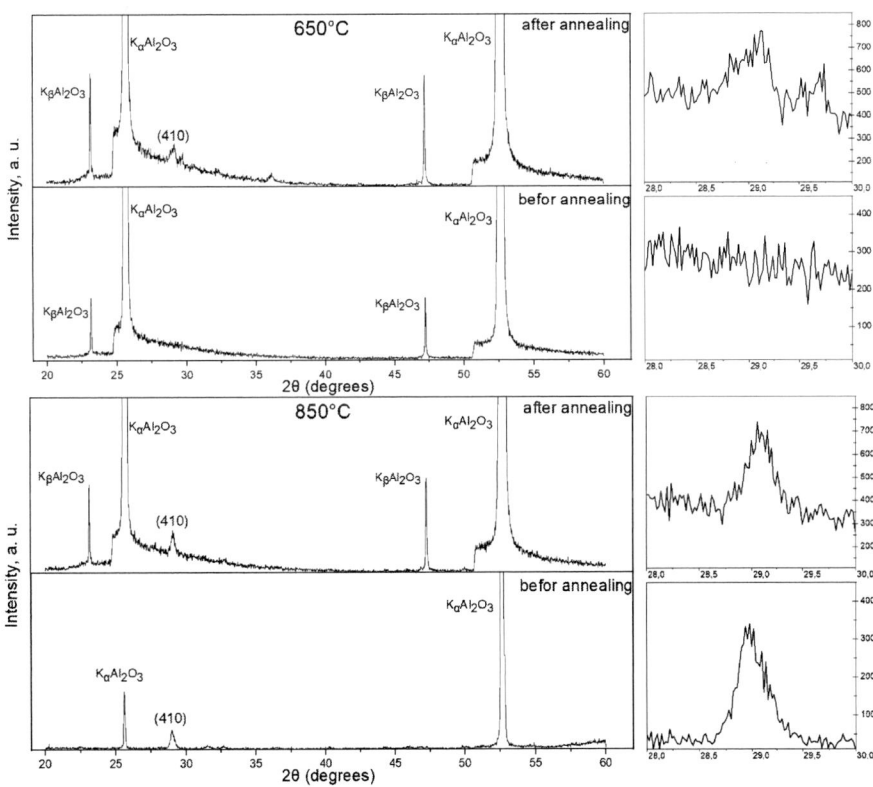

Fig. 1. Diffractograms of SBN thin films before annealing on a monocrystalline sapphire substrate.

Figures 2 and 3 show diffractograms of the mentioned films subjected to high-temperature annealing in air. Thin films obtained at a deposition temperature of 650°C and 850°C were annealed at a temperature of 1000°C for half an hour. The samples obtained at a substrate temperature of 750°C and 880°C were annealed at a temperature of 1100°C for an hour.

Fig. 2. Diffractograms of SBN thin films obtained at a temperature of 650°C and 850°C, respectively, before and after annealing at 1000°C.

979-8-3503-8358-4/23 $31.00 © 2023 IEEE

On the diffractograms after annealing (Figure 2), a clear demonstration of the reflex corresponding to the reflection from the plane (410) is noticeable, and its amplitude has almost doubled. Structural analysis of the thin film deposited at the lowest temperature also showed the appearance of peaks from secondary dielectric phases, negatively affecting the ferroelectric properties of the SBN film. The formation of the SBN phase for a low-temperature sample after annealing is unusual, but this fact may also be traced in [2]. The diffractogram of a thin-film structure obtained at a temperature of 850°C and subjected to annealing differs in the absence of reflexes from secondary phases.

Diffractograms of films subjected to post-growth high-temperature treatment at a temperature of 1100°C (Figure 3) also show an increase in the intensity of the reflex (410). For a film obtained at a temperature of 750°C, after annealing, the greatest increase in the amplitude of the SBN reflex is recorded with an increase in peak intensity by more than 3 times, which indicates the best results in a series of experiments.

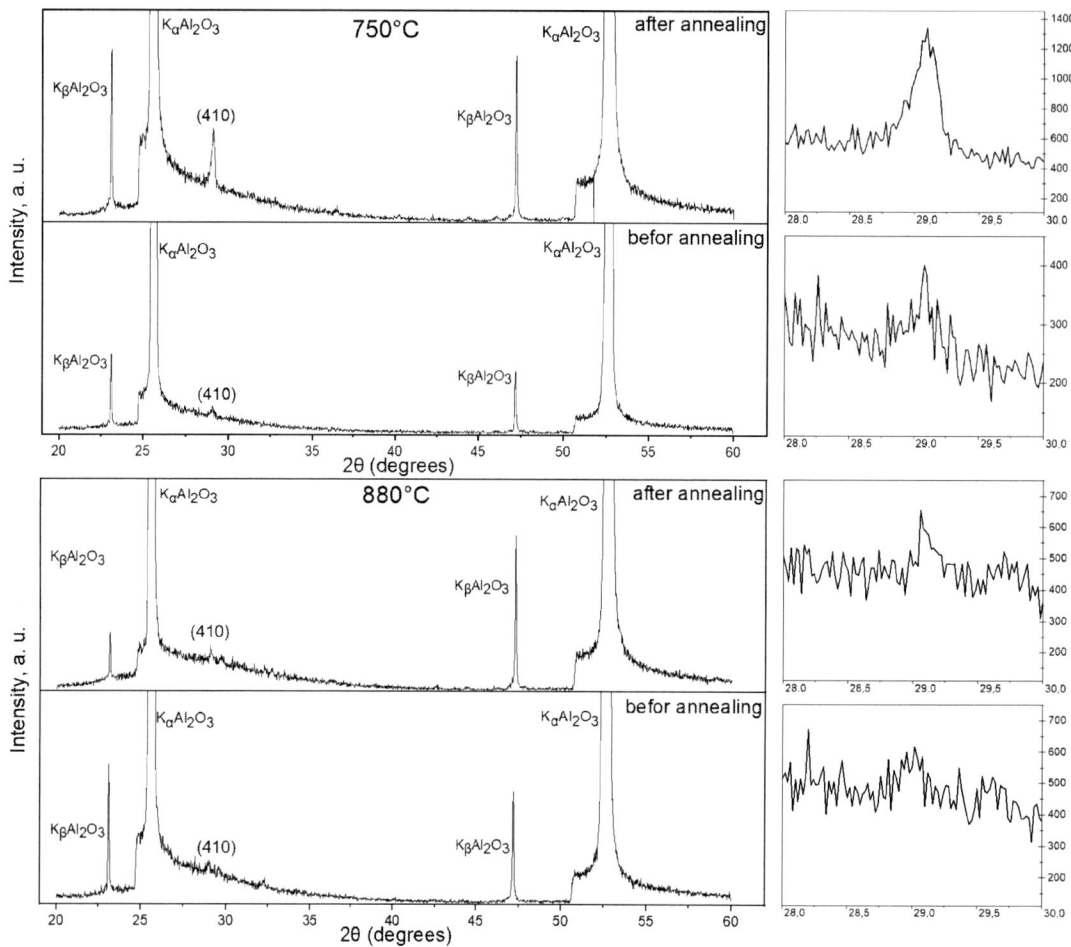

Fig. 3. Diffractograms of SBN thin films obtained at a temperature of 750°C and 880°C, respectively, before and after annealing at 1100°C.

IV. CONCLUSION

The possibility of the growth of thin films $Sr_{0.75}Ba_{0.25}Nb_2O_6$ with characteristic planes (410) and (311) parallel to the surface of substrate of sapphire is demonstrated. It is noted that only at a substrate temperature of 880°C growth is observed in both of these directions. In order to improve the structural composition of the film, high-temperature annealing in air at temperatures of 1000-1100°C was applied, as a result of which there was a sharp increase in the intensity of reflection from the plane (410) (more than 3 times). This fact gives reason to believe that the selected conditions for the synthesis of thin films are close to optimal.

ACKNOWLEDGMENT

This work was supported by the Ministry of Science and Higher Education of Russian Federation under Grant № 075-01438-22-07 - FSEE-2022-0015.

REFERENCES

[1] G.N. Tolmachev, A.P. Kovtun, I.N. Zakharchenko, I.M. Aliyev, A.V. Pavlenko, L.A. Reznichenko, I.A. Verbenko. Synthesis, structure and optical characteristics of barium-strontium niobate thin films. Solid State Physics, 2015, Volume 57, issue 10. (in Russian)

[2] Hsiu-Fung Cheng, Gong-Shing Chiou, Kou-Shung Liu, I-Nan Lin. Ferroelectric properties of (Sr0,5Ba0,5) Nb2O6 thin films synthesized by pulsed laser deposition. Elsevier Science B.V. 1997, 113/114. DOI: /10.1016/S0169-4332(96)00794-5

[3] J. de Los S. Guerra, R. G. Mendes, and J. A. Eiras, I. A. Santos, E. B. Araújo. Investigation of nonlinear dielectric properties in Sr0.75Ba0.25Nb2O6 relaxor ferroelectric thin films. American Institute of Physics, 2008, Vol. 103, DOI:10.1088/0953-8984/20/13/135209.

[4] SEUNG EON MOON, MIN HWAN KWAK, YOUNG-TAE KIM, HAN-CHEOL RYU, SU-JAE LEE, KWANG-YONG KANG. Measurement of Microwave Dielectric Properties of (Sr,Ba)Nb2O6 Thin Films. Integrated Ferroelectrics, 2004, Vol.66, pp. 275–281. DOI: /10.1080/10584580490895680.

[5] M. Melo, E.B. Araujo, A.P. Turygin, V.Ya. Shur & A.L. Kholkin. Physical properties of strontium barium niobate thin films prepared by polymeric chemical method. Ferroelectrics, №1, Vol. 496, pp. 177-186. DOI: /10.1080/00150193.2016.1155035.

[6] Boulay, M. Cuniot-Ponsard, J. M. Desvignes, A. Bellemain. Dielectric and Ferroelectric Properties of SrxBa1−xNb2O6

(SBN:x) Thin Films. Ferroelectrics, 2007, №1, Vol. 353, pp.10–20. DOI: /10.1080/00150190701367010

[7] A.V. Pavlenko, A.P. Kovtun, S.P. Zinchenko, D.V. Stryukov Structure, dielectric and optical properties of SBN-50 c-oriented films grown on a Pt/Al2O3 substrate. Letters in ZhTF. 2018, № 6, Vol. 11. pp. 31-37. (in Russian). DOI: /10.1134/S1063785018060068

[8] Alamanda V. Prasadarao, Ulagaraj Selvaraj, and Sridhar Komarnen. J. Fabrication of Sr2Nb2O7 thin films by sol-gel processing. Mater. Res., 1995, No. 3, Vol. 10, pp. 704-707. DOI: /10.1557/JMR.1995.0704

[9] Ilya M. Beskin, Sunah Kwon, Agham B. Posadas, Moon J. Kim, Alexander A. Demkov. Growth and Structure of Strong Pockels Material Strontium Barium Niobate on SrTiO3 and Si by Molecular\Beam Epitaxy. Adv. Photonics Res. 2021, №.10, Vol. 2, pp.2100111. DOI: /10.1002/adpr.202100111.

Research of Microwave Radiation Absorption of FDM Printing Materials

Ilya Chemerev
Institute of Nano- and Microsystem Technology
Moscow Institute of Electronic Technology "MIET"
Zelenograd, Russia
ichemerev@yandex.ru

Abstract— **This paper presents the results of microwave radiation absorption research for such FDM 3D printing materials as ABS, PLA, PETG. Samples for measurements were made using the FDM 3D-printer "FlyingBear Ghost 4S". Anechoic chamber AMS-8500 was used for performing the measurements in. A "П6-140-3" microwave antenna and a high frequency generator were used to generate microwave radiation. Radiation absorption was measured on frequencies ranging from 18 to 24 GHz. Also the influence of infill percentage on a microwave radiation absorption was studied during the experiments. Data gained during the research will be used in design and production of radio transparent enclosures for electronic devices via FDM 3D-printing.**

Keywords—microwave radiation, radio transparency, FDM, 3D printing, PLA, ABS, PETG

I. INTRODUCTION

Thanks to the growing popularity of 3D printing as a fast and cheap way to produce prototypes and small series of products, the variety of 3D printed cases has expanded. For example, it can be used for microwave radio electronic equipment, which includes a radiating microwave antenna.

While working on one of these devices, it was found that the device enclosure prototypes made of such filaments as acrylonitrile butadiene styrene (ABS), polylactide (PLA) and polyethylene terephthalate glycol (PETG) have different radio transparency.

In the course of studying scientific articles on the topic, it was found that radio transparency is affected by the dielectric constant of the material ε and the dielectric loss tangent tgδ [1]. However, these values may vary depending on the radiation frequency range. The values of these parameters are presented in Table 1 [2], [3].

For a more detailed research, it was decided to make samples from the mentioned filaments and compare their effect on microwave radiation in the K-band.

TABLE I. DIELECTRIC PARAMETERS OF FILAMENTS

Filament	Band type	ε	tgδ
PLA	Ka-band	2,27	$6{,}2 \cdot 10^{-3}$
PETG	Ka-band	2,47	$1{,}6 \cdot 10^{-3}$
ABS	S-band	2,45	$4{.}15 \cdot 10^{-3}$

II. MATERIALS AND METHODS

3 experiments were carried out:

Experiment 1. Determination of the most radio transparent filament among the studied

Experiment 2. Determining the effect of sample thickness on radio transparency

Experiment 3 Determination of the influence of the infill volume (in %) on the sample radio transparency

The filament materials chosen for the study were PLA, PETG and ABS. All of these filaments are available on the market, can be printed by most FDM 3D printers, and are widely used in enclosure prototyping. Also, ABS plastic has been used for many years for the mass production of enclosures for electronic devices by injection molding.

Samples for research are plates 113x113 mm, its thickness varies from 1.5 to 4.5 mm and infill from 5 to 100%, the infill pattern is rectilinear. Files for printing were prepared using Prusa Slicer 2.4.2.

The samples were printed on a FlyingBear Ghost 4S FDM 3D printer with a 0.6mm nozzle and a 0.3 mm layer height. The print speed, airflow and temperature settings were adjusted according to the recommendations for printing with PLA, PETG and ABS filaments.

A total of 5 samples were printed for each filament:

1) *Thickness 1.5mm, 100% infill*
2) *Thickness 3mm, 100% infill*
3) *Thickness 4.5mm, 100% infill*
4) *Thickness 3mm, 5% infill*
5) *Thickness 3mm, 10% infill*

The determination of the most radio transparent filament was carried out on samples of 3 mm thickness, because most enclosures for electronic devices available for purchase have a 3mm thick wall.

Result of measurements in the first approximation are the values of the peak gain of the antenna, and in the second approximation - the radiation pattern of the antenna.

The study of radio transparency was carried out in an anechoic chamber AMS-8500. The source of microwave radiation was the antenna «П6-140-3», with a frequency range of 18-26.5 GHz, which corresponds to the K-band. Measurements were made at frequencies of 18, 20 and 24 GHz

For measurements, the test sample was mounted on the antenna using paper adhesive tape, after which the antenna rotated in the ground plane with a 1 degree step. The scheme of the measuring setup is shown in Fig.1.

979-8-3503-8358-4/23 $31.00 © 2023 IEEE

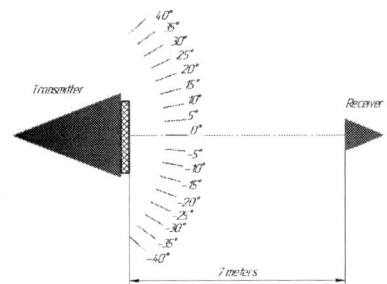

Fig. 1. Scheme of measurement setup. Top view.

III. EXPERIMENT RESULTS

As a result of the measurements, tables of maximum peak gain values (Tables 2–4) and microwave radiation patterns were obtained. Allowable deviation of the peak gain of the antenna with a plate is not more than -1 dB at each frequency.

A. Experiment 1. Determining the most radio transparent filament among the studied

TABLE II.   RESULTS OF EXPERIMENT 1

Filament	Sample	Frequency, GHz		
		18	20	24
		Maximum gain, dB		
-	No sample	-25.78	-27.02	-25.87
PLA	3mm 100% inf	-26.89	-27.66	-25.78
PETG	3mm 100% inf	-27.10	-27.71	-25.72
ABS	3mm 100% inf	-26.69	-27.20	-25.61

The table shows that at a frequency of 18 GHz, ABS has the best radio transparency, at frequencies of 20 and 24 GHz the peak gain does not change by more than -1 dB, and to determine the most radio transparent material for these frequencies it is necessary to look at the radiation patterns (Fig. 2, 3). The closer the pattern of the antenna with a samle plate to the pattern of the antenna without a sample, the higher the radio transparency.

It can be seen from the plots that for a frequency of 20 GHz, ABS is the most radio transparent filament over the entire gap, and for 24 GHz, all plastics have similar radio transparency in the range of [-12; +13] degrees, beyond which all filaments significantly degrade the signal.

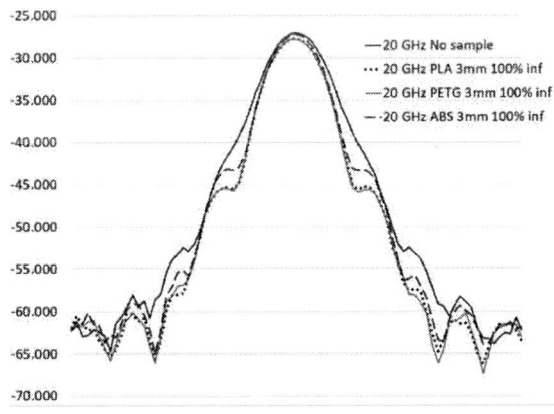

Fig. 2. Pattern of the microwave radiation of the antenna at a frequency of 20 GHz.

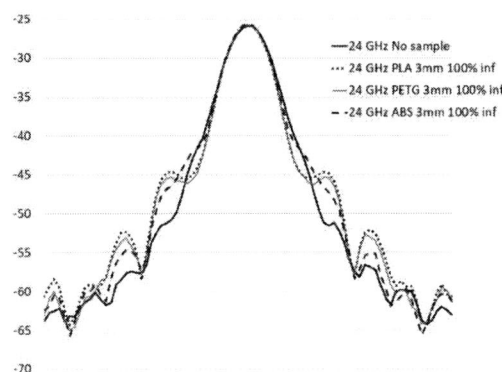

Fig. 3. Pattern of the microwave radiation of the antenna at a frequency of 24 GHz.

B. Experiment 2. Determining the effect of sample thickness on radio transparency

TABLE III.   RESULTS OF EXPERIMENT 2

Filament	Sample	Frequency, GHz		
		18	20	24
		Maximum gain, dB		
-	No sample	-25.78	-27.02	-25.87
PLA	1.5mm 100% inf	-26.27	-27.71	-26.37
	3mm 100% inf	-26.89	-27.66	-25.78
	4.5mm 100% inf	-25.81	-26.98	-26.28
PETG	1.5mm 100% inf	-26.64	-27.79	-26.75
	3mm 100% inf	-27.10	-27.71	-25.72
	4.5mm 100% inf	-25.97	-27.00	-26.44
ABS	1.5mm 100% inf	-26.48	-27.60	-26.28
	3mm 100% inf	-26.69	-27.20	-25.61
	4.5mm 100% inf	-25.89	-26.96	-26.19

The table shows that at a frequency of 18 GHz, there is a decrease in gain by more than -1 dB for PLA and PETG samples with a thickness of 3 mm, while ABS satisfies the tolerance over the entire thickness range. At 20 and 24 GHz, all materials across the entire thickness range meet tolerance, but at 20 GHz, each 4.5 mm thick sample and at 24 GHz, each 3 mm thick sample increases output gain by 0.02 to 0.26 dB. For a deeper analysis, the radiation patterns of the microwave radiation of the antenna with plates made of the studied filaments at frequencies of 20 and 24 GHz were plotted, Fig. 4-9.

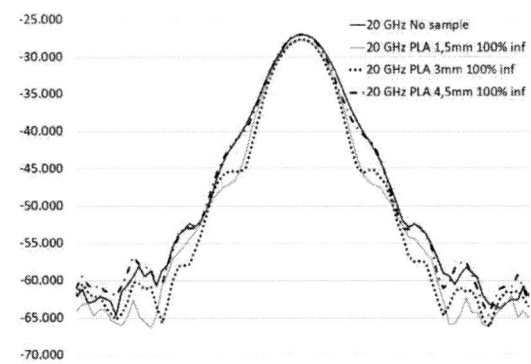

Fig. 4. Pattern of the microwave radiation of the antenna with a PLA plate at a frequency of 20 GHz

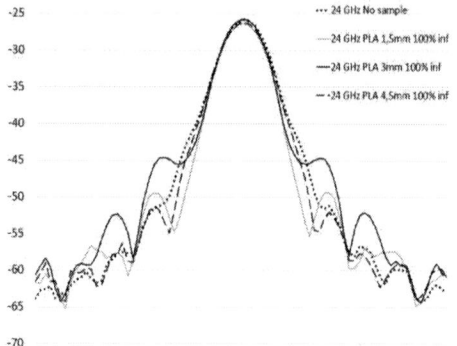

Fig. 5. Pattern of the microwave radiation of the antenna with a PLA plate at a frequency of 24 GHz

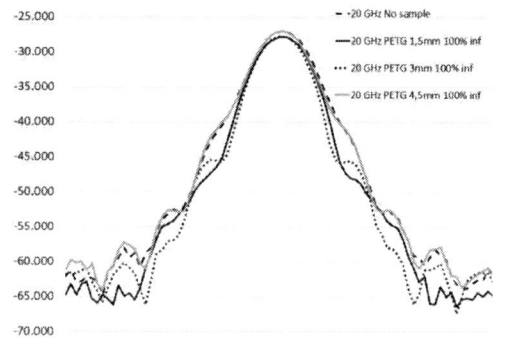

Fig. 6. Pattern of the microwave radiation of the antenna with a PETG plate at a frequency of 20 GHz

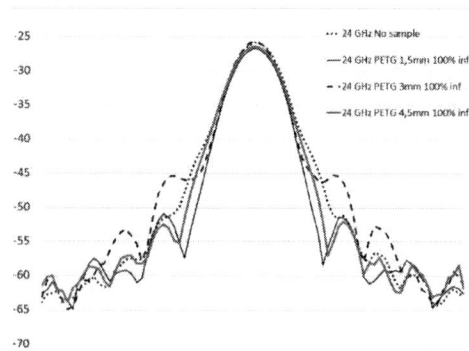

Fig. 7. Pattern of the microwave radiation of the antenna with a PETG plate at a frequency of 24 GHz

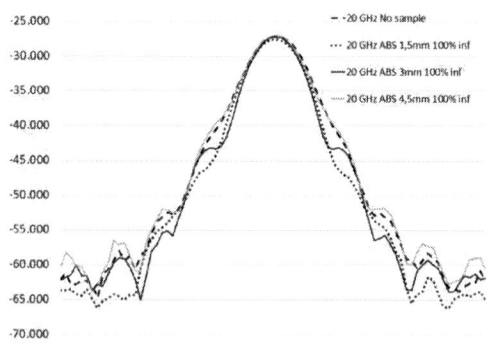

Fig. 8. Pattern of the microwave radiation of the antenna with a ABS plate at a frequency of 20 GHz

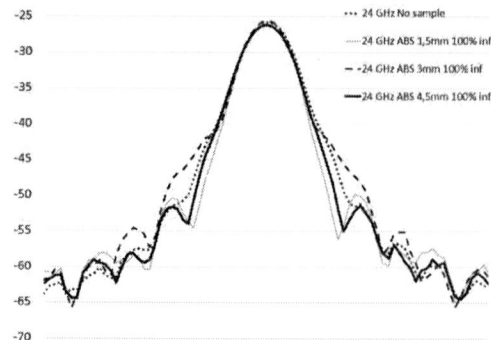

Fig. 9. Pattern of the microwave radiation of the antenna with a ABS plate at a frequency of 24 GHz

For PLA filament, it can be seen that at a frequency of 20 GHz plate with a thickness of 4.5 mm, and at a frequency of 24 GHz plate with a thickness of 3 mm narrows the radiation pattern, which increases the maximum peak radiation gain

For PETG filament, it can be seen that at a frequency of 20 GHz, a 4.5 mm thick plate narrows the radiation pattern, and at a frequency of 24 GHz, a 3 mm thick plate increases the radiation gain at peak points.

For ABS filament, it can be seen that at a frequency of 20 GHz, a 4.5 mm thick plate narrows the radiation pattern and increases the radiation gain at some peak points. At 24 GHz, the 3 mm thick plate widens the angle of the antenna beam and increases the beam power at peak points.

C. Experiment 3 Determining the effect of infill percentage of the sample on a radio transparency

TABLE IV. RESULTS OF EXPERIMENT 3

Filament	Sample	Frequency, GHz		
		18	20	24
		Maximum gain, dB		
-	No sample	-25.78	-27.02	-25.87
PLA	3mm 5% inf	-25.88	-27.02	-25.81
	3mm 10% inf	-25.99	-27.08	-25.87
	3mm 100% inf	-26.89	-27.66	-25.78
PETG	3mm 5% inf	-26.06	-27.03	-25.80
	3mm 10% inf	-26.09	-27.05	-25.78
	3mm 100% inf	-27.10	-27.71	-25.72
ABS	3mm 5% inf	-26.03	-27.09	-25.89
	3mm 10% inf	-26.03	-27.01	-25.80
	3mm 100% inf	-26.69	-27.20	-25.61

Prior to experiment 3, it was hypothesized that a higher percentage of infill would reduce the peak output gain of the antenna.

As a result of measurements, the hypothesis was fully confirmed for all filaments at a 18 GHz frequency, for PLA and PETG at a 20 GHz frequency. For the ABS at a 20 GHz frequency the hypothesis was only partially confirmed, and at a 24 GHz frequency the hypothesis was not confirmed for any filament, fig. 10. For a deeper analysis, the radiation patterns of the microwave radiation of the antenna with plates of the filaments under research with different infill amount at a 24 GHz frequency were plotted, Fig. 11 – 13.

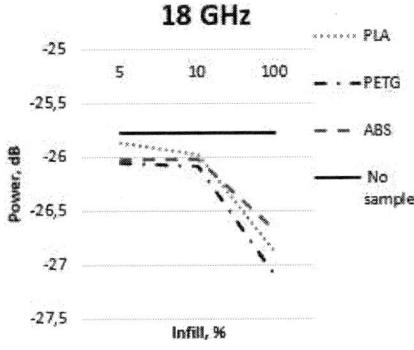

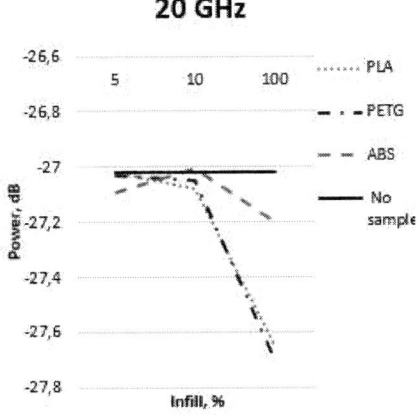

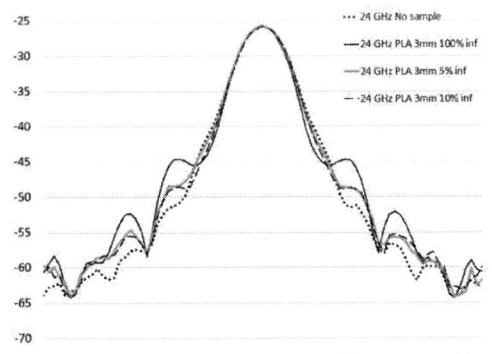

Fig. 10. Effect of infill quantity on peak power

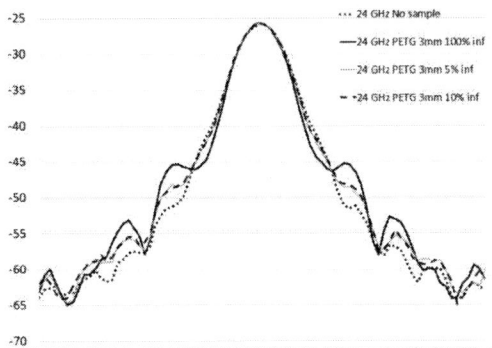

Fig. 12. Pattern of the microwave radiation of the antenna with a PETG plates with different infill % at a frequency of 24 GHz

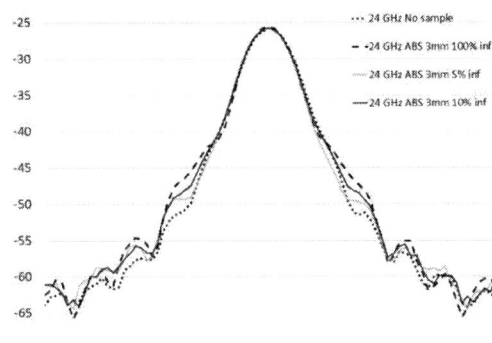

Fig. 13. Pattern of the microwave radiation of the antenna with an ABS plates with different infill % at a frequency of 24 GHz

In a more detailed examination of the radiation patterns for the frequency of 24 GHz, it was determined that at 5 and 10% infill of the samples, the resulting pattern was closest to the original pattern, while fully filled samples have strong deviations (about 5 dB) at the peak points of the patterns.

This is probably connected to an increase in the dielectric constant of the samples with increasing filling. In [4], it is said that with an increase in the infill % of the model for ABS plastic, an increase in the dielectric constant occurs, which can lead to gain losses.

Also, for all the work done, one general conclusion can be drawn that the test samples are microwave lenses that are capable of changing the shape of the antenna radiation. It is shown by the obtained radiation patterns for all three experiments.

Using the obtained data, data from early studies in this area [5], and the known features of FDM 3D printing, it can be assumed that in order to obtain the most efficient microwave lenses, printing with several materials should be used to obtain a multilayer flat lens with different dielectric constants in the layers. Such a lens will be able to focus the power of the antenna at the certain point, or vice versa, dissipate power and as a result it will increase the radiation angle.

ACKNOWLEDGMENT

Special thanks to Konstantin Knyazev for measurement assistance

REFERENCES

[1] N. Korobova, "Piezoelectric MEMS technologies" in the book Advanced Piezoelectric Materials. Science and Technologies. Second ed. Kenji Uchino. Elsevier. 2017. Pp.533-574.

Fig. 11. Pattern of the microwave radiation of the antenna with a PLA plates with different infill % at a frequency of 24 GHz

[2] Sobolev D.I., Proyavin M.D., Parshin V.V., Belousov V.I., Ryabov A.V. 3D-printed wideband windows with low reflection for high frequency radiation. VserossiyskayaNauchno-TechnicheskayaKonferencia "ElectronikaImikiroelectronikaSVCH" [Russian scientifical and technichal conference "Electronics and Microelectronics of High Frequencies"], Saint Petersburg, 2021, pp. 52-56.(In Russian)

[3] Demidemko E.V., Kuzmin S.V., Kirik D.I. 3D printing of antenna-feeder devices using polymer materials. VserossiyskayaNauchno-TechnicheskayaKonferencia "ElectronikaImikiroelectronikaSVCH" [Russian scientifical and technichal conference "Electronics and Microelectronics of High Frequencies"], Saint Petersburg, 2018, pp. 491-495.(In Russian)

[4]

Balashov A.Yu., Gyul'magomedov N. H.; Ermilov A.S. The use of 3D printing in the development of structures withradio transparent properties. VIIMezhdunarodnayakonferenciya"Additivnyetekhnologii:nastoya-shchee i budushchee" [VII International Conference "Additive technologies: present and future"], Moscow, 2021, pp. 180-189. (In Russian)

[5] S. Zhang, R. K. Arya, S. Pandey, Y. Vardaxoglou, W. Whittow, R. Mittra. 3D-printed planar graded index lenses. IET Microwaves, Antennas & Propagation, 2016, V. 10, Iss. 13, pp. 1411-1419.

Design and Fabrication of Normally-on and Normally-off Ultrathin AlN-based Transistors

Olga B. Chukanova
Chair of Quantum Physics and Nanoelectronics
National Research University of Electronic Technology
Moscow, Russia
kukhtuaeva@mail.ru

Vladimir I. Egorkin
Chair of Quantum Physics and Nanoelectronics
National Research University of Electronic Technology
Moscow, Russia
egorkinvi1962@mail.ru

Valery E. Zemlyakov
Chair of Quantum Physics and Nanoelectronics
National Research University of Electronic Technology
Moscow, Russia
vzml@rambler.ru

Maxim N. Zhuravlev
Chair of Quantum Physics and Nanoelectronics
National Research University of Electronic Technology
Moscow, Russia
maxim@org.miet.ru

Abstract— This article demonstrates simulation of normally-off and normally-on GaN-based transistors with ultrathin AlN layer. Influence of the heterostructure parameters on the transistor's characteristics has been researched. Verification of modeling parameters by experimental data has been carried out. It's shown that polarization charge at the AlN/GaN interface and the transistor characteristics strongly depend on AlN thickness and the heterostructure growth conditions. The device with 3 nm AlN layer is normally-off and has the maximum drain current about 430 mA/mm at 0 V on the gate electrode. At the same time the normally-on one with 7 nm AlN layer has about 900 mA/mm maximum drain current at 3 V gate voltage.

Keywords— *gallium nitride, amplifier, HEMT, AlN/GaN heterostructure*

I. INTRODUCTION

GaN technology are widely used today. AlGaN/GaN transistors are excellent candidates for millimeter wave power amplifiers. Next problems should be decided to increase the frequency range: to reduce the geometric transistor dimensions and to increase maximum drain current of transistor and breakdown voltage. These tasks can be decided using AlGaN/GaN FET (field effect transistor) technology. But growth of AlN layer less then 10 nm leads to heavily degradation of channel carrier density and accordingly to poor the device characteristics. Therefore, the task to ensure high channel carrier density, high breakdown voltages and high frequencies at the same time is difficult. Moreover, these devices should be inexpensive. i.e., they need be formed on heterostructure based on silicon wafer. This article demonstrates transistor based on ultrathin heterointerface AlN/GaN to solve aforementioned problems.

II. SIMULATION FEATURES OF ALN/GAN HETEROSTRUCTURES

In these heterostructures channel is on heterointerface AlN/GaN. The huge bandgap of the materials and high interface charge density σ leads to channel with high charge density (about a few units 10^{13} cm^{-2}) without barrier layer doping. The channel carrier density can be changed varying the barrier layer thickness. Therefore, it is important to simulate the dependence of the channel carrier concentration on the barrier layer thickness $Nch(h)$. It is necessary to solve quantum-mechanical task to calculate the charge concentration, but [1] shows that the same result can be obtained if we restrict ourselves to using the simplest classical diffusion-drift model. Fig.1 demonstrates calculation of channel charge density on barrier layer thickness for different mole fractions of AlN obtained as result of a numerical solution in Sentaurus Technology Computer-Aided Design (TCAD) (solid line) and as a result of calculations in "mathematics" according to (1).

$$\sigma(x) = 0.0483x \cdot$$
$$\cdot \left\{ \left[\frac{5x + 103}{-32x + 405} \right] \cdot (0.73x + 0.73) + (0.11x + 0.49) \right\} + \tag{1}$$
$$+0.052x$$

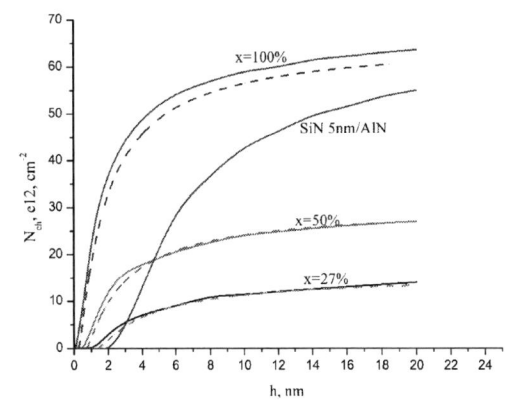

Fig. 1. Dependence of channel carrier density *Nch* on barrier thickness *h* for Al$_x$Ga$_{1-x}$N/GaN heterostructure obtained as a result of a numerical solution in Sentaurus TCAD (solid line) and as a result of calculations in "mathematics" according to (1) for Al mole fraction *x* 0.27, 0.5 and 1.0

Fig.1 demonstratrates that we can get channel carrier density in AlN/GaN heteristructures about $6 \cdot 10^{13}$ cm^{-2}. Actually, it's not like that. Increase of barrier layer Al mole fraction results in lattice mismatch increasing, formation of deformation in barrier layer and decreasing of mobility μ. [2] points out that thikcness of AlN 7 nm is critical and thicker layer has a lot of dislocation and cracks that are visible to the naked eye (Fig.2). It is important for this structure to find "window" of high mobility and low sheet resistance. The electron mobility on the interface AlN/GaN is limited by optical and acoustic phonons scattering [3]. This scattering is integral process and the only way to increase the mobility is to decrease sheet resistance. Consequently, it is necessary to adapt of the structure growth conditions so that the surface

979-8-3503-8358-4/23 $31.00 © 2023 IEEE

roughness is low, in particular to reduce the growth temperature. This feature can be taking into account at simulation with parametr "Polarisation Activation" on the interface AlN/GaN, which is equal 1 in Sentaurus TCAD by default. This parametr is difficulty calculated so it can be choosen experimentally woth TCAD. However it is necessary to take the structures of the same manufacturers as this parameter depends on the growth conditions.

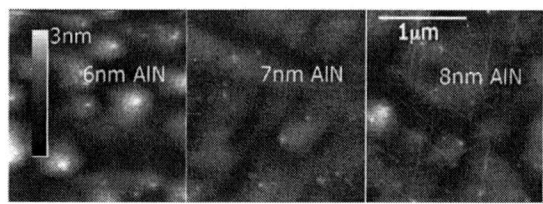

Fig. 2. AFM images showing scans of the AlN surfaces after growth [2].

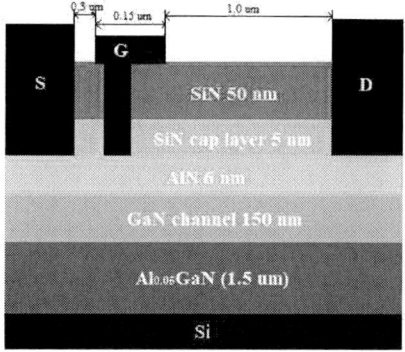

Fig. 3. Cross section of the simulated AlN/GaN-on-Si HEMT.

First of all structure from [4] have been considered. [4] demonstrates French scientists experiment results, selected parameters of simulation and the influence of the parameters on the characteristics of the device.The structure includes next layers: PECVD SiN 50 nm, In-situ SiN cap layer 5nm, AlN 6 nm, buffer GaN 1 um, high-resistance substrate Si. The gate length is 0.15 um, the gate-to-drain length is 0.3 um and the gate-to-source length is 1.0 um. The basic parameter "Polarisation Activation" is 0.35 in this case.

The spontaneous and piesoelectric polarization in GaN structures leads to the presence of an interface charge even at boundary interface between electrodes. Passivation layer compensates this charge. Accordingly to technology passivation can be incomplete, i.e. uncompensated charge, which will affect on channel charge density, remains on the interface after passivation. Passivation can be simulated with different ways. First way is describing of states concentration and energy in the bandgap. The second way, more simplified, is describing a fixed charge on the interface. The dependence of current-voltage characteristics on fixed charge has been researched with simulation of structure. For the researched structure a fixed charge 8.6e13 cm^{-2} corresponds complete passivation. The channel disappears between electrodes at 6e13 cm^{-2} fixed charge and a parasitic channel forms at 10e13 cm^{-2}. For our case we use 7.5e13 cm^{-2} fixed charge.

As well influence of AlN thickness on the device type and its characteristics have been researched (Fig.4). To form normally-off transistor based on this structure AlN layer should be less then 3 nm [5]. To estimate a critical AlN

thickness a statistical one-dimensional problem based on the solution of the Poisson equation has been solved. In case 6 nm AlN layer the transistor is normally-on and has 2.5 A/mm drain current at the 4 V gate voltage, cutoff voltage about -3V, slope of 641 mS/mm at the work point (drain voltage is 2 V, gate voltage is -2.37 V), which doesn't contradict the results presented in the article [4].

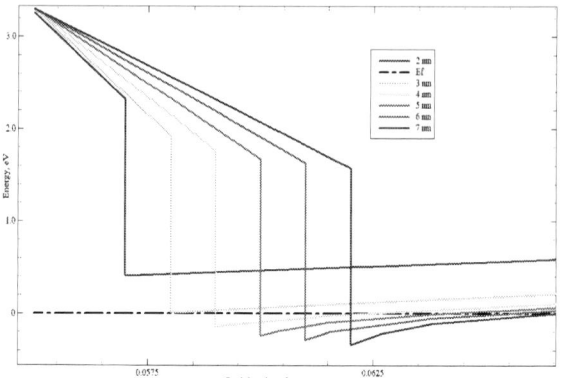

Fig. 4. Influence of AlN thickness on the heterostructure band diagram (solid – conduction band, dotted – Fermi level): blue – 2 nm, green – 3 nm, light blue – 4 nm, pink – 5 nm, grey – 6 nm and purple – 7 nm. Blue and green lines correspond to normally-off transistors.

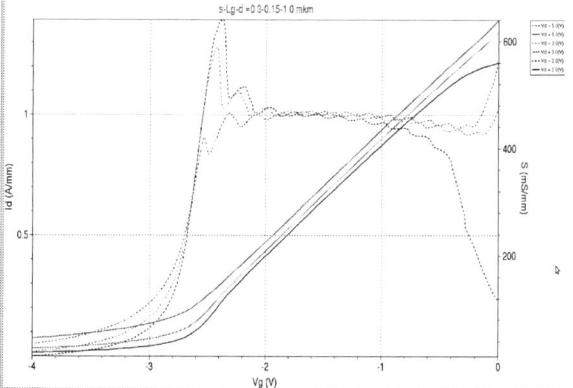

Fig. 5. Transfer characteristics (solid) and transconductance (dotted) of transistor (Fig.3) at various drain voltages: blue – 5V, red – 3V and black – 2V. The transconductance is 641 mS/mm at the work point gate voltage of -2.37 V and drain voltage of 2V.

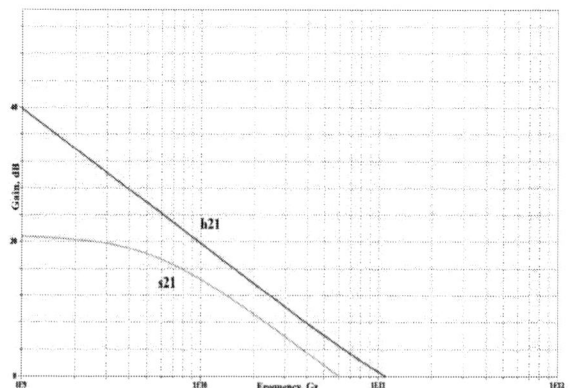

Fig. 6. Small signal analysis of structure (Fig.3): red – h21, pink – s21. Cutoff frequency is 96 GHz and gain is 20 dB.

Also small-signal analysis and calculation of microwave characteristics have been calculated, namely, s21 and h21.At the selected operating point the cutoff frequency is 96 GHz and gain is about 20 dB (Fig.6), which also doesn't contradict the article result.

III. Experiment and Verification of Model

For our research GaN-on-Si and GaN-on-sapphire substrates have been grown. The device formation technology is standard for GaN. First of all, an inter-device isolation is formed, then Ti/Al/Ni/Au ohmic contacts are formed on AlN barrier layer for which Si_3N_4 is first etched. Then ohmic contacts are get rapid thermal annealing at temperature 850°C. Next, Si_3N_4 passivation is deposited. It is bled off in the required place for the gate leg, and the Ni/Au gate metal is deposited. So, metallization of first and second levels are formed.

Transistor with gate length of 1 um, gate-to-source length 1.5 um and gate-to-drain length 3.5 um has been formed on the PECVD SiN 50 nm/in-situ SiN cap layer 5 nm/AlN 3 nm/buffer $Al_{0.05}GaN$ 1,5 um / high-resistive Si substrate. It is normally-off. Also, we have managed to form normally-on transistor with gate length of 0.15 um, gate-to-source length 0.3 um and gate-to-drain length 1 um based on GaN-on-sapphire heterostructure with 7 nm AlN layer and normally-off one with the same design based on GaN-on-sapphire heterostructure with 2.5 nm AlN layer.

Having processed experimental data a verification and calibration of the models have been carried out. "Polarization Activation" parameter for our structures is equal to 0.22 and the fixed charge on the interface of passivation - 8.25e13 cm-2. According to the simulation the transistor based on heterostructure with 7 nm AlN layer is normally-on and has maximum drain current in opened state of 640 mA/mm at 3 V on the gate electrode and cutoff voltage of -2V (Fig.7). The device based on the heterostructure with 3 nm AlN is normally-off and has maximum drain current of 400 mA/mm at 2V gate voltage and cutoff voltage of 0.2 V. Also 2.5 nm AlN transistor is normally-off and has 300 mA/mm maximum drain current at 2V gate voltage and 0.2 V cutoff voltage(Fig.8). The simulation results are different from experimental data by no more than 16% (Fig.9). But our mathematical model doesn't taking into account temperature since the inclusion of it deteriorates the convergence of the problem and the result will directly depend on a type of thermal contacts which difficulty compared with the experiment. However the simulation parametrs can be choosen exactly by verification of simulation with the experimental data so the simulation and experiment results will differ by no more than 16%.

IV. Conclusoin

Thus, the numerical simulation of normally-on and normally-off devices based on ultra-thin AlN/GaN heterointerfaces has been carried out. Their current-voltage characteristics have been calculated. We have showed that theoretical calculation of channel carrier density can be heavily differ from reality as conditions of structure growth aren't taking into account, namely, ultra-thin AlN layer. Due to the strong difference between the band gaps of AlN and GaN in such structures grown AlN is amorphous with many defects and at a thickness of more than 7 nm it can even crack. All this leads to the presence of roughness at the AlN/GaN heterojunction and high sheet resistance, as a result, to a low concentration of charge carriers in the channel

compared to the possible maximum. The generated devices have good characteristics, but, as has been shown, such devices can have much higher drain currents, transconductance, and microwave characteristics. Therefore, it is necessary to work out the conditions for the growth of structures, to aim a larger main parameter "Polarization Activation" and to develop ways to reduce the sheet resistance in such devices.

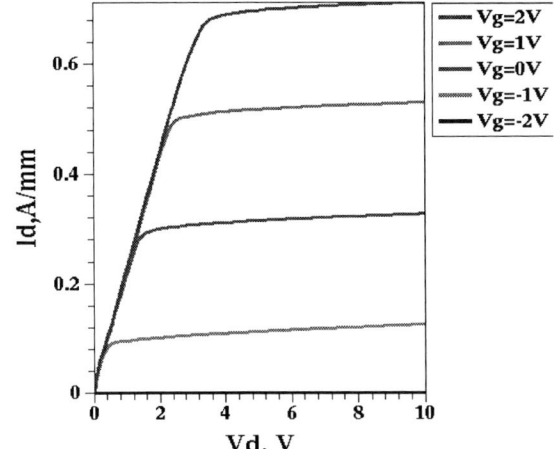

Fig. 7. Simulated current-voltage characteristics of transistor with 7 nm AlN

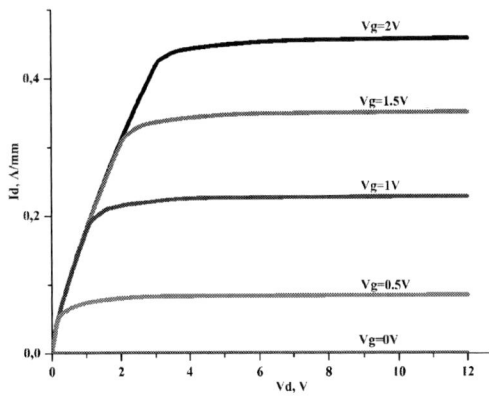

Fig. 8. Simulated current-voltage characteristics of transistor with 3 nm AlN

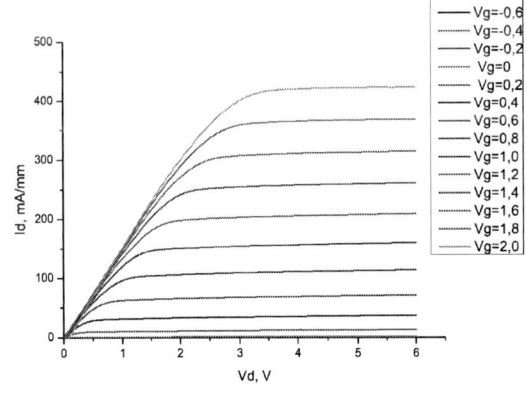

Fig. 9. Experimental current-voltage characteristics of transistor with 3 nm AlN: results differ by no more 16%

REFERENCES

[1] Lee K.S., Yoon D.H., Bae S.B., Park M.R., Kim G.H. Self-Consistent Subband Calculations of AlGaN/GaN Singl Heterojunctions. *ETRI Journal*, 2002, no. 4 (24), pp. 270-278.

[2] Cao Y., Jena D. High-mobility window for two-dimensional electron gases at ultrathin AlN/GaN heterojunctions. *Applied Physics Letters*, 2007, V. 90, p. 182112.

[3] Cao Y., Wang K., Orlov A., Xing H., Jena D. Very low sheet resistance and Shubnikov-de-Haas oscillations in two-dimensional electron gases at ultrathin binary AlN/GaN heterojunctions. *Applied Physics Letters*, 2008, V. 92, p. 152112.

[4] Medjdoub F., Zegaoui M., Waldhoff N., Grimbert B., Rolland N.,Rolland P. Above 600 mS/mm Transconductance with 2.3 A/mm Drain Current Density AlN/GaN High-Electron-Mobility Transistors Grown on Silicon. *Applied Physics Express 4*, 2011, V. 4, p.064106.

[5] Harrouche K., Kabouche R., Okada E., Medjdoub F. High Power AlN/GaN HEMTs with record power-added-efficiency >70% at 40 GHz. IEEE/MTT-S International Microwave Symposium (IMS). Los Angeles, CA, USA, 2020, pp. 285-288.

Investigation of Mechanical Characteristics of Pressuring Materials of Generators Stator Winding in Thermal Aging Process

Nikita A. Fedotov
Institute of Energy
Peter the Great St. Petersburg
Polytechnic University
St. Petersburg, Russia
nifed4@gmail.com

Tatyana M. Shikova
Institute of Energy
Peter the Great St. Petersburg
Polytechnic University
St. Petersburg, Russia
t_shikova@mail.ru

Victor O. Belko
Institute of Energy
Peter the Great St. Petersburg
Polytechnic University
St. Petersburg, Russia
vobelko@gmail.com

Andrey M. Kostelov
Power machines – ZTL, LMZ,
Electrosila, Energomachexport
St. Petersburg, Russia
a.m.kosteliov@gmail.com

Emil R. Mannanov
Power machines – ZTL, LMZ,
Electrosila, Energomachexport
St. Petersburg, Russia
emil-mannanov@mail.ru

Dmitriy A. Chernyshov
Peter the Great St. Petersburg
Polytechnic University
St. Petersburg, Russia
mr.chernyshev.dima@mail.ru

Abstract - During the exploitation of the generator, forces in the radial and tangential direction are applied to the stator winding of the generator. These forces can lead to vibration of the stator winding. Vibration is one of the main conditions to occurrence Vibration Sparking (VS). VS is failure mechanism of the stator insulation system therefore stator winding has to effectively consolidate by pressure springs to avoid vibration. While operating of generator pressuring materials are exposed to temperature around 120 °C. This temperature can reduce their mechanical properties due to thermal aging process. Consequently, the method for determining mechanical characteristics of pressuring materials in the thermal aging process is developed. Thermal aging was carried out at temperatures 130,155 and 180°C .As a result the trend of changes mechanical properties of pressuring springs in thermal aging process is obtained.

Keywords – Insulation system, slot partial discharge, vibration sparking, pressuring ripple springs, mechanical characteristics, thermal ageing

I. INTRODUCTION

Insulation system of high-voltage turbogenerators conditionally is divided into two components including groundwall insulation of stator winding and corona protection system (CPS) consisting of conductive coating and pressuring side ripple springs. CPS defends groundwall insulation against electrical failure mechanism such as Slot Partial Discharge (SPD) and Vibration sparkling (VS). SPD occurs when electric field strength between the groundwall insulation and stator core is more than electrical strength of air. VS takes place with vibration of stator winding when the surface of groundwall insulation lose points of contact with stator core whereby arc ignites [1,2]. VS is aggressive failure mechanism because of temperature of arc reaches 2000 °C and it can burn out conductive coating in the arc area (Fig. 1) [1-6].

During the operation of the turbogenerator, the stator winding is exposed to vibrations due to electromagnetic forces in the radial and tangential directions [7,8]. This indicates that the occurrence of vibrations of the stator core winding remains a problem demanding for study and requiring a solution for more reliable performance of high-voltage insulation.

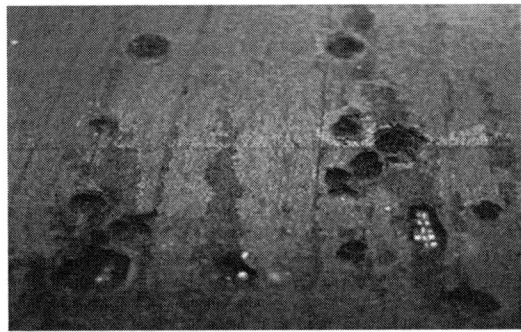

Fig. 1. Local outburning of conductive coating by VS [2].

In the work of Dombrowski et al. [7] describes the forces acting on the stator bar in the radial and tangential directions. The forces acting on the stator bar in the radial direction arise due to the flow of electric current through the conductor of the stator rod winding. Under the influence of the ampere force, depending on the direction of the current, the conductor can start moving in two directions: either in the direction down the slot, or in the direction towards the wedge. The stator bars moving is a real cause of radial direction vibration. The forces acting on the stator winding in the tangential direction appear because of the displacement of part of the stator magnetic flux due to the steel teeth saturation. This force can press the stator winding in the direction of the rotor rotation and has a pulsating character, which can cause vibrations in the tangential direction. [8] shows using finite element modeling (FEM) to determine the radial and tangential forces induced in a real generator. The forces arising in the radial and tangential directions don't provide such values that will cause vibrations of the stator winding in the slot when using pressuring materials in the initial state.

Consequently, construction of stator slot includes the pressuring side ripple springs to eliminate the vibration of stator windings.

979-8-3503-8358-4/23 $31.00 © 2023 IEEE

These materials consist of fiberglass and epoxy-resin, thickness h=0,4-1 mm, and have a form of wave as shown in the Fig.2

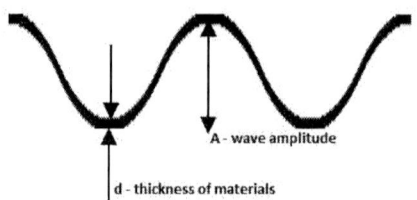

Fig. 2. Construction of pressuring materials.

Pressuring materials reliably consolidate stator winding and don't give the opportunity to creating vibration. However, operating temperature of stator slot is usually equal to 120 °C. Therefore, mechanical characteristics of pressuring materials can change while exploitation of generator. Strong decrease in mechanical characteristics can lead to VS. Nowadays, we don't have enough literature data concerning change in the mechanical characteristic of pressuring materials in thermal aging process. And it's aim of our work.

II. MATERIALS AND METHODS

A. Materials

The investigations were conducted on manufacturing samples of stator bar pressuring conductive materials various thickness (0,5-1 mm).

There are four mechanical characteristics that fully describe the performance of pressuring materials are as follows specific compressive force (SCF), wave amplitude, bending strength (BS), modulus of elasticity (ME). SCF are determined by compressive loads, and the BS and ME by bending loads. The wave amplitude parameter investigated separately.

B. Compressive load

Investigations of mechanical characteristics by compressive load allows to define SCF parameter. A test sample (150*40 mm) must be placed between two metal plates (150*60 mm) then sample subjected to a compression process to determine the UUR on a MARK-10 universal bursting machine. During the sample compression, a deformation curve is recorded on the computer. Deformation curve used to determine the SCF of the material. The test scheme is shown in Fig.3.

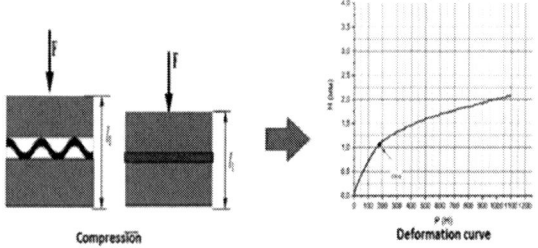

Fig. 3. Test scheme of compression.

The full compressive force is defined in deformation curve. So, SCF parameter can be calculated by:

$$\sigma_f = \frac{F}{S},\qquad(1)$$

where F is a full compressive force, N, S – Surface area, m^2.

C. Determining of wave height

The parameter wave height is determined in MARK-10 universal bursting machine, the sample is clamped between two metal plates with a force of 2.5 N. The distance remaining between the two plates shows the wave height of the material. (Fig.4) Wave height includes the amplitude of the wave and the thickness of the material as it's shown in (2). The test scheme is presented in Fig.4

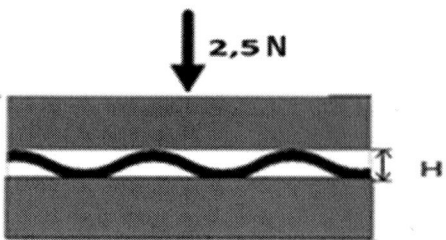

Fig. 4. Test scheme for determining WH parameter.

So, SCF parameter can be calculated by, where A is a Wave amplitude, d is a material thickness, H is a wave height:

$$A = H - d\qquad(2)$$

D. Bending load

Study of mechanical characteristics by bending load are presented in [9] allows to determine BS and ME parameters. Experimental investigations were conducted using tree-point loading (Fig. 5) system in follow conditions: $L_v = 22$ mm is distance between supports, $b = 10$ mm is samples width, $h = 0,5-1$ mm is samples thickness, $V = 10$ mm/s is moving speed of loading tip. Stress-strain curves, i.e., dependencies of the bending load F on the deflection distance, were determined for each experimental sample.

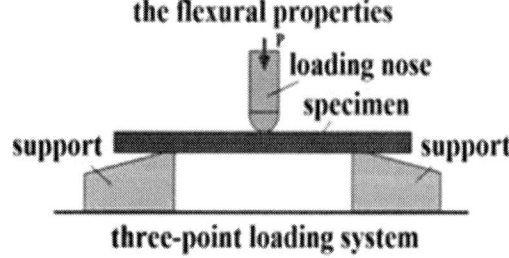

Fig. 5. Test scheme of three-point loading system

Bending stress was calculated by this expression (3):

$$\sigma_f = \frac{FL_v}{2bh^2}\qquad(3)$$

The ultimate bending strength was determined in maximum load value. Elastic modulus E_f was calculated with tangent of deformation curve inclination. To determine the mechanical characteristics the universal testing machine MARK - 10 was used.

The test samples are anisotropic materials; thus, their bending characteristics may differ depending on the direction of force application. Therefore, it is necessary to study the characteristics of bending loads in two directions: along the wave and across the wave.

E. Thermal ageing

Research of thermal ageing were conducted for various temperature depending on the loads. In case of compressive load thermal ageing temperature was selected 130, 155 and 180 °C and samples was fully clamped between the metal plates as it's shown at Fig. 6.

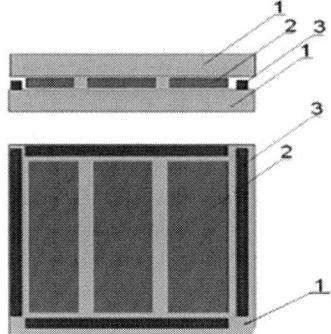

Fig. 6. Clamping device with samples.1 – metal plates; 2- samples; 3 – limited clamps.

In case of bending load thermal ageing was carried out at three temperatures including 155, 180 and 200°C in a free position.

III. RESULTS AND DISCUSSION

A. Results of compressive load

It has been experimentally established that thermal aging in the clamped state sharply reduces the SCF index aging for all materials as shown in the Fig. 7. It is obvious that in the initial state, the thicker sample has a high force value (d=1mm, $\sigma = 225 \cdot 10^3$ N/m^2) than the thinner sample (d=0,6mm, $\sigma = 37,5 \cdot 10^3$ N/m^2). Therefore, as a percentage, Sample 1 SCF indicator decreased much more and became equal to 2-4% relative to the initial state. The force of the sample 2 is reduced to 13-20% and for the sample 3 is 23-36%.

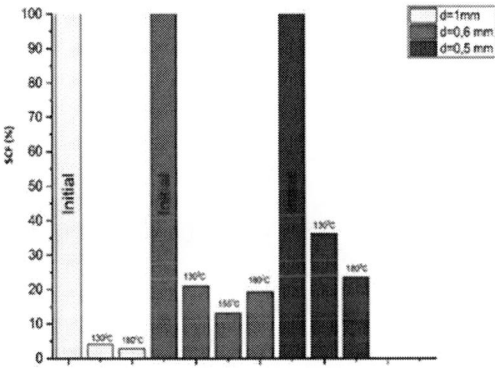

Fig. 7. Changing SCF parameters during thermal aging.

But even considering the large loss, the pressuring material has not lost its operability according to literary sources in the case of all samples. Therefore, the pressuring materials will eliminate vibration of stator bar.

Similarly, the characteristic of the wave amplitude for all samples is reduced as presented in the Fig 8. In the initial state, the wave amplitude values are approximately the same for all samples (4-5 mm). As a result of thermal aging, this indicator decreases below 20%. However, the side pressuring materials will maintain the contact points between the groundwall

insulation and the stator core even at such low values. Consequently, VS doesn't occur.

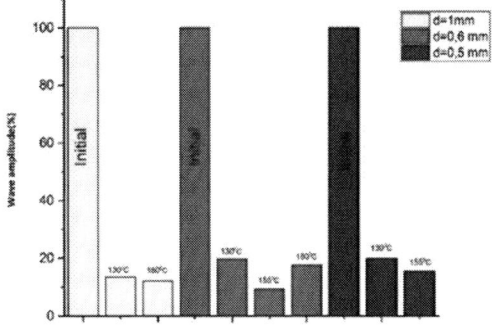

Fig. 8. Changing wave amplitude during thermal aging.

B. Results of bending load.

Studies of mechanical characteristics under bending loads don't show significant changes during thermal aging. The determination of the BS wasn't lower than 80% even after 1000 hours of aging. In the case of the elastic modulus, this indicator didn't decrease below 65% during thermal aging. A critical reduction can be considered the loss of 50% of the initial state during such tests.

IV. CONCLUSION

It is established that, pressuring materials lose their mechanical characteristics during prolonged thermal aging. If the pressuring material of the insulation system doesn't fix the stator bar and doesn't hold the contact points between the groundwall surface and the stator core, then VS will eliminate the insulation system. Now, the decrease in the characteristics of materials hasn't reached critical values capable of destroying the insulation system. Consequently, mechanical characteristics investigation is continued.

REFERENCES

[1] M. Liese and M. Brown, "Design-dependent slot discharge and vibration sparking on high voltage windings," in IEEE Transactions on Dielectrics and Electrical Insulation, vol. 15, no. 4, pp. 927-932, August 2008

[2] G. C. Stone, C. V. Maughan, D. Nelson and R. P. Schultz, "Impact of Slot Discharges and Vibration Sparking on Stator Winding Life in Large Generators," in IEEE Electrical Insulation Magazine, vol. 24, no. 5, pp. 14-21, September-October 2008, doi: 10.1109/MEI.2008.4635657.

[3] W. G. Moore, A. Khazanov, "Insulation degradation in generator stator bars due to spark erosion and partial discharge damage," IEEE International Symposium on Electrical Insulation, San Diego, CA, 2010, pp. 1-5.

[4] V. O. Belko, Y. K. Petrenya, A. M. Andreev, A. M. Kosteliov and M. B. Roitgarz, "Numerical Simulation of Discharge Activity in HV Rotating Machine Insulation," 2019 IEEE Conference of Russian Young Researchers in Electrical and Electronic Engineering (EIConRus), Saint Petersburg and Moscow, Russia, 2019, pp. 800-802.

[5] V. Belko, A. Plotnikov, Y. Petrenya and T. Shikova, "Study of Characteristics of Vibration Sparking in HV Rotating Machine Insulation," 2019 16th Conference on Electrical Machines, Drives and Power Systems (ELMA), Varna, Bulgaria, 2019, pp. 1-4.

[6] A.S. Reznic, I.O. Ivanov, A.M.Andreev, E.R.Mananov, T. M. Shikova, " The Effect of Discharge Activity on the Performance of Corona-Protective Semiconducting Coatings of the Stator Bar Insulation," 2021 IEEE Conference of Russian Young Researchers in Electrical and Electronic Engineering (EIConRus), Saint Petersburg and Moscow, Russia, 2021, pp. 1228 - 1231.

[7] I.A. Glebov, V.V. Dombrovskii, A.A. Dukshtau et al. Gigrogeneratory[Hydrogenerators],Leningrad, Energoizdat Publ., 1982, 368 p.

[8] B. Sanosian, P. Wendling, T. Pham and W. Akaishi, "Electromagnetic Forces On Coils And Bars Inside The Slot of Hydro-Generator," 2019

IEEE Energy Conversion Congress and Exposition (ECCE), 2019, pp. 1754-1760, doi: 10.1109/ECCE.2019.8913254

[9] I. O. Ivanov, A. S. Reznik, E. G. Feklistov, T. M. Shikova, N. A. Fedotov and Y. K. Petrenya, "Mechanical Characteristics Investigation of Mica-Containing Insulation of High Voltage Rotating Machines," *2022 Conference of Russian Young Researchers in Electrical and Electronic Engineering (ElConRus)*, 2022, pp. 1017-1021, doi: 10.1109/ElConRus54750.2022.9755593.

Causes of Resistance Decrease of Corona Protection Materials in Thermal Aging Process

Nikita A. Fedotov
Institute of Energy
Peter the Great St. Petersburg
Polytechnic University
St. Petersburg, Russia
nifed4@gmail.com

Tatyana M. Shikova
Institute of Energy
Peter the Great St. Petersburg
Polytechnic University
St. Petersburg, Russia
t_shikova@mail.ru

Victor O. Belko
Institute of Energy
Peter the Great St. Petersburg
Polytechnic University
St. Petersburg, Russia
vobelko@gmail.com

Alexandr S. Reznik
Institute of Energy
Peter the Great St. Petersburg
Polytechnic University
St. Petersburg, Russia
alexxxandr2803@mail.ru

Andrey M. Kostelov
Power machines – ZTL, LMZ,
Electrosila, Energomachexport
St. Petersburg, Russia
a.m.kosteliov@gmail.com

Emil R. Mannanov
Power machines – ZTL, LMZ,
Electrosila, Energomachexport
St. Petersburg, Russia
emil-mannanov@mail.ru

Abstract—During the operation of the high-voltage electrical machine partial discharge (PD) may occur in the slot of stator winding. A resistance conductive coating based on graphite is applied to the surface of wall insulation to eliminate PD. The coating resistance decreases rapidly when high temperature impact to insulation system. Low value of resistance can lead to the appearance of vibration sparking (VS). In our work, we consider a different factor affected to quality of conductive coating which include the chemical nature of the polymer matrix, the degree of curing of the binder, the degree of grinding of the filler (graphite) and the technological process of manufacturing enamel for a conducting coating. This paper shown that the unfinished process of forming the polymer matrix structure is the most probably cause of resistance instability.

Keywords— rotating machines; stator slot; anti-corona system; slot partial discharge, vibration sparking, coating resistance

I. INTRODUCTION

One of the main failure mechanisms of the insulation of the stator winding of high-voltage generators are discharges in the slot, which include partial discharge (PD) and vibration sparking (VS) [1--5]. A low-resistance coating is applied to the surface of the mica-containing insulation to eliminate the PD. It's a composite material with graphite powder as a filler. If there are enough points of electrical contact of the coating with a stator core, the current in the coating will shunt the air between the insulation surface and the stator core. Consequently, the discharge doesn't occur. But loss of contact points due to the vibration of the stator core leads to a vibration spark, destroying the insulation in the early stages of operation [5].

To eliminate these processes, restrictions have been introduced on the resistance value of the coating, which must be in a certain range during the whole operational life of high-voltage generators (5 – 500 kOhm) [4-8]. [8] shows there is a sharp decrease resistance and output beyond the permissible range already at the first stages of accelerated thermal aging (Fig. 1). This can lead to the appearance of vibration sparking. Therefore, one of the important problems include the determining the reasons of sharp decrease in the resistance of the conductive coating and finding methods to remove them.

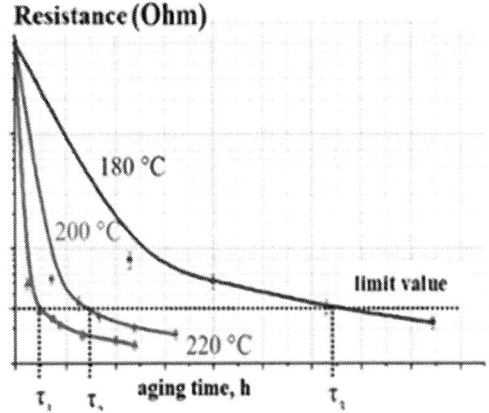

Fig. 1. Coating resistance change during thermal aging [8]

The coating is a composite material consisting of a polymer matrix and an inorganic filler - powdered graphite, which determines the conductive properties of the coating. It is known that the characteristics of such a composite material depend on the characteristics of the matrix and filler, and their quantitative content. In addition, factors such as the size and uniformity of the distribution of filler particles, the surface activity of graphite, the chemical nature, and the degree of curing of the matrix are interested in our work. At the same time, the change in the quantitative content of the components wasn't considered [8 - 12].

II. METHODS

A. Processing

Studies were carried out for two types of enamel coating, where graphite powder was used as a filler, and polymer resins with a hardener were used as a matrix: Sample 1 - polyester, Sample 2 – polyurethane resin. The enamel samples were made in laboratory conditions according to the production recipe. The enamel components were mixed manually. However, the composition was carried out additional (more intensive) mixing by a magnetic stirrer for 15-20 minutes to study the effect of the uniformity of particle distribution and to ensure the stability of conducting clusters. After manufacturing, the enamel coating was applied to the surface of mica-containing insulation plates with a size of 50*100mm or to the surface of fiberglass plates with foil electrodes. On such samples, the resistance was measured using the RLC-meter APPA 703 at room temperature.

B. Filler pre-heating

The surface of the filler was activated by pre-heating the filler in a muffle furnace at temperatures of 350, 600, 800 and 1000°C to increase the adhesion strength of filler particles and the matrix and stabilize the location of filler particles. Subsequently, the graphite powder was cooled to room temperature and used to the preparation of enamel.

C. Filler griding

Evaluation of the effect of particle size on resistance stability was researched using additional grinding of graphite powder in a ball mill. At the same time, the granulometric composition of the leading filler was determined before and after grinding using a laser analyzer of microparticles "Laska".

D. TGA analysis

The study of the thermal resistance of coatings was studied using thermal analysis methods on devices from NETZSCH (Germany). Thermogravimetric analysis (TGA), on the TG 209 F1 device, in the temperature range from 30 to 500 °C at a heating rate of 10 °/min in an oxidizing medium (air). The weight of the samples is 1-2 mg. The thermal stability of the studied samples was evaluated by the values of the thermal stability index t5, i.e., by the temperature corresponding to a 5% mass loss under TGA test conditions.

E. Monitoring of curing process

The kinetics of curing of the enamel coating was evaluated by monitoring the tangent of the dielectric loss angle (tgδ) at a frequency of 1 kHz (device) of the coating matrix applied to fiberglass in a stationary temperature regime at temperatures of 120 - 180°C.

III. RESULT AND DISCUSSION

A. Investigation of the thermal resistance of coatings by TGA methods

The TGA results show that the temperature resistance indices of the samples are in the range from 188 to 354 °C. For the coating sample 1, the lowest value of t5 is 188.4 °C and the residual mass is 38.92%. So, sample 1 has a low temperature resistance – no more than 190°C, whereas for sample 2 this characteristic is 100°C higher. Consequently, if the temperature on the insulation surface is reached, even at certain points, the value of 155°C, the operational life of the sample 1 will significantly reduce.

Based on this, we can assume that a sharp decrease in resistance during accelerated thermal aging for sample 1 is associated with thermal degradation processes, and other mechanisms operate for sample 2. However, the values of these coatings in the initial state have similar values i.e., changing the matrix doesn't affect the final resistance value. with a constant amount of filler.

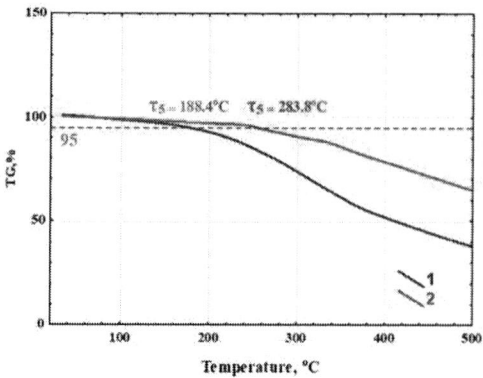

Fig. 2. Curves of TGA coatings 1 and 2

B. The effect of additional mixing of the composition

Insufficient mixing of the composition leads to uneven distribution of the filler, formation of aggregates or unstable connection of filler particles with the matrix. This can reduce the stability of the coating resistance under temperature exposure. However, a comparison of samples 1 made with manual stirring and more intensive stirring with a magnetic stirrer didn't show a difference between these options. During the drying process at a temperature of 100°C, the resistance of both coating options decreases from 190 – 200kOhm to 8-10 kOhm. In case of sample 2 (Fig. 3), the difference between the mixing options practically disappears, starting from a temperature of 100 - 120°C i.e., additional mixing doesn't have a significantly influence on the values of the coating resistance.

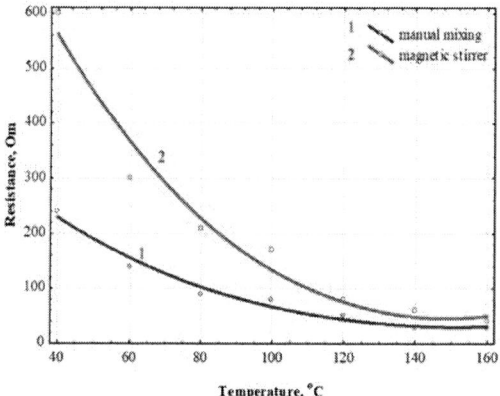

Fig. 3. The effect of the mixing method on the resistance of the coating 2 (1- manual mixing, 2 – magnetic mixing)

C. Evaluation of the effect of particle size by additional grinding

The probability of formation of conducting clusters in composite materials consisting of a matrix and a powder filler is directly related to the size of the filler particles with constant volume content [11, 12]. The average particle size changed slightly – from 10.06 to 9.44 microns during the grinding process. The resistance of the sample 2 based on the initial and grinding fillers showed no difference in values both in the initial state and in the process of thermal aging (Fig.4). Moreover, it is shown that a sharp decrease in resistance occurs already in the first 5 hours of temperature exposure. So, we can assume that this is directly related to the stabilization of the matrix – the end of the curing process.

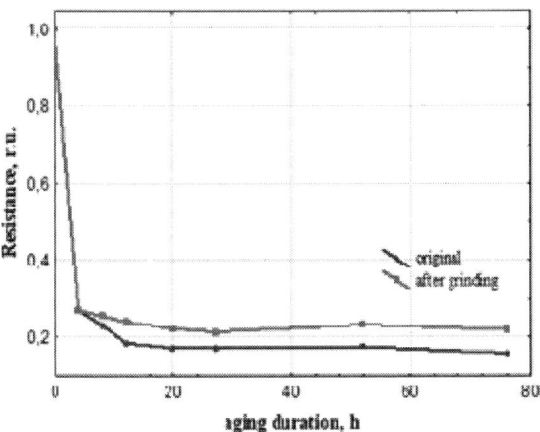

Fig. 4. The effect of additional grinding of the filler on the resistance of the coating 2 in the process of thermal aging at temperature

D. Influence of filler heat treatment

According to the literature data [11, 12, 14], pretreatment of the filler surface by calcining filled compounds and compositions containing a colloidal graphite mixture at elevated temperature is used in the manufacture. Calcining increases the strength of the components interaction and stabilizes the position of the filler particles. Calcination of the filler at temperatures of 600, 800, 1000°C didn't lead to a significant change in the resistance of the coating. Regardless of the calcination temperature, the resistance value decreases 5-10 times during the subsequent thermal aging of all samples at a temperature of 100 °C (Fig. 5) and 10-20 times at a temperature of 200°C. Thus, calcination of the filler isn't a way to stabilize the resistance of the coating.

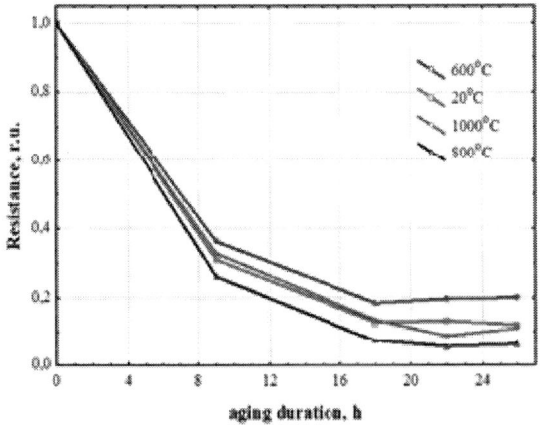

Fig. 5. Resistance changes of coating samples 2 after calcination of the filler (during heat treatment at 100°C)

It is necessary to note the low rate of resistance change during the holding of samples at room temperature and the absence of a resistance stabilization area (Fig. 6). This fact shows that the curing process doesn't conduct at the room temperature. Consequently, coating resistance can considerably decrease at the thermal aging. The kinetics of curing process is important issue to assess the effect of temperature on the coating condition.

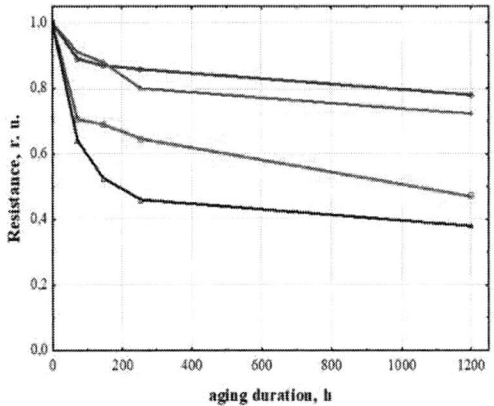

Fig. 6. Resistance changes of coating samples 2 after calcination of the filler during exposure at room temperature (notation from Fig.6)

E. Control of the curing process

The completeness of the matrix curing leads to the final fixation of filler particles and stabilization of the coating resistance. In our study, a cold-curing compound is used as a matrix. But even in this case, the final formation of the structure doesn't occur at room temperature, as indicated above (Fig.6). The curing process of the coating matrix of sample 2 is monitored in the stationary temperature mode - 120, 140, 160, 180°C (Fig. 7). At the same time, the temperature of 120°C corresponds to the minimum operating temperature. Received data show that ending of curing process conducts after 4 hours process at a temperature below 120°C. During this time, the properties of the coating may stabilize, and the sharp area decrease resistance will disappear with thermal aging.

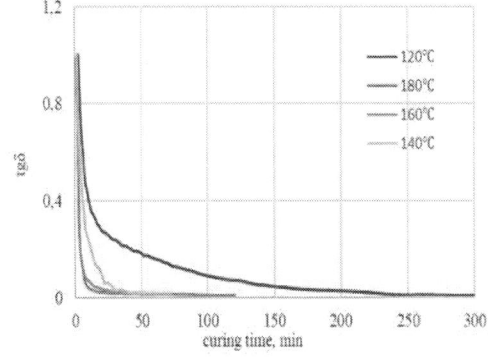

Fig. 7. Change of tgδ during the curing of the coating matrix 2 in stationary mode

IV. CONCLUSION

As a result of studies shown that additional mixing of the composition, changing the size of graphite particles and additional treatment of their surface by calcination are not the cause of a sharp decrease in the resistance of coatings at the first stage of thermal aging. The most probable issue of the instability of resistance is the incomplete process of forming the structure of the polymer matrix during thermal aging. Ending of this process can be completed at the initial stage of operation of the winding at operating temperature for 4-5 hours. The evaluation of the resistance of the coatings must be carried out after heat treatment at a temperature of 100-130°C during manufacture. The duration of heat treatment must be previously established by determining the dependence of the resistance on the time of heat treatment.

REFERENCES

[1] W. G. Moore, A. Khazanov, "Insulation degradation in generator stator bars due to spark erosion and partial discharge damage," IEEE International Symposium on Electrical Insulation, San Diego, CA, 2010, pp. 1-5.

[2] G. C. Stone and C. Maughan, "Vibration Sparking and Slot Discharge in Stator Windings," Conference Record of the 2008 IEEE International Symposium on Electrical Insulation, Vancouver, BC, 2008, pp. 148-152.

[3] G.C. Stone et al, "Electrical Insulation for Rotating Machines", IEEE Press-Wiley, 2004.

[4] M. Liese and M. Brown, "Design-dependent slot discharge and vibration sparking on high voltage windings," in IEEE Transactions on Dielectrics and Electrical Insulation, vol. 15, no. 4, pp. 927-932, August 2008.

[5] Y. Xia, Z. Jia, Z. Guan, L. Wang, and Z. Zhang, "Principles and characteristics of vibration sparking in high voltage stator slots," IEEE Trans. Dielectr. Electr. Insul., vol. 20, no. 1, pp. 42–53, 2013, DOI: 10.1109/TDEI.2013.6451340

[6] V. O. Belko, Y. K. Petrenya, A. M. Andreev, A. M. Kosteliov and M. B. Roitgarz, "Numerical Simulation of Discharge Activity in HV Rotating Machine Insulation," 2019 IEEE Conference of Russian Young Researchers in Electrical and Electronic Engineering (EIConRus), Saint Petersburg and Moscow, Russia, 2019, pp. 800-802.

[7] V. Belko, A. Plotnikov, Y. Petrenya and T. Shikova, "Study of Characteristics of Vibration Sparking in HV Rotating Machine Insulation," 2019 16th Conference on Electrical Machines, Drives and Power Systems (ELMA), Varna, Bulgaria, 2019, pp. 1-4.

[8] A.S. Reznic, I.O. Ivanov, A.M.Andreev, E.R.Mananov, T. M. Shikova, " The Effect of Discharge Activity on the Performance of Corona-Protective Semiconducting Coatings of the Stator Bar Insulation," 2021 IEEE Conference of Russian Young Researchers in Electrical and Electronic Engineering (EIConRus), Saint Petersburg and Moscow, Russia, 2021, pp. 1228 - 1231.

[9] T. Tanaka, "Aging of polymeric and composite insulating materials. Aspects of interfacial performance in aging," IEEE Trans. Dielectr. Electr. Insul., vol. 9, no. 5, pp. 704–716, 2002, DOI: 10.1109/TDEI.2002.1038658.

[10] J. A. Brydson, Plastics materials. Elsevier, 1999

[11] Blythe T. Electrical properties of polymers / T.Blythe, D.Bloor. - 2nd ed. - Cambridge: Cambridge University Press, 2005 (2008). - xi, 480 p

[12] Mihajlin YU.A. Termoustojchivye polimery i polimernye materialy.— SPb.: Professiya, 2006.—624 s., il

[13] M. Shikova, "The improvement of insulation systems made from resin-rich materials for large-scale electric machines," Russ. Electr. Eng., vol. 78, no. 3, pp. 113–117, 2007, DOI: 10.3103/S1068371207030030.

[14] A. Bezborodov and T. Shikova, "Technological factors of insulation systems for large rotating machines: Influence on thermal and electrical properties," in 2020 International Conference on Diagnostics in Electrical Engineering (Diagnostika), 2020, pp. 1–4, DOI: 10.1109/Diagnostika49114.2020.9214657

Optimization Analysis of the Main Insulation Structure of the Large Generator Stator Bar

Feng Shengxi
Institute of Energy of SPBPU
Peter the Great St.Petersburg Polytechnic University
Saint Petersburg, Russia
fengshengxi01@gmail.com

Victor O. Belko
Higher School of High Voltage Power Engineering of SPBPU
Peter the Great St.Petersburg Polytechnic University
Saint Petersburg, Russia
vobelko@gmail.com

Abstract—**Currently, the most important task is to ensure the operational reliability of generators. As the power and voltage of generators increase, the requirements for the insulation properties of generators also become higher. Therefore, it is of great significance to carry out the optimization analysis of the main insulation structure of the stator bar of a large generator to improve the electric field distribution in the stator bar insulation system. Using PTC Creo3.0 and COMSOL Multiphysics connected through the Livelink interface an automatic simulation system is created for estimating the electric field distribution in stator bar insulation. The parameterized modeling of the slot area of the generator stator bar is of great guiding significance for the optimization of the insulation structure of large machines. It can provide a theoretical basis for optimization of the insulation system design of the stator bars.**

Keywords—*insulation structure optimization, stator bar, parametric modeling, simulation systeminsert*

I. INTRODUCTION

The important factor limiting the capacity of the generator is the insulating part of the motor. The insulating part of the motor is mainly the insulating part of the stator coil.In order to develop a higher voltage level generator, the insulation material or insulation structure must be optimized.

Generally, the stator bars of AC machine are composed of several parallel strands. Due to the existence of leakage magnetic field in the slot, the induction electromotive force is generated in the strands of stator bar, and the circulating current between the strands will cause a series of problems, such as uneven temperature rise of the strands. It is a serious threat to the insulation of stator bars and the service life of motors. The electric field of a large hydroelectric generator is mainly concentrated at the end of the stator wire rod, so partial discharge is prone to occur at the end of the stator wire rod.Partial discharge is the main cause of insulation damage to the stator wire rod of the generator.In order to optimize the groove portion of the stator wire rod, parameterized modeling and simulation analysis of the groove portion of the stator wire rod are carried out to reduce the occurrence of partial discharge of the stator wire rod.

Using PTC Creo3.0, the linear part and the transposition part of the groove part of the stator wire rod of a large generator are constructed separately, and then assembled to complete the construction of a three-dimensional solid parameterized model of the transposition of the groove part of a large generator. And then comsol and PTC Creo are connected through the Livelink for PTC Creo module. Finally, an automatic simulation system is created for the electric field distribution in slot bar by adopting APP developer in the comsol software.

The system greatly improves the research convenient an accessible method for the structural design of the stator bar. When using the simulation system, automatic modeling and simulation calculation can be realized by inputting the relevant parameters such as the structural size, material properties and electromotive force of the generator bar. The parameterized modeling of the groove portion of the generator stator wire rod is of great guiding significance for the optimization of the insulation structure of large motors and the improvement of the electric field of the groove portion of the wire rod. It can provide a theoretical basis for the transposition method, transposition optimization, strand elevation, insulation material and insulation thickness design of the groove portion of the wire rod.

II. MODEL OF STATOR IN SLOT

A. The structure of stator in slot

The stator wire rod of the generator is composed of many strands arranged in a row.Due to the different positions of the strands in the magnetic field, a potential difference will form between the strands, resulting in circulation.Under the influence of circulation, the average temperature of the wire rod will increase, reducing the efficiency of the generator.In order to reduce the impact of circulation, the stator wire rod transposition technology must be used.

The transposition model of the groove part of the stator wire rod is shown in Figure 1 .

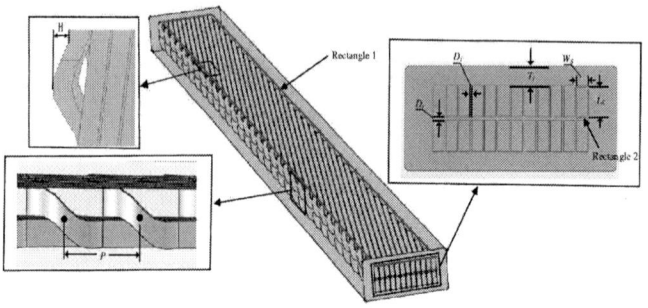

Fig. 1. The structure of stator in slot.

The control parameters of the stator wire rod transposition model are shown in Table 1

TABLE I. CONTROL PARAMETERS OF THE MODEL

Control parameters of the model					
Para meter	Name	Para meter	Name	Parame ter	Name
L_S	Section length	W_S	Section length	N	Number of strands
α	Transposition angle	P	Transposition pitch	H	Transposition height
D_r	Distance between rows	D_i	Distance between strands	T_i	Bond thickness

α is the transposition angle of the stator wire rod when the strands are transposed. N is the number of strands of a single stator wire rod, and N is generally even in the parameterized model.

B. Model of stator in slot

Parameterized modeling of the groove part of the stator wire rod through the parameters in Table 1.

The modeling of the linear part of a single row of strands of a wire rod adopts the method of array, and a sketch is constructed on the "FRONT reference plane". The cross-section of a single strand is defined as a rectangular plane 1, and the length and width of the matrix are constrained by equation 1 and 2.

$$W_i= (N/2-1) \cdot (W_S+D_i) \tag{1}$$
$$L_i =2N \cdot P \tag{2}$$

Construct sketch 2, choose to use the previous plane drawing, select the right and bottom sides of "Rectangle 1" as the reference, create a reference point, use this reference point as the reference, draw "line segment 1", the relationship between line segment 1 is determined by equation 3 and 4.

$$X_i = L_S + W_S \tag{3}$$
$$Y_i =P \tag{4}$$

Taking the right and lower sides of "Rectangle 1" as a reference, draw a line segment to the origin and name it "Line segment S1".Create a scan, the relationship between the scan cross-section is as follows equation 5 and 6.

$$X_{ii} = L_S \tag{5}$$

$$Y_{ii} = W_S \tag{6}$$

The scanned part of the single wire rod is shown in Figure 2.a. Take "Scan" 1" as the base, select the array, and array up and left respectively. The resulting array diagram is shown in Figure 2.b

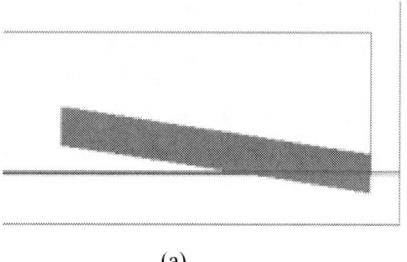

(a)

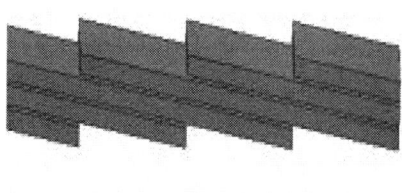

(b)

Fig. 2. Single-row model

Select "Line Segment 2" to be drawn in the reference plane "RIGHT plane". Line segment 2 takes the origin as the reference point and draws to the left. Create a reference plane "DTM1".Select the left end point of "Line Segment 2" as the reference, and select parallel to the "FRONT plane"; create "Sketch 5", select "DTM1 plane" as the reference plane, adjust the viewing angle, take the end point of "Line Segment 2" as the reference, and draw "Rectangle 2". Using the same method to construct a model of the second row of non-transposed strands, the resulting model can be mirrored to obtain the four-row linear transposition model part. The second row linear model and the fourth row linear transposition model are shown in Figure 3.

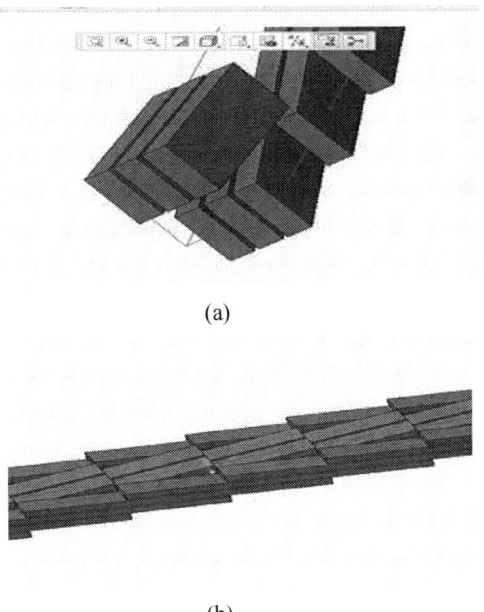

(a)

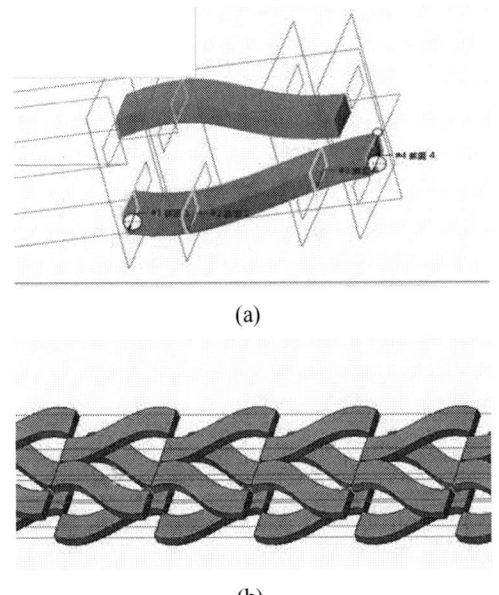

(b)

Fig. 3.   Double-row model and four-row model

Create four reference planes within the unit intercept, the middle two reference planes are the raised strand cross-sections, and the corresponding strand cross-sections are mixed and scanned, and we can get a parameterized transposition model of pitch. model.Array and scan the resulting single-pitch transposition model to obtain a complete four-row strand transposition model. The transposition model within a single pitch and the complete four-row strand transposition model are shown in Figure 4.

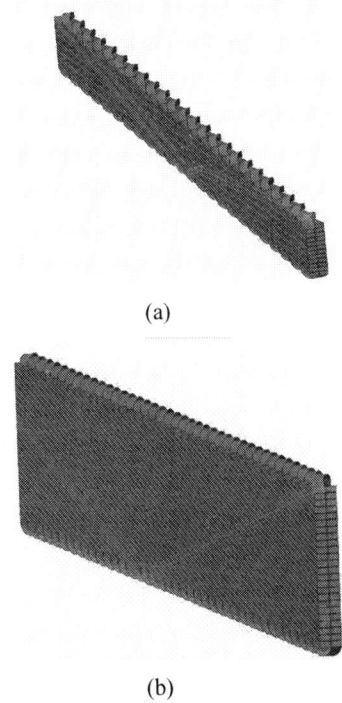

(a)

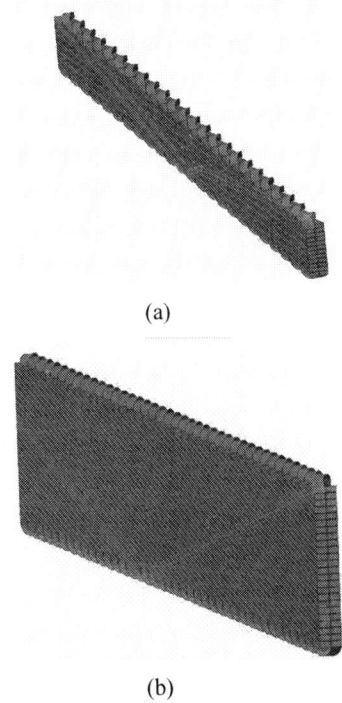

(b)

Fig. 4.   Strand transposition model

Select a reference point of the truncated plane of the strand of the transposition model, select a point where the straight line part coincides with it, assemble the two models, and increase the coincidence constraints.You can get a complete model of the groove of the stator wire rod.

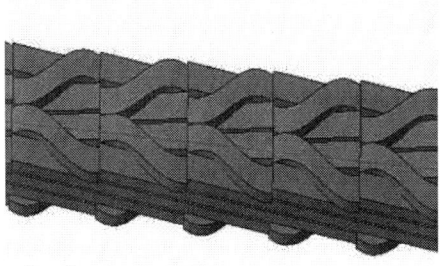

Fig. 5.   Complete model of the groove of the stator wire rod

Through the above steps, a parameterized model of the stator wire rod can be established.Figure 6.a is a model of a stator wire rod with 24 strands in 360-degree transposition, Figure 6.b is a stator wire rod model with 48 conductors in 300-degree transposition.

(a)

(b)

Fig. 6.   The model of stator bar conductive core.

At this point, the parameterized model of the transposition part of the groove part of the stator wire rod of a large generator is completed.The input data will automatically change the size of the model, and the changes in the groove of the generator stator wire rod can be seen more intuitively through parameterized modeling.

III. MODEL SIMULATION

Using PTC Creo3.0, the linear part and the transposition part of the groove part of the stator wire rod of a large generator are constructed separately, and then assembled to complete the construction of a three-dimensional solid parameterized model of the transposition of the groove part of a large generator. And then comsol and PTC Creo are connected through the Livelink for PTC Creo module. Finally, an automatic simulation system is created for the electric field distribution in slot bar by adopting APP developer in the comsol software.

A. Five models of electric field distrbution

Using the stator wire rod automatic simulation system, a 360-degree full transposition model of 56 strands was constructed.The model is simplified in COMSOL, and the electric field of the unshielded structure and the shielding structure of the semiconductor laminate, the semiconductor putty structure, the semiconductor hat structure, and the fully shielded structure are analyzed.

(a)

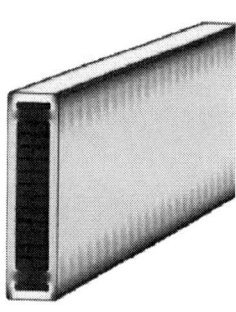

(b)

(c)

(d)

(e)

Fig. 7.   Five models of electric field distribution

As can be seen from Figure 6, the electric field of the wire rod treated with the shielding structure is mainly concentrated on the narrow surface of the outer rounded part of the semiconductor, the wire rod. All four structures effectively improve the concentration of electric field in the transposition part of the wire rod.

The maximum degree of optimization is ranked from high to low, the fully shielded structure, the semiconductor hat structure, the semiconductor putty structure, and the semiconductor platen structure.The semiconductor platen structure is not as good as several other structures for improving the electric field distribution.Mainly, the insertion of the semiconductor laminate into the insulation causes the insulation thickness to become thinner, and the optimization of the electric field effect cannot effectively offset the increase in the electric field caused by the decrease in the insulation thickness.

IV.   CONCLUSION

The three-dimensional modeling software Creo 3.0 is used for parameterized modeling of the groove structure of the stator wire rod, and the parameterized control of many transpositions is realized. You can enter the corresponding data in the parameter interface, and Creo 3.0 will automatically construct a model of the groove of the stator wire rod under this parameter.The parameters that can be controlled are the cross-section length of the strands, the cross-section width of the strands, the longitudinal spacing between the strands, the spacing between the strands, the height of the strand transposition height, the number of strands, the angle of the wire rod transposition and the pitch of the wire rod transposition and other parameters.Through the three-dimensional model, the electric and magnetic fields in the stator rod groove of the generator can be analyzed and calculated, so as to optimize the parameters in the stator groove, which is of great guiding significance for the design of high-efficiency motors.

ACKNOWLEDGMENT

This thesis, from the selection of topics, to the late application software, the use of Creo 3.0 software for the establishment of parameterized models, program design, the writing of the thesis, and the revision and review of the thesis, was completed under the guidance of Teacher Hu Haitao. Teacher Hu is kind, rigorous in teaching, and serious and responsible guidance. I admire it.Not only did it enable me to better master knowledge in my studies, but it also taught me a lot in living as a human being, enabling me to think better independently, be down-to-earth and serious, and better master knowledge. It has been a great inspiration and help.It is my honor to meet such a serious and responsible instructor as Teacher Hu at the moment of graduation.Thank you for having such a good teacher on the way to study!

REFERENCES

[1] Zhang X., Cao T., Zhang B. A parameterized modeling method for transposition structure of stator wire rod of large motor. Master of Engineering Thesis. Heilongjiang, China, 2018, pp. 17-27.

[2] Liang Y., Bian X., Yu., Yang Y. Analytic Algorithm for Strand Slot Leakage Reactance of the Transposition Bar in an AC Machine. IEEE TRANSACTIONS ON INDUSTRIAL ELECTRONICS. Chongqing, China, 2014, pp. 5232-5237.

[3] Gu J., Zhang Y., Wei Y., Li X. Research of Internal Shielding Structure for Stator Coil of 1000 MW Turbogenerator. Shanghai Jiaotong University. Shanghai, 2011, pp. 2-5.

[4] Popova A.N. Numerical study of discharge activity in the groin area of the stator winding of high-voltage electric machines. [Peter the Great St.Petersburg Polytechnic University], 2018, no. 3, pp. 23-64. (in Russian)

[5] Ivanovsky A.A. Development of a mathematical model for calculating electrophysical processes in the isolation system of stator windings of turbo generators. [Power Machine], 2011, no. 3, pp. 2-6. (in Russian)

[6] Krzheminsky T.B. Investigation of physical and mechanical characteristics of materials of elastic grooved sealing of generator stator winding rods. [Peter the Great St.Petersburg Polytechnic University], 2021, no. 3, pp. 5-64. (in Russian)

[7] Liang Y, Bian X, Yang L, et al. Numerical calculation of circulating current losses in stator transposition bar of large hydro-generato. IET Science. Heilongjiang, China, 2014, pp. 485-490.

[8] Staubach C, Kempen S, Pohlmann F. Calculation of electric field distribution and temperature profile of end corona protection systems on large rotating machines by use of finite element mode. IEEE Conference Record of the 2010 IEEE International Symposium on Electrical Insulation. 2010: 1-6.

979-8-3503-8358-4/23 $31.00 © 2023 IEEE

Influence of Laser Radiation Power on Raman Spectra of Nanostructured Silicon Wires with Crystallographic Orientation (111)

Alena Yu. Gagarina
Department of Micro- and Nanoelectronics
Saint Petersburg Electrotechnical University "LETI"
Saint Petersburg, Russia
gagarina.au@gmail.com

Alexey Kuznetsov
Department of Nanoheterostructures Physics and Technology
Alferov Federal State Budgetary Institution of Higher Education and Science Saint Petersburg National Research Academic University of the Russian Academy of Sciences
Saint Petersburg, Russia
leshiy2698@mail.ru

Yulia M. Spivak
Department of Micro- and Nanoelectronics
Saint Petersburg Electrotechnical University "LETI"
Saint Petersburg, Russia
ymkanageeva@yandex.ru

Aleksey D. Bolshakov
Department of Nanoheterostructures Physics and Technology
Alferov Federal State Budgetary Institution of Higher Education and Science Saint Petersburg National Research Academic University of the Russian Academy of Sciences
Saint Petersburg, Russia
bolshakov@live.com

Vyacheslav A. Moshnikov.
Department of Micro- and Nanoelectronics
Saint Petersburg Electrotechnical University "LETI"
Saint Petersburg, Russia
vamoshnikov@mail.ru

Abstract—In this paper an experimental study of laser irradiation power effect on the Raman spectra of nanostructured porous silicon wires (NPSW) obtained on n-Si(111) is reported. NPSW were obtained by modified metal-assisted electrochemical etching of c-Si. The laser radiation power varied from 0.064 mW to 5.94 mW. Analysis of the spectra showed the following at low power density the received signal is poorly distinguishable against the background of the main mode for Si at 520 cm^{-1}. As the power increases, one can observe characteristic shifts of the entire Raman spectrum towards low frequencies up to 483 cm^{-1} for the fundamental optical phonon mode. Repeated studies at lower powers did not reveal significant structural changes in the NPSW. Based on the data obtained, new possibilities of the technique for characterizing of NPSW by the Raman scattering are discussed.

Keywords— porous silicon, porous silicon nanowires, Raman spectroscopy, nanomaterials

I. INTRODUCTION

Nowadays porous silicon (PS) is widely used in many areas of semiconductor electronics. One of the promising forms of PS is nanostructured porous silicon wires (NPSWs). Due to the possibility of controlling the morphology and properties of the nanostructure at the stage of synthesis by varying the technological conditions of synthesis and functionalization, NPSWs are actively studied as a gas-sensitive material for detection of ultra-low concentrations of industrial gases [1-5]. Thus, the possibility of detecting ammonia in liquid solutions with a detection limit of 80 ppb and a sensitivity of 0.2% μmol–1·L was demonstrated in [1]. In [2], it was proposed to use structures based on nanostructured silicon as a chemoresistive device to detect hydrocarbon vapors of different masses in oil reservoirs. Another topical direction is the development of photoluminescent biosensors based on NPSWs using optical

signal quenching, which makes it possible to detect biomolecules with a sensitivity up to the femtomolar range without the use of traditional fluorescent labels [10–11]. Such systems work on the principle of cooperative hybridization for selective DNA capture and optical emission of quantum-limited charge carriers in NPSWs, the quenching of which is used as a detection mechanism. Using the hepatitis B virus (HBV) genome as an example, it was demonstrated that it is possible to achieve a detection limit of 2 copies/reaction for a synthetic genome and 20 copies/reaction for a genome isolated from human blood [10]. In addition, as photoluminescence in the visible range is actively used to detect biological objects. For example, in [11] photoluminescence of nanostructured silicon wires was used as a cut-off signal for selective detection of C-reactive protein in human serum.

One of the methods for controlling NPSWs is the Raman scattering spectroscopy, which allows estimating the phase and elemental composition, as well as the presence and magnitude of mechanical stresses in NPSWs. It is known that high power of laser radiation can have a significant influence on the structure of nanostructured silicon. Thus, because of local heating, amorphization of the samples and increased quantity of structural defects are possible. Therefore, there is a need to determine the optimal modes for obtaining informative and undistorted Raman spectra of NPSWs.

In this work, we present an experimental study of the effect of laser irradiation power on the Raman spectra of NPSWs obtained on an n-Si with crystallographic orientation (111).

II. MATERIALS AND METHODS

NPSWs were obtained by two-stage modified metal-assisted electrochemical etching (MAECE) of single-crystalline n-type Si with crystallographic orientation (111). The first stage of MAECE involved electrochemical etching of c-Si in aqueous solution of hydrogen fluoride, isopropanol and 0.01M silver nitrate for 5 min at anodization current density of 15 mA/cm2. At the second stage, electrochemical etching of the resulting composite was carried out in aqueous solution of hydrogen fluoride and isopropanol at anodizing current density of 180 mA/cm2 for 20 minutes. Afterwards the sample was post-treated in aqueous solution of nitric acid and isopropanol and dried at 20°C.

Raman spectra were measured at room temperature using the Horiba LabRAM HR 800 spectrometer. Laser power was varied from 0.064 mW to 5.94 mW. The excitation wavelength was 532 nm.

III. RESULTS AND DISCUSSION

Figure 1a shows the Raman spectra for NPSWs at different laser irradiation powers in the range 460-560 cm^{-1}. In this case, for the first-order optical phonon mode a strong shift to the low-frequency range is observed with increasing laser irradiation power. For an irradiation power of 5.94 mW the Raman shift of the main silicon mode reaches 36.8 cm-1, for 3.26 mW 31.5 cm-1, for 1.95 mW 11.1 cm-1 and 2.9 cm-1 at 0.45 mW respectively.

It is assumed that the high dependence of the Raman shift on the laser irradiation power is related to local heating of the nanostructure and the subsequent occurrence of mechanical stresses at the heating point, as indicated by the high shifts to the lower energy range. This feature can be associated with the lower thermal conductivity of nanostructures silicon wires as compared to c-Si due to the complex phase composition and porous structure.

In order to assess the effect of laser irradiation power on the crystal structure of sample and to determine the possibility of recrystallization processes, it was proposed to compare the Raman spectra of the sample at the point previously irradiated at 0.064 mW, at which the minimum deviation from the classical c-Si Raman spectrum is observed (Figure 1b). At the same time for all points a slight shift (the maximum shift was 1.21 cm^{-1} after irradiation with 5.94 mW laser power) of the main silicon peak was observed, which can be associated either with the occurrence of mechanical stresses characteristic of porous silicon structure or extremely minor recrystallization processes occurring in the subsurface layers of the nanostructure.

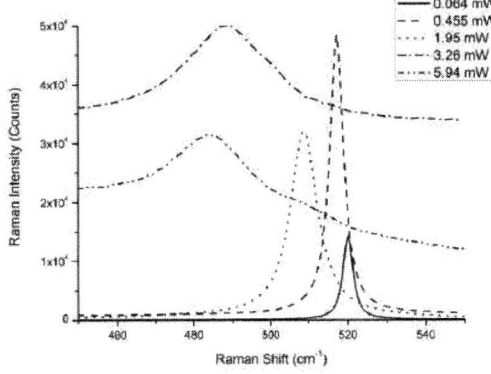

a

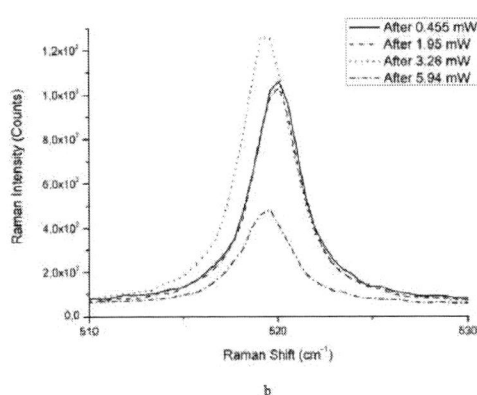

b

Fig. 1. Raman shift spectra of NPSWs: (a) - after exposure of the sample to different laser powers; (b) - after irradiation of the sample in the same point with 0.064 mW laser power

In addition, Figure 2 shows that NPSWs have amorphous component, as evidenced by asymmetric broadening of the peaks at 480 and 960 cm^{-1}. For NPSWs can be observed an oxide phase, which is indicated by the occurrence of a band at 400 cm^{-1}, associated with vibrations of the Si-O-Si bond.

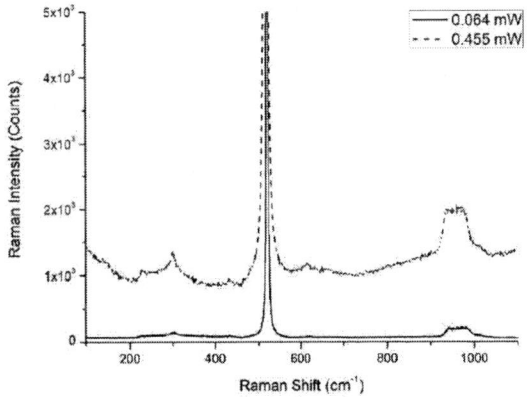

Fig. 2. NPSWs Raman shift spectra obtained at 0.455 mW and 0.064 mW

IV. CONCLUSION

Raman spectroscopy investigation of the nanostructured porous silicon wires n-Si(111) demonstrated that NPSWs obtained by modified metal-assisted electrochemical etching represent an array of wires with a crystalline core covered by amorphous and oxidized silicon layers. The formation of silicon nanocrystals occurs as a result of the

979-8-3503-8358-4/23 $31.00 © 2023 IEEE

disproportionation reaction near the pore surface. This is evidenced by the shift of the Raman shift spectra towards lower energies and the broadening of the main silicon peak associated with the deformation of the Si crystal lattice and the localization of optical phonons in Si nanocrystals.

REFERENCES

[1] Kondratev V.M, Morozov I.A., Vyacheslavova E.A., Kirilenko D.A., Kuznetsov A., Kadinskaya S.A., Nalimova S.S., Moshnikov V.A., Gudovskikh A.S., Bolshakov A.D. Silicon Nanowire-Based Room-Temperature Multi-environment Ammonia Detection. *ACS Applied Nano Materials*, 2022, vol. 5, no. 7, pp. 9940-9949.

[2] Jeribi M., Nafie N., Boujmil M. F., Bouaicha M. (2021). Response modulation of silicon nanowires-based sensor to carbon number in petroleum vapor detection. *Fuel*, 2021, vol. 304, pp. 121260.

[3] Naama S., Hadjersi T., Keffous A., Nezzal G. (2015). CO2 gas sensor based on silicon nanowires modified with metal nanoparticles. *Materials Science in Semiconductor Processing*, 2015, vol. 38, pp. 367-372.

[4] Bobkov A., Luchinin V., Moshnikov V., Nalimova S., Spivak Y. Impedance Spectroscopy of Hierarchical Porous Nanomaterials Based on por-Si, por-Si Incorporated by Ni and Metal Oxides for Gas Sensors. *Sensors*, 2022, vol. 22, no. 4, pp. 1530.

[5] Gagarina A.Yu., Bogoslovskaya L.S., Spivak Y.M., Moshnikov V.A., Bobkov A.A., Shumilo M.V., Kondratev V.M., Khalugarova K. Impedance spectroscopy of hybrid structures based on nanostructured silicon impregnated with Au and NiO. *Vestnik Moskovskogo Universiteta*, 2022, vol. 4, pp. 2241402 (in Russian).

[6] Korany F. M., Hameed M. F. O., Hussein M., Mubarak R., Eladawy M. I., Obayya S. S. A. (2018). Conical structures for highly efficient solar cell applications. *Journal of Nanophotonics*, 2018, vol. 12, no. 1, pp. 016019.

[7] Spivak Y.M., Belorus A.O., Panevin A.A., Zhuravsky S.G., Moshnikov V.A., Bespalova K.,. Somov P.A, Zhukov Yu., Komolov A.S., Chistyakova L.V., Grigorieva N.Yu. Porous silicon as a nanomaterial for targeted drug delivery systems to the inner ear. *Technical Physics*, 2018, vol. 88, no. 9, pp. 1394-1403 (in Russian).

[8] Spivak Y. M., Mjakin S. V., Moshnikov V. A., Panov M. F., Belorus A. O., Bobkov A. A. Surface functionality features of porous silicon prepared and treated in different conditions. *Journal of Nanomaterials*, 2016, vol. 2016, pp. 2629582.

[9] Smerdov R., Spivak Y., Bizyaev I., Somov P., Gerasimov V., Mustafaev A., Moshnikov V. Advances in Novel Low-Macroscopic Field Emission Electrode Design Based on Fullerene-Doped Porous Silicon. *Electronics*, 2020, vol. 10, no. 1, pp. 42.

[10] Leonardi A.A.; Lo Faro M.J.; Petralia S.; Fazio B.; Musumeci P.; Conoci S.; Irrera A.; Priolo F. Ultrasensitive label- and PCR-free genome detection based on cooperative hybridization of silicon nanowires optical biosensors. *ACS Sensors*, 2018, vol. 3, no. 9, pp. 1690–1697.

[11] Irrera A.; Leonardi A.A.; Di Franco C.; Lo Faro M.J.; Palazzo G.; D'Andrea C.; Manoli K.; Franzò G.; Musumeci P.; Fazio B.; Torsi T., Priolo F. New generation of ultrasensitive label-free optical si nanowire-based biosensors. *ACS Photonics*, 2018, vol. 5, no. 2, pp. 471–479.

Electroplasma Technologies for Cleaning, Polishing and Welding of Metals

Fivzat M. Gaysin
Kazan National Research Technical University named after A.N. Tupolev–KAI
Kazan, Russia
FMGaysin@kai.ru
0000-0002-4720-3775

Liliya N. Bagautdinova
Kazan National Research Technical University named after A.N. Tupolev–KAI
Kazan, Russia
LNBagautdinova@kai.ru
0000-0002-5300-9880

Almaz F. Gaisin
Kazan National Research Technical University named after A.N. Tupolev–KAI
Kazan, Russia
AFGaysin@kai.ru

Azat F. Gaisin
Kazan State Power Engineering University
Kazan, Russia

Karim Sh. Mastyukov
Kazan National Research Technical University named after A.N. Tupolev–KAI
Kazan, Russia

Dzhaudat U. Zakirov1
Kazan National Research Technical University named after A.N. Tupolev–KAI
Kazan, Russia

Abstract—The purpose of this work is to use new physical phenomena and processes for the development and creation of a technological complex of in- stallations. Technological installations of various capacities have been devel- oped and created on the basis of multichannel discharge: 100, 60 and 25 kW. Installations for cleaning and polishing with the use of electric discharge plasma with liquid electrodes allow: to process parts of all sizes and shapes without de- formation and force impact; to process surfaces of hard-to-reach places of com- plex parts; to exclude acids and alkalis from the technological process; to use safe reusable electrolytes (active metal salts); to remove foreign inclusions - corrosion centers; to have productivity and overall efficiency 3-5 times higher (compared to machining); electricity consumption is higher, however, the total operating costs are 3 times lower than with electrochemical processing; the equipment used is versatile - suitable for cleaning, polishing and welding. An example of the successful application of processing in installations is the removal of dielectric coatings (resins, paints, polymers), polishing of metal surfaces, welding of thin-walled metals. Experimental studies have shown that a multichannel discharge burns in the pressure range P = 105-104 Pa, with a further decrease in pressure; the multi-channel discharge turns into a volumetric discharge. Technological processes of cleaning and polishing are realized on the basis of multichannel discharge. On the basis of the volume discharge, the welding process in the electrolyte is realized. The technology for obtaining the required metal surface microrelief and welding of homogeneous and dissimilar metals has been developed. Pilot plants for cleaning, polishing and welding in electrolyte with a capacity of 100, 60 and 25 kW have been designed and built. Technological tests of the units were carried out, which showed the efficiency of the applied technology.

Keywords—electrical discharge; cleaning polishing; welding technologies.

I. INTRODUCTION

One of the ways to obtain a low-temperature plasma is to use an electric discharge between a metal and electrolytic electrodes [1-9]. These discharges are used in elec-troplasma technology [10-17] for cleaning, polishing and welding [18], as well as for heating, cutting metals, coating metal and polymer coatings [19, 20].

The development of modern technology makes ever higher demands on the quality of metals and alloys. Currently, the possibilities of cleaning metals and alloys using traditional processing methods—mechanical, chemical—are practically exhausted. These methods have a number of disadvantages: low productivity, poor quality and accuracy of surface treatment; increased energy consumption and environmental harmfulness of the technology; the need to apply special measures for waste disposal; high cost of consumables, which leads to a decrease in the competitiveness of prod- ucts. In this regard, the problem arises of developing new technological processes for cleaning the surface of metals and alloys. A promising direction is the use of high- energy methods: laser [21], plasma and electron-beam, which allow saving raw mate- rials and reagents, increasing labor productivity, improving the surface quality of the processed material and making it possible to obtain materials with new physical and mechanical properties. One of the promising technologies in the field of cleaning the surface of metals and alloys is its treatment with low-temperature plasma of an elec- tric discharge. The interest in plasma discharges with a liquid electrode for use in technological purposes lies in the fact that it combines the properties of two technolo- gies: chemical and plasma [22-24]. The use of a non-equilibrium electrolyte plasma of a multichannel discharge [25-26] often provides an increase in the efficiency of many technological processes, such as the plasma-chemical formation of surfaces with desired properties on various materials.

Multichannel discharge is an electrical discharge at atmospheric pressure, consist- ing of microchannels. A discharge can burn between a liquid cathode and a solid an- ode (fig.1a), and also between the cooled metal cathode and the flowing electrolytic anode (fig.1b). In the second case, without cooling the metal cathode, the multichan- nel discharge transforms into a contracted discharge from the cathode side.

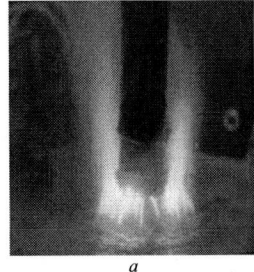 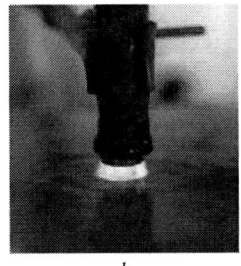

a   *b*

Fig. 1. Types of multichannel discharge with a liquid cathode (a) and a liquid anode (b).

979-8-3503-8358-4/23 $31.00 © 2023 IEEE

II. Materials and Methods

On the basis of a multichannel discharge, technological installations of various capac- ities have been developed and created.

Figure 2a shows an installation with a power of 100 kW. The installation consists of a control panel with a power supply from a three-phase network and a discharge chamber with a matrix system for fixing the processed elements. Such a matrix sys- tem is capable of simultaneously cleaning and polishing many elements. From the control panel, the treatment mode (voltage) is selected, the treatment time is set, as well as automatic immersion and lifting.

Figure 2b shows an installation with a power of 60 kW, which allows processing of parts of smaller sizes.

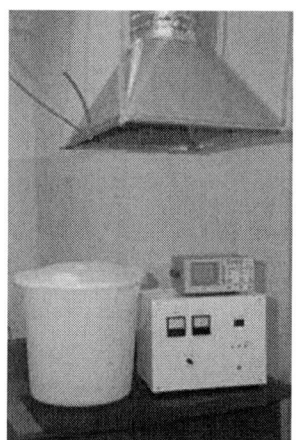 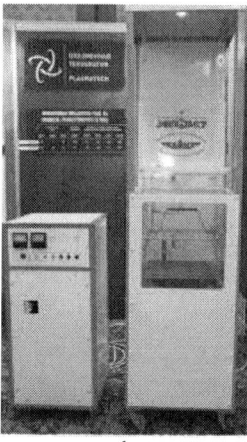

a   *b*

c

Fig. 2. Technological complexes for cleaning, polishing and welding metals in electrolyte of various power: a - 100 kW, b - 60 kW, c - 25 kW.

Installations for cleaning and polishing using an electric discharge plasma with liquid electrodes allow:

- to process parts of all sizes and shapes without deformation and force im- pact;
- to process surfaces of hard-to-reach places of complex parts;
- to exclude acids and alkalis in the technological process;
- to use safe electrolytes of reusable use (salts of active metals);
- to remove foreign inclusions - corrosion centers;
- to have productivity and total efficiency 3-5 times higher (compared to me- chanical processing);

- electricity consumption is higher, however, the total operating costs are 3 times lower than with electrochemical processing;
- the equipment used is versatile - suitable for cleaning, polishing and welding.

III. Results and Discussion

An example of a successful application of treatment on a 50 kW installation is the removal of dielectric coatings (resins, paints, polymers) shown in fig. 3.

Fig. 3. Removing the dielectric coating: part before and after processing.

The next example is the descaling of metals formed after quenching. Descaling re- duces power losses, eliminates sparks, fires, reduces erosion and wear of parts.

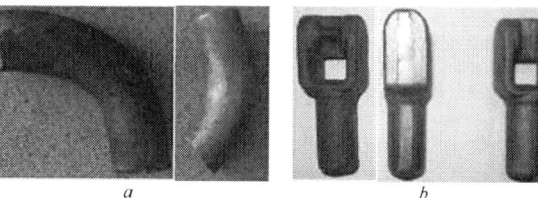

a   *b*

Fig. 4. Descaling: details before and after processing (material - steel 12XH2P).

Figure 5 shows the result of cleaning a mold (rod) with a multichannel discharge to form tablets. Cleaning occurs when a part is immersed in an electrolytic cathode and a positive potential is applied to it, and processing takes place within 5-60 seconds, depending on the area of the treated surface. As a result of cleaning, the adhesion of the rod is reduced due to the polishing of the surface.

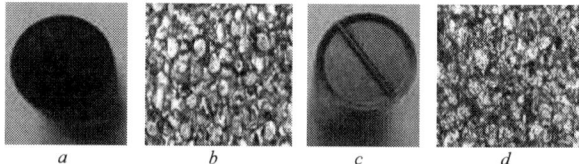

a   *b*   *c*   *d*

Fig. 5. Photos of untreated (a, b) and treated (c, d) surface of a steel rod (steel 01T31507). Photos b and d magnification 20 times (inverted microscope).

Multichannel discharge allows polishing the surface of a metal product by immer- sion in an electrolyte. The exposure time depends on the treated area. Each material has its own composition and concentration of electrolyte. Non-aggressive salt solu- tions are used as electrolytes. Fig. 6-8 show examples of cleaning and polishing various products.

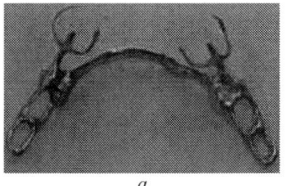 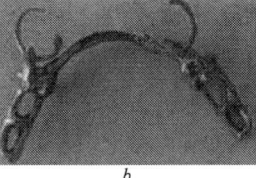

a   *b*

Fig. 6. Photos of untreated (a) and treated (b) surfaces of the denture (steel).

Figure 7 shows the results of processing medical needles. The photographs in Fig. 7 b and c, taken with a confocal scanning microscope, show a smoothing of the sur- face relief and a decrease in the surface roughness of medical needles with a maxi- mum roughness factor of Ra3.281 μm for an untreated needle (fig. 7b) and Ra0.243 for a machined needle (fig. 7c), which corresponds to the 9th grade of roughness.

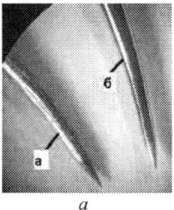 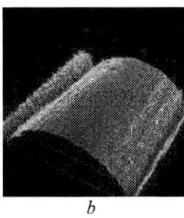 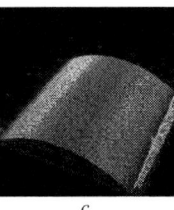

a   *b*   *c*

Fig. 7. Photos of medical needles (a), untreated (b) and treated (c) surface of an atraumatic needle. Image captured with a confocal scanning microscope.

Fig. 8 shows a variant of cleaning and polishing for structural alloy steels.

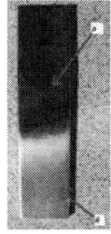 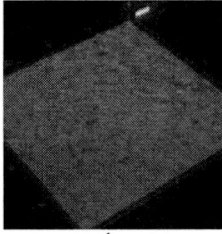 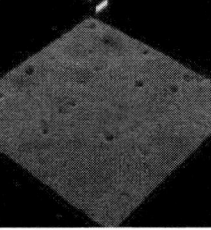

a   *b*   *c*

Fig. 8. Photographs of a steel sample (a), untreated (a) and treated (b) surfaces of steel grade 12XH2P.

The possibility of welding homogeneous and dissimilar metals using an electric discharge plasma in a liquid on a 100 kW installation was investigated. For thin- walled metals (from 1 to 5 mm thick), welding is performed in the voltage mode $150 \leqslant U \leqslant 500$ V, discharge current $0,5 \leqslant I \leqslant 200$ A, welding time 1-60 s.

Figure 9 shows samples of welding of D16 duralumin (fig. 9a), steel 3311 (fig. 9b), copper grade M4 (fig. 9c), steel with copper (fig. 9d). For example, copper with a thickness of 1 mm and a length of 100 mm is welded within 1 second.

a – Duralumin welding samp   *b – Welding steel sample*

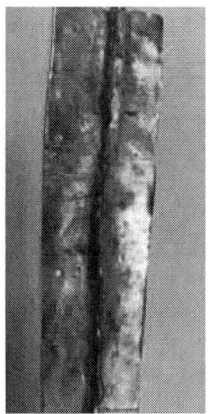 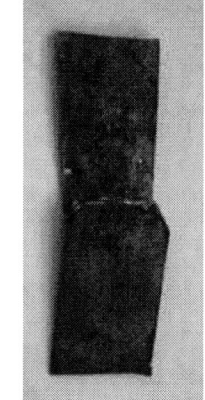

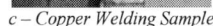

c – Copper Welding Sample   *d – Sample of welding copper to steel*

Fig. 9. Photos of welding samples of various metals.

IV. CONCLUSION

Experimental studies have shown that a multichannel discharge burns in the pressure range $P = 10^5\text{-}10^4$ Pa; with a further decrease in pressure, the multichannel discharge transforms into a volumetric discharge. On the basis of a multichannel discharge, technological processes of cleaning and polishing are implemented. Based on the volumetric discharge, the electrolyte welding process is realized.

Technologies have been developed for obtaining the required microrelief of the surface of metals and welding of homogeneous and dissimilar metals.

Pilot industrial installations for cleaning, polishing and welding in electrolyte with a capacity of 100, 60 and 25 kW have been developed and built. Technological tests were carried out, which showed the effectiveness of the technological complex and the applied technology.

REFERENCES

[1] Bruggman P.J., et al: Plasma-liquid interactions: A review and roadmap. Plasma Sources Sci.and Technol., 053002 (2016).

[2] Gaisin, A.F., Kashapov, N.F., Kuputdinova, A.I., Mukhametov, R.A.: Discharge between Liquid Jet and Metallic Electrodes. Technical Physics 63(5), 695–699 (2018).

[3] Gaisin, A.F., Sadriev, R.S., Bagautdinova, L.N., Nasibullin, R.T., Gaisin, F.M., Mast- yukov, S.C.: Low-Power Electric Discharges with Metallic, Dielectric, and Electrolytic Electrodes at Low Frequencies and Atmospheric Pressure. High Temperature 58(6), 777– 780 (2020).

[4] Takseitov, R.R., Galimova, R.K., Yakupov, Za.Y.: Calculation of portable properties of some real gas mixtures at high temperatures. Journal of Physics: Conference Series 1588(1). 012065 (2020).

[5] Gajsin F.M., Galimova R.K., Khakimov R.G.: Vapor-gas discharge with nontraditional electrodes. Elektronnaya Obrabotka Materialov (5), 27–29 (1994).

[6] Bagautdinova, L.N., Sadriev, R.S., Gaysin, A.F., Nasibullin, R.T., Gaysin, F.M., Mast- yukov, S.C.: Features of a Low-Power, Low-Frequency, AC Arc between Solid and Liquid Electrodes. High Temperature 58(3), 441–443 (2020).

[7] Bagautdinova, L.N., Sadriev, R.S., Gaysin, A.F., Mastyukov, S.C., Gaysin, F.M., Fakh- rutdinova, I.T., Leushka, M.A., Gaysin, A.F.: Some Features of Dielectric Barrier Dis- charge with Liquid and Solid Electrodes . High Temperature 57(6), 944–947 (2019)

[8] Sadriev, R.S., Son, E.E., Bagautdinova, L.N., Gaisin, A.F., Gaisin, F.M.: Experimental study of an impulse electric discharge with liquid electrodes. High Temperature 55(2), 310–311(2017).

[9] Gaisin, A.F., Basyrov, R.S., Son, E.E.: Model of glow discharge between an electrolytic anode and a metal cathode. High Temperature 53(2), 188–192 (2015)

[10] Yang, Y., Cho, Y. I., Fridman, A., Plasma discharge in liquid: Water Treatment and Ap- plication, CRC Press (2012).

[11] Фортов В.Е. и др.: Плазменные технологии. Изд-во МФТИ, Москва (2006).

[12] Frolov, V.Ya., Ivanov, D.V. Calculation of a plasma composition and its thermophysical properties in cases of maintaining or quenching of electric arcs 2018 Journal of Physics: Conference Series, 1058(1), 012040.

[13] Murashov, I., Frolov, V., Kadyrov, A. Development of the arc plasma torch operation mathematical model for spheroidization of fine-dispersed powders 2018 Journal of Phys- ics: Conference Series, 1058(1), 012024.

[14] Murashov, I., Frolov, V., Ivanov, D. Numerical simulation of DC air plasma torch modes and plasma jet instability for spraying technology, Proceedings of the 2016 IEEE North West Russia Section Young Researchers in Electrical and Electronic Engineering Confer- ence, EIConRusNW 2016, 7448261, с. 625-628.

[15] Frolov, V.Y., Toropchin, A.I. The influence of arc plasma parameters on the form of a welding pool , 2015 Technical Physics Letters, 41(7), с. 635-637.

[16] Frolov, V., Murashov, I., Ivanov, D. Special aspects of DC air plasma torch's operating modes under turbulent flow conditions , 21st Symposium on Physics of Switching Arc 2015, FSO 2015, с. 129-133.

[17] Tsareva, A.M., Makaeva, R.K., Safina, D.M., Galimova, R.K.: Diagnostics of Fracture of the Blower Impeller of Gas Turbine Engine Core by Using the Holographic Interferome- try. Russian Aeronautics. 63(2), 362–365 (2020).

[18] Bagautdinova, L.N., Basyrov, R.Sh., Galimzyanov, I.I., Gaysin, Al.F., Gaysin, A.Z.F., Gaysin, F.M., Fakhrutdinova, I.T.: New technology for welding aluminum and its alloys. Materials Today: Proceedings, 19, 2566–2567 (2019).

[19] Mardanov R. R., Kayumov R. R., Akhatov M. F., Gaisin A. F.: A review of use of barrier discharge for modification of surface of polymers. Journal of Physics: Conference Series, LTP Coatings 2019, 1588 012029 (2020).

[20] Mardanov R. R., Kayumov R.R., Akhatov M.F., Gazizova A.I., Gaisin A.F.: Modification of the surface of polyethylene by low-temperature plasma and liquid electrodes. Journal of Physics: Conference Series, LTP Coatings 2019, 1588 012028 (2020).

[21] Giliazov M.R., Nagulin K.Y., Gilmutdinov A.K.: The laser beam positioning and focusing system for surface treatment of products. Optics & Laser Technology. 119, 105624 (2019).

[22] Khlyustova A.V., Maximov A.I.: Physical chemistry of plasma-solution systems High Energy Chemistry 43(3), 149 (2019).

[23] Sirotkin N.A., Khlyustova A.V., Titov V.A.: Chemical composition and processes in the dc discharge plasma of atmospheric pressure with a liquid electrolyte cathode Plasma. Chemistry and Plasma Processing. 40(1), 187-205 (2020).

[24] Gaisin, Al.F., Abdullin, I.Sh., Basyrov, R.Sh., Petriakov, S.Y.: Research of radio- frequency electric discharge in electrolyte in the processes of surface treatment. Journal of Physics: Conference Series. 567(1), 012014 (2014).

[25] Gaisin, A.F., Bagautdinova, L.N., Gaisin, A.F., Sadriev, R.S., Gaisin, F.M., Galimzyanov, I.I., Gilmutdinov, A.H., Shakirova, E.F.: Thermograms of High-Frequency Capacitive Discharge between Solid and Liquid Electrodes. High Temperature. 56(5), 821–823 (2018).

[26] Galimzyanov, I.I., Gaisin, A.F., Fakhrutdinova, I.T., Shakirova, E.F., Akhatov, M.F., Kayumov, R.R.: Characteristics of the Development of Electric Discharge between Jet Anode and Liquid Cathode. High Temperature. 56(2), 296–298 (2018).

Modeling of the Magnetic Field and Current Density Distributions in HTS SMES Systems

Vladislav M. Govor
Peter the Great St. Petersburg Polytechnic University
Saint-Petersburg, Russia
pwnway@gmail.com

Alexander G. Kalimov
Peter the Great St. Petersburg Polytechnic University
Saint-Petersburg, Russia

Evgenii N. Kobzar
Peter the Great St. Petersburg Polytechnic University
Saint-Petersburg, Russia

Abstract—Ensuring high performance of superconductor magnetic energy storages (SMES) requires conducting calculations to establish operational limits. This paper proposes an example of such calculations. The T-A formulation in conjunction with the homogenization approach is applied for high temperature superconducting (HTS) pancake coil modeling. The T-A formulation is relevant for calculation of tapes with a thin superconducting layer. And the homogenization approach simplifies problem formulation. Kim critical state model and the continuous E-J power law describe the material properties of the coil. Given problem formulation allows to gain distributions of the magnetic flux density and the current density at the coil cross-section. The dependence of hysteresis losses on the transport current amplitude and frequency are obtained. Based on these dependences, the value of critical current at 50 Hz is determined for the considered coil configuration. The current value, at which pancake coil remains in a superconducting state over the entire frequency range from $5 \cdot 10^{-5}$ Hz to 50 Hz, is calculated.

Keywords— HTS modeling, SMES, superconducting 2G tapes, coil homogenization, T-A formulation, hysteresis losses.

I. INTRODUCTION

Renewable power sources such as wind and solar power plants have an issue that needs to be solved before they would play a vital role in systems of power generation. This issue is related to power stability and to inconsistency of power producing. Renewable power sources operate in conjunction with energy storage systems to overcome periods of power fluctuations [1]. SMES systems meet all requirements applied to these systems.

During periods of instability, SMES systems provide an utility of power management, thereby granting additional stability to a system [2]. In cases when such as a loss of a transmission line aligns with a high load or one of major power sources, a power system could suffer from low-frequency oscillations or voltage fluctuations. SMES systems are capable of providing sufficient active and reactive power to stabilize a power system in this situation [3]. At the same time, they are able to accumulate great amount of energy and to charge and discharge at very high speed (Fig. 1). Moreover, SMES could withstand multiple recharge cycles without efficiency degradation [4, 5].

One of the SMES main parts is an HTS coil. SMES consists of a stack of pancake coils, which are wounded with a 2G HTS tape. These tapes have several layers. The set of layers differs from one tape to another. However, all current flows only through the HTS layer while it is in superconducting state. But surpassing the critical value of magnetic field or current density or increased heating because of losses during transient process could lead to the HTS transition to a normal state.

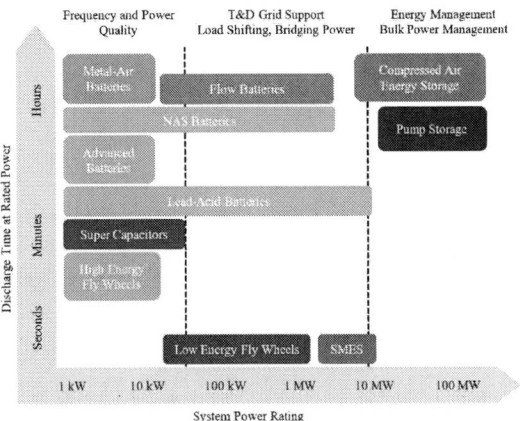

Fig. 1. Comparison of discharge time and power ratings for different energy storage systems

A magnetic field penetrates into type-II superconductors when its strength exceeds the first critical value. Current vortices, which appeared on a superconductor surface, are attracted to pinning centers located at crystal lattice defects. Further vortices penetration into deeper superconductor layers requires a Lorentz force exceeding the attracting force of pinning centers. Critical state models (CSM) provide the macroscopic description of these electromagnetic processes.

The two main critical state models are Bean-London CSM [6] and Kim CSM [7]. The assumption of the first CSM is that the critical current density remains constant throughout a superconductor. Kim's critical state model sets the dependence of current density form magnetic flux density as follows

$$\frac{\alpha}{\mathbf{J}} = B_0 + \mathbf{B},$$

where α and B_0 are the empirical constants dependent on a critical temperature and a type of superconductor materials. Bean-London model is useful for an estimation of some quantities, but in general Kim model and its modifications are applied for calculations of superconductors [8].

Whereas Bean-London CSM establishes a step change of the electric field at a moment of a superconductor transition to a normal state, Kim model assumes a continuous function [9]. The E-J power law [10] sets such a relation between the current density and the electric field strength.

$$\mathbf{E}(\mathbf{J}) = E_0 \cdot \frac{\mathbf{J}}{J_c} \cdot \left| \frac{\mathbf{J}}{J_c} \right|^{n-1},$$

979-8-3503-8358-4/23 $31.00 © 2023 IEEE

where $E_0 = 1$ μV/cm is the standard electric field criterion, J_c is the critical current density, $n = 20 - 30$ (for 2G HTS) is the empirical power factor.

Losses in superconductor occur in cases when the transport current or the applied magnetic field changes over time. There are two types of losses in composite superconductor: the hysteresis losses that occur inside a superconducting part of tapes and the coupling losses inside metal parts of tape [11].

2G HTS tapes are relatively expensive, so it becomes a vital issue to ensure the SMES maximum performance [12]. Nowadays a determination of the certain limit for SMES normal operation could be performed with FEM calculations. FEM models of HTS tapes use different formulations of Maxwell's equations [13]. In this paper the *T-A* formulation was implemented with COMSOL Multiphysics. This type of formulation proved its efficiency for a thin superconductor modeling that becomes more relevant for pancake coils made of flat HTS tapes [14].

This paper is organized as follows: Section 2 gives theoretical description of the problem formulation. Section 3 provides the overview of results that were obtained. Section 4 summarizes conclusions of the study.

II. PROBLEM FORMULATION

One of the earliest T-A formulations is introduced in [15]. The modified version of this formulation could be found in [16]. Due to its simplicity the T-A formulation coupled with model homogenization [17] becomes a powerful tool to conduct calculations of pancake coils. The task of modeling of a pancake superconducting coil is a non-trivial problem due to the set of several factors. The first one is a nonlinear material property, which makes solution convergence more difficult. The second one is the high value of aspect ratio because of the great difference between width and thickness of 2G HTS tapes. Model homogenization easily deals with the second factor. The main idea is that a superconducting coil could be represented in a model as a bulk with homogeneous structure and anisotropic properties [18]. This technique allows reducing the number of elements necessary for coil modeling without accuracy degradation. Visualization of the homogenized model and differences from a full one could be observed in Fig. 2.

The calculation is performed with respect to two quantities: the magnetic vector potential **A** and the current vector potential **T**. The **A** calculation is performed for the entire model, and the **T** calculation only for the area of a coil cross-section. Other quantities could be described by a set of equations. The two main of them are

$$\nabla \times \left(\frac{1}{\mu} \nabla \times \mathbf{A} \right) = \mathbf{J},$$

$$\nabla \times \rho \nabla \times \mathbf{T} = -\frac{\partial \mathbf{B}}{\partial t},$$

where μ is the magnetic permeability, ρ is the resistivity, **B** is the magnetic flux density.

The current density is defined as

$$\mathbf{J} = \nabla \times \mathbf{T}.$$

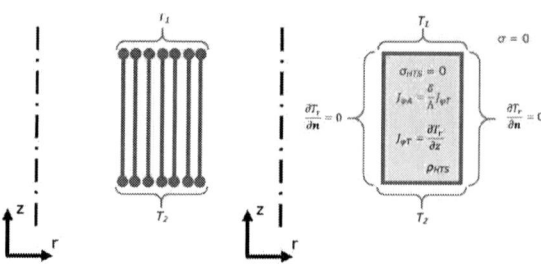

Fig. 2. Transformation of the full coil model (left) to the homogenized coil model (right). The red colored text is related to the A formulation and the black colored text is related to the T formulation

The HTS layers are much thinner than the full tape. The current is concentrated in superconducting layer, when the current density is lower than the critical value. This fact allows assuming that **T** vector consists of the only T_r component and the current density consists of the only J_φ component as well:

$$J_{\varphi T} = \frac{\partial T_r}{\partial z}.$$

The second of two main equations becomes

$$\frac{\partial}{\partial z} \left(\rho \frac{\partial T_r}{\partial z} \right) = \frac{\partial B_r}{\partial z},$$

where B_r could be derived from $\mathbf{B} = \nabla \times \mathbf{A}$.

The first type boundary conditions are obtained from the equation for calculating current through the cross-section S of a superconducting layer.

$$I = \int_S \mathbf{J} dS = \int_S \nabla \times \mathbf{T} dS = \oint_{\partial l} \mathbf{T} dl,$$

where ∂l is the contour of this cross-section.

Considering T_z component equal to zero, equation for current evaluation turns to

$$I = (T_1 - T_2)\delta,$$

where δ is the superconducting layer thickness. Then the boundary conditions applied to top and bottom sides of the coil set up to provide the current vector potential difference between them. Also, the Newman boundary conditions set up at the vertical sides of the coil are

$$\frac{\partial T_z}{\partial \mathbf{n}} = 0,$$

where **n** is the normal unit vector of this sides.

The coil area resistivity equals to ρ_{HTS} derived from the E-J power law.

The current density for the *A* formulation is recalculated as

$$J_{\varphi A} = \frac{\delta}{\Lambda} J_{\varphi T},$$

where Λ is the tape thickness which includes insulation, $J_{\varphi T}$ is the φ-component of current density obtained with the *T* formulation. This new value of the current density is set up as an external current source for the entire coil cross-section. The zero-conductivity property is applied both for the coil and the surrounding area in the *A* formulation.

The HTS electromagnetic properties are set with the E-J power law ($n = 25$) and Kim critical state model. Its parameters were taken from [19] at the temperature equal to 77 K.

$$J_c(\boldsymbol{B}) = J_{c0}\frac{B_0}{B_0 + |B_r|},$$

where $J_{c0} = 1{,}11 \cdot 10^{10}\frac{A}{m^2}$, $B_0 = 120$ mT.

Following equation is used for the evaluation of instantaneous losses

$$\xi = \iint_{HTS} \boldsymbol{E} \cdot \boldsymbol{J} \cdot r\,dr\,dz.$$

Since there is no flowing current inside the coil at the very beginning of simulation, the hysteresis losses are different throughout two halves of the first period. However, the pattern of losses curve of the second part of the first period is further reproduced over all consequential periods. That is, instantaneous losses time dependence becomes steady by the beginning of the second half of the first cycle. Thus, the instantaneous losses integrated from $T/2$ to T and multiplied by two represent the steady-state value of losses occurred in the coil during one period.

$$W = 2 \cdot \int_{\frac{T}{2}}^{T} \iint_{HTS} \boldsymbol{E} \cdot \boldsymbol{J} \cdot r\,dr\,dz\,dt.$$

III. RESULTS

The problem of critical current calculation was solved for the pancake coil wounded with 2G HTS tape in a state of carrying sinusoidal alternating transport current. All dimensional parameters of the model are represented in Table 1. The specification of the SCS 12050 tape manufactured by SuperPower [20] is considered in a model design. The tape wide side is oriented along the vertical axis.

TABLE I.   PARAMETERS OF THE MODELED COIL

Parameter	Value
Tape width	12 mm
Tape thickness	93 μm
Thickness of superconducting layer	1 μm
Thickness of insulation between turns	200 μm
Coil inner radius	100 mm
Number of turns	20
Single tape critical current	300 A

The considered cases are the transport current flowing with variable amplitude and frequency. The transport current changes its instantaneous value by the sinusoidal law.

A. Current variation

There is constant frequency of 50 Hz and variable amplitude from 10% to 100% of critical value in the first case. Losses are obtained for each current amplitude value.

Fig.3 shows the relation between the hysteresis losses per cycle and the amplitude of transport current. The value of the hysteresis losses instantly rises, when the transport current exceeds 80% of critical current. The hysteresis losses are considered more accurately at the interval of transport current from 70% to 80% of the critical value. The slightly higher increase could be noticed after 77% value. The instantaneous losses during the period for the two amplitudes closest to 77% are presented in Fig. 4.

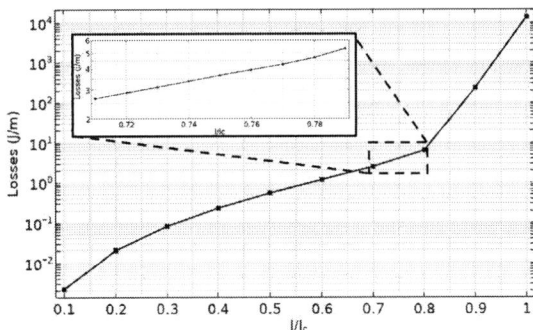

Fig. 3. Dependence of hysteresis losses per period on the amplitude of transport current

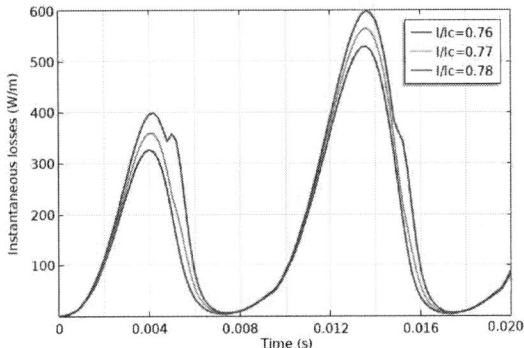

Fig. 4. Instantaneous losses of pancake coil

Obvious distortion of the instantaneous losses curve could be seen at 78% value. An appearance of the local maximums means the current surpassing critical value. The other two cases require consideration that is more detailed. Their magnetic flux density and current distributions for the second half of cycle are presented in Fig.5 and Fig. 6.

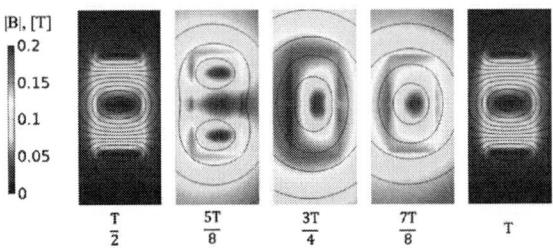

Fig. 5. Magnetic flux density distribution at the transport current equal to 76% of critical current

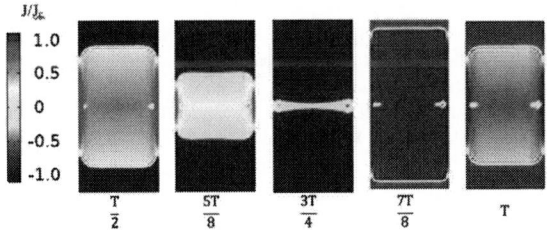

Fig. 6. Current density distribution over the coil cross-section at the transport current equal to 76% of critical current

The magnetic flux density distribution of the two cases has no significant difference. It could be seen that at zero transport current there are three zones with zero field. They are located at the center of the coil and at the top and bottom sides. The central zero field zone remains throughout the

979-8-3503-8358-4/23 $31.00 © 2023 IEEE   49

entire cycle, but the side zones are superseded to the center, and the magnetic field along the coil perimeter grows as the transport current increases. These three zones merge into one at the moment of maximal transport current.

The current density distributes deeper into coil turns as the transport current increases. In case of 76% transport current areas of zero current density exist at every moment. The superconducting coil remains incompletely penetrated. But in the case of 77% transport current areas of zero current density disappear after the transport current reaches its maximum value. There is no space for redistribution of the current density inside coil turns after complete penetration of coil, and further change of applied current leads to a local excess of the current density over the critical value. It causes the increase in the electric field strength according to the E-J power law and the corresponding rise of losses.

The established value of critical current at 50 Hz for presented coil configuration is 76% of the single tape critical current value, and it equals to 228 A. The exploitation of SMES proceeds with current value about 70 – 80% of coil critical value to provide maximum performance and sufficient reliability. Taking into account the fact that SMES usually consist of several HTS pancake coils, the actual value of the operating current during a transient state should be even lower than 160 A.

B. Frequency variation

The constant values of the transport current and its frequency variating from $5 \cdot 10^{-5}$ Hz to 50 Hz are considered in this case. The calculation of the hysteresis losses are performed for three current values (50%, 55% and 70% of the single tape critical current).

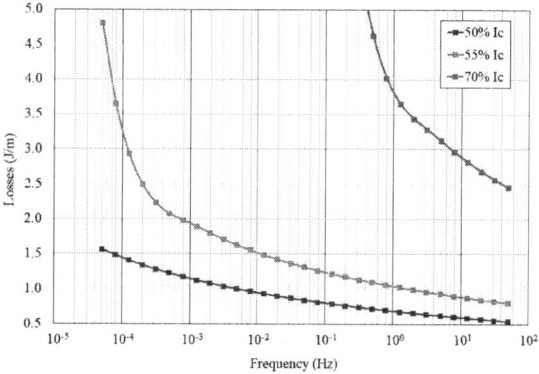

Fig. 7. Hysteresis losses per cycle of pancake coil as a function of the transport current frequency

The losses rise is observed at two values of the transport current. The hysteresis losses significantly increase at frequencies lower than $5 \cdot 10^{-4}$ Hz and 2 Hz at 55% and 70% current, respectively. The instantaneous losses time dependences for the first frequency and its adjacent value are presented in Fig. 8.

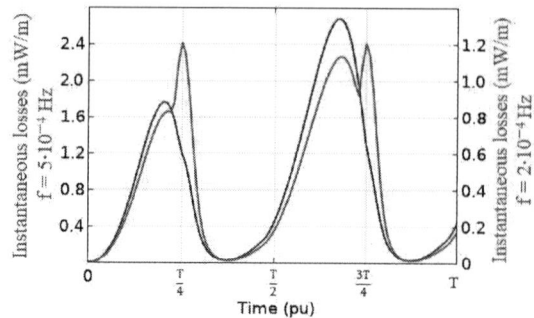

Fig. 8. Instantaneous losses of pancake coil at two frequencies of the transport current ($I/I_c = 0.55$)

Closer look at these dependences demonstrates the presence of a curve pike on the lower frequency. The pike presence evidences the local transition of the tape to a normal state. In case of 50% transport current, the coil completely remains in superconducting state over entire frequency range. The critical current of the coil decreases with the period increasing.

Despite the lower value of the instantaneous losses, the hysteresis losses per cycle are higher for low frequencies. The losses per cycle are an integral quantity, and even relatively low instantaneous value could lead to significant overall losses.

IV. CONCLUSION

The 2G HTS pancake coil was considered in this study. The T-A formulation of problem in conjunction with the homogenization method were applied to evaluate the coil critical current and to investigate its dependence on frequency. Kim critical state model and the E-J power law were implemented to describe material properties of the superconductor.

The combination of mentioned approaches allows defining the limits of the coil operation and estimating the hysteresis losses in superconductor. The critical current of the considered HTS coil configuration is 76% of the single tape critical current value at 50 Hz. Moreover, modeling with variable frequency represents the coil critical current decrease and the hysteresis losses rise at low frequencies. That is, a determined critical current from one frequency would be over critical at another. Thus, the considered coil configuration remains in superconducting state at 50% transport current over entire frequency range from $5 \cdot 10^{-5}$ Hz to 50 Hz.

So, ensuring the maximum performance of SMES systems requires considering every operational state of the system and the critical current calculations for each of them, because these values could differ drastically.

REFERENCES

[1] Shaaban, E.F., Hassan, A.E.W., Mansour, D.E.A.: Using SMES for Voltage Stabilization of PMSG Based Wind Energy System. In: IEEE Conference on Power Electronics and Renewable Energy, CPERE 2019 (2019). https://doi.org/10.1109/CPERE45374.2019.8980257.

[2] Xue, X.D., Cheng, K.W.E., Sutanto, D.: A study of the status and future of superconducting magnetic energy storage in power systems. Supercond. Sci. Technol. 19, (2006). https://doi.org/10.1088/0953-2048/19/6/R01.

[3] Ngamroo, I.: An optimization of superconducting coil installed in an hvdc-wind farm for alleviating power fluctuation and limiting fault current. IEEE Trans. Appl. Supercond. 29, (2019). https://doi.org/10.1109/TASC.2018.2881993.

[4] Akinyele, D.O., Rayudu, R.K.: Review of energy storage technologies for sustainable power networks. Sustain. Energy Technol. Assessments. 8, (2014). https://doi.org/10.1016/j.seta.2014.07.004.

[5] Indu, P.S., Jayan, M. V.: Frequency regulation of an isolated hybrid power system with Superconducting Magnetic Energy Storage. In: Proceedings of 2015 IEEE International Conference on Power, Instrumentation, Control and Computing, PICC 2015 (2016). https://doi.org/10.1109/PICC.2015.7455752.

[6] Bean, C.P.: Magnetization of high-field superconductors. Rev. Mod. Phys. 36, (1964). https://doi.org/10.1103/RevModPhys.36.31.

[7] Kim, Y.B., Hempstead, C.F., Strnad, A.R.: Magnetization and critical supercurrents. Phys. Rev. 129, (1963). https://doi.org/10.1103/PhysRev.129.528.

[8] Robert, B.C., Fareed, M.U., Ruiz, H.S.: How to Choose the Superconducting Material Law for the Modelling of 2G-HTS Coils. Materials (Basel). 12, 2679 (2019). https://doi.org/10.3390/ma12172679.

[9] Morandi, A.: 2D electromagnetic modelling of superconductors. Supercond. Sci. Technol. 25, (2012). https://doi.org/10.1088/0953-2048/25/10/104003.

[10] Rhyner, J.: Magnetic properties and AC-losses of superconductors with power law current-voltage characteristics. Phys. C Supercond. its Appl. 212, (1993). https://doi.org/10.1016/0921-4534(93)90592-E.

[11] Grilli, F., Pardo, E., Stenvall, A., Nguyen, D.N., Yuan, W., Gomory, F.: Computation of losses in HTS under the action of varying magnetic fields and currents. IEEE Trans. Appl. Supercond. 24, (2014). https://doi.org/10.1109/TASC.2013.2259827.

[12] Dechanupaprittha, S., Sakamoto, N., Hongesombut, K., Watanabe, M., Mitani, Y., Ngamroo, I.: Design and analysis of robust SMES controller for stability enhancement of interconnected power system taking coil size into consideration. In: IEEE Transactions on Applied Superconductivity (2009). https://doi.org/10.1109/TASC.2009.2018492.

[13] Grilli, F.: Numerical modeling of HTS applications. IEEE Trans. Appl. Supercond. 26, (2016). https://doi.org/10.1109/TASC.2016.2520083.

[14] Liang, F., Venuturumilli, S., Zhang, H., Zhang, M., Kvitkovic, J., Pamidi, S., Wang, Y., Yuan, W.: A finite element model for simulating second generation high temperature superconducting coils/stacks with large number of turns. J. Appl. Phys. 122, (2017). https://doi.org/10.1063/1.4995802.

[15] Zhang, H., Zhang, M., Yuan, W.: An efficient 3D finite element method model based on the T-A formulation for superconducting coated conductors. Supercond. Sci. Technol. 30, (2017). https://doi.org/10.1088/1361-6668/30/2/024005.

[16] Berrospe-Juarez, E., Zermeño, V.M.R., Trillaud, F., Grilli, F.: Real-time simulation of large-scale HTS systems: Multi-scale and homogeneous models using the T-A formulation. Supercond. Sci. Technol. 32, (2019). https://doi.org/10.1088/1361-6668/ab0d66.

[17] Clem, J.R., Claassen, J.H., Mawatari, Y.: AC losses in a finite Z stack using an anisotropic homogeneous-medium approximation. Supercond. Sci. Technol. 20, (2007). https://doi.org/10.1088/0953-2048/20/12/008.

[18] Zermeno, V.M.R., Abrahamsen, A.B., Mijatovic, N., Jensen, B.B., Sørensen, M.P.: Calculation of alternating current losses in stacks and coils made of second generation high temperature superconducting tapes for large scale applications. J. Appl. Phys. 114, (2013). https://doi.org/10.1063/1.4827375.

[19] Yuan, W.: Second-Generation High-Temperature Superconducting Coils and Their Applications for Energy Storage. Springer London, London (2011). https://doi.org/10.1007/978-0-85729-742-6.

[20] SuperPower Inc.: 2G HTS Wire Specification, http://www.superpower-inc.com, last accessed 2020/12/03.

The Effect of Liquid Dielectric Impregnation on the Self-Healing Characteristics

Ivan O. Ivanov
Higher School of High Voltage Engineering
Peter the Great St.Petersburg Polytechnic University
Saint Petersburg, Russia
ivanov.eicc@yandex.ru

Efrem G. Feklistov
Higher School of High Voltage Engineering
Peter the Great St.Petersburg Polytechnic University
Saint Petersburg, Russia
efrem.feklistov@mail.ru

Abstract — Self-healing processes in PP, PET and PPS metalized films in dry and impregnated by different liquid dielectrics states were investigated. It was determined that liquid dielectric impregnation increases electric strength and self-healing energy. In the cases of impregnated polymer films the value of breakdown voltage and self-healing energy have increased by 4 – 12 % and 7 – 24 % respectively, and self-healing duration remained constant. All types of used liquid dielectrics have led to increasing of dissipated energy during self-healing process. It was shown that non-polar liquid dielectrics with low value of dielectric permittivity have a less effect on dissipated energy increasing.

Keywords — *metallized film capacitor, polymer film, liquid dielectric, self-healing characteristics*

I. Introduction

Metallized film capacitors (MFC) are the key components of different high voltage pulse power equipment such as impulse voltage and current generators, switching impulse generators, laser pumping systems, particle accelerators, and others [1-5]. There are various options for the design of present capacitors type. The main part is capacitance element made of polymer film covered by a thin metal layer with thickness of several tens of nanometers which is used as electrodes. MFCs replaced capacitors with foil electrodes and combine dielectric consisted on polymer film and capacitor paper impregnated by liquid dielectric. In the capacitors industry mineral (petroleum-based) oil and different aromatic hydrocarbons are used as a liquid dielectric [6-9]. The advantages of MFCs are high operating electric field (more than 200 kV/mm) and high reliability due to self-healing (SH) ability. SH process consists in isolating of breakdown channel after the dielectric breakdown and following restoring of operating ability [10-12].

The use of liquid dielectric impregnation is associated with the necessity to exclude air gaps inside the capacitive element in order to increase the operating electric field. The improvement of MFCs manufacturing technology made it possible to abandon the use of liquid dielectrics and switch to the production of dry-type high-voltage capacitors. The technology of liquid dielectric impregnation of capacitance elements was improved and that made it possible not to use a paper in capacitors' design. Nevertheless, the industry has not completely replaced impregnated capacitors with dry MFCs. Particularly, liquid dielectrics are still used in low voltage capacitors.

The main reason of impregnated capacitance elements using is the need to increase volumetric energy density. It is known that volumetric energy density W_V is proportional to the value of relative permittivity ε and the square of electric field E

$$W_V = \frac{\varepsilon_r \varepsilon_0 E^2}{2} k. \qquad (1)$$

In this equation ε_0 is dielectric constant and k is a packaging factor. Using of liquid dielectric in high voltage capacitors allows to increase the values of relative permittivity and electric field. In present days leading manufacturers of high-voltage pulse capacitors produce both dry and impregnated MFCs.

The purpose of this study was to investigate the effect of the liquid dielectric impregnation on the SH characteristics. In accordance with this, the main tasks included the development of a methodology and an experimental setup for studying SH processes on model samples.

II. Materials and Methods

Experimental investigations of SH processes were carried out using model samples of capacitance element. Various combinations of polymer films and liquid dielectrics were studied. Polypropylene (PP), polyethylene terephthalate (PET) and polyphenylene sulfide (PPS) films in dry state and in combination with liquid dielectrics: mineral capacitors oil and dioctyl phthalate (DOP) were investigated. Polymer films were 6 um thickness and had metallization layer with 20 nm thickness.

The view of model sample of capacitance element is shown in Fig. 1. Polyimide mask with 20 um thickness was placed between investigated films to localize the breakdown zone and to exclude surface discharges. The model samples were fixed in a hydraulic press that created pressure ~ 3 atm. on the films. Block diagram of the experimental setup is shown in Fig. 2. The main elements of the experimental setup are high voltage direct current source (HVDC), supply capacitor $C_0 = 47$ nF, current resistance $R_0 = 1$ Ω, R-C voltage divider and digital oscilloscope. To study the SH process, a linear rise of the applied voltage at a rate of $dV/dt = 50$ V/s was used. After electrical breakdown of the polymer film HVDC source was switched of and voltage and current waveforms were saved.

979-8-3503-8358-4/23 $31.00 © 2023 IEEE

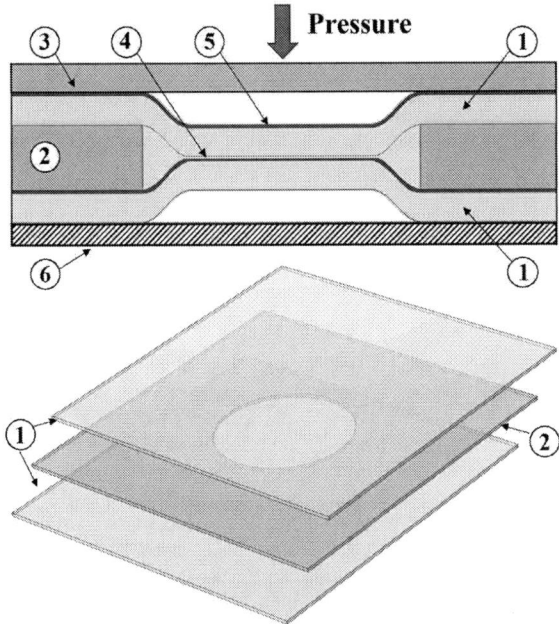

Fig. 1. The view of model sample of capacitance element: 1– investigated polymer film, 2 – polyimide mask, 3 – cover polyimide film, 4 – liquid dielectric, 5 – metallization.

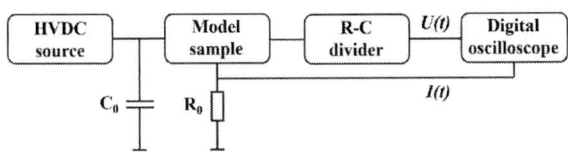

Fig. 2. Block diagram of the experimental setup.

III. RESULTS AND DISCUSSION

More SH energy is one of the most important characteristics of SH process [13]. SH energy is a part of stored energy which dissipated in the breakdown channel. SH energy values were used to compare experimental samples types and to determine the effect of impregnation with a liquid dielectric. SH energy is determined by equation

$$W_{SH} = \int_{0}^{\tau_{SH}} U(t) \cdot I(t) dt \,, \qquad (2)$$

where $U(t)$ is voltage, $I(t)$ is current and τ_{SH} is SH duration. Typical voltage and current waveforms are shown in Fig. 3.

It is well known that impregnation of capacitance elements by liquid dielectric leads to increasing its electric strength that allows to increase capacitor's operating electric field. This effect is achieved by filling the air gaps with a liquid dielectric which has higher values of electric strength relative permittivity. Experimental investigations showed that impregnation of investigated films by mineral capacitors oil leads to electric strength increasing by 4 – 6 % and SH energy increasing by 8 – 12 %. In the case of impregnation by DOP electric strength increasing by 7 – 9 % and SH energy increasing by 17 – 24 % was observed. Weibull distributions of electric strength and SH energy of dry and impregnated PP films are shown in Fig. 4 and 5. The same distributions were obtained for PET and PPS films.

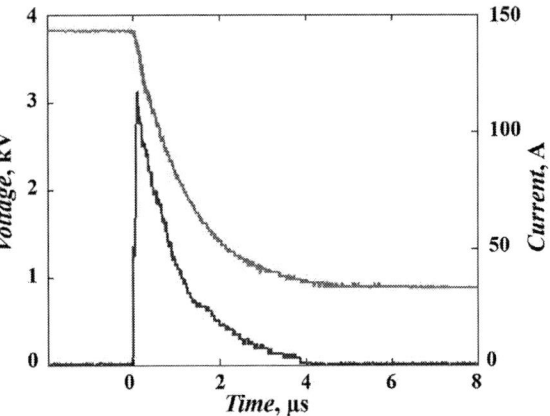

Fig. 3. Typical voltage and current waveforms. The case of dry PP.

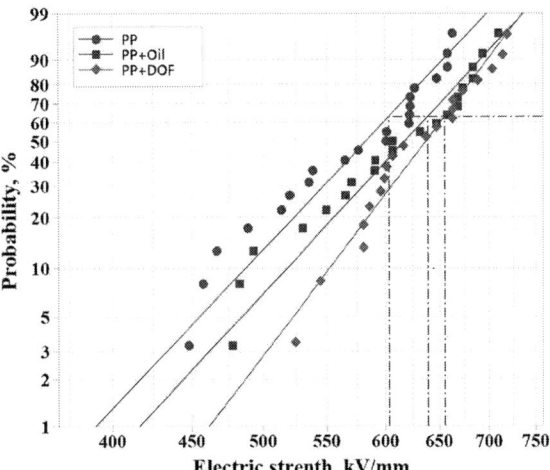

Fig. 4. Weibull distribution of electric strength of dry and impregnated PP films.

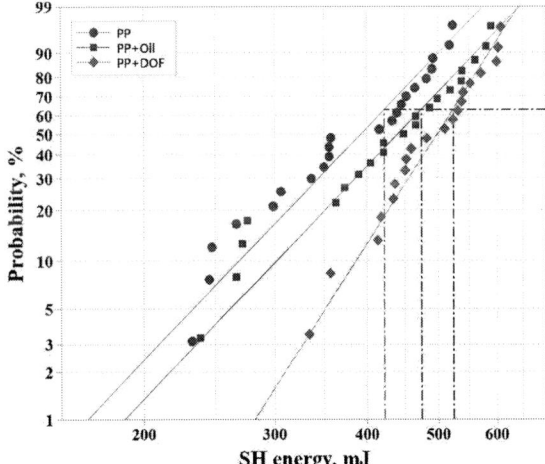

Fig. 5. Weibull distribution of SH energy of dry and impregnated PP films.

The obtained voltage and current waveforms made it possible to determine the dependences of breakdown channel resistance $R_{CH}(t)$ during SH acts and the duration of SH acts. It was established that minimal value of the $R_{CH}(t)$ dependence takes value of 15 – 60 Ω and the duration of SH

979-8-3503-8358-4/23 $31.00 © 2023 IEEE

acts varies in the range of 2.5 – 8 μs. The obtained values do not depend on the types of polymer film and liquid dielectric. Typical dependance of breakdown channel resistance during SH act is shown in Fig 6.

In the presence of a liquid dielectric between polymer films, it was expected to detect its effect on evolution of microarc discharge in the breakdown channel. It was assumed that the duration of SH events would be shorter in impregnated samples compared to dry experimental samples. This was motivated by the assumption of a possible increase in the heat flux from the breakdown channel and a decrease in the duration of the microarc discharge. As a consequence, the shorter duration of SH process should also lead to a decrease in the SH energy. However, based on the obtained results, it can be concluded that presence of a liquid dielectric does not affect on the development of a microarc discharge in the breakdown channel due to short duration of the process (units of microseconds).

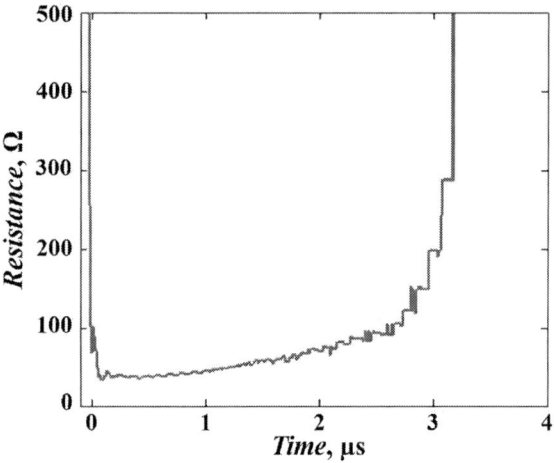

Fig. 6. Typical dependance of breakdown channel resistance during SH act. The case of dry PP.

TABLE I. SELF-HEALING CHARACTERISTICS OF INVESTIGATED POLYMER FILMS

Samples type	E_{BD}, kV/mm	W_{SH}, mJ	τ_{SH}, μs
PP	604	423	3.6
PP + Oil	639	475	3.9
PP + DOP	656	524	4.1
PET	585	410	5.5
PET + Oil	604	446	5.9
PET + DOP	621	487	6.3
PPS	434	220	3.4
PPS + Oil	460	244	3.5
PPS + DOP	482	260	3.6

The average values of electric strength, SH energy and SH duration of investigated dry and impregnated polymer films are presented in the Table I. It is seen that PPS film has a slightly lower electric strength than PP and PET films. These results are consistent with the literature data [6]. Due to lower breakdown voltage PPS film has significantly less SH energy in comparison with other studied films. As a

result, lower SH energy provides higher reliability of electrical capacitor.

Despite the fact that mineral capacitors oil and DOP have approximately the same electric strength (~ 20 kV/mm) experimental samples with DOP impregnation have more higher electric strength. This result may be associated with a higher value of relative permittivity of DOP ($\varepsilon \approx 5.5$). Mineral capacitors oil is non-polar liquid dielectric with lower value of relative permittivity ($\varepsilon \approx 2.2 - 2.4$).

IV. CONCLUSION

The influence of liquid dielectric impregnation on self-healing characteristics were studded. Experimental investigations were carried out using model samples of capacitance element. The self-healing parameters: electric strength, dissipated energy, duration of the process and the dependencies of breakdown channel resistance were determined. As expected, impregnation of polymer films by liquid dielectrics leads to increasing of electric strength. The highest value of electric strength was observed when DOP was used as impregnating structure. This result may be associated with a higher value of relative permittivity of DOP. On the other hand, higher value of electric strength leads to higher value of dissipated energy during self-healing process that is a negative factor for capacitor reliability. The significant increasing of self-healing duration was not found. The obtained results can be used in numerical simulation of thermal fluctuations in metallized film capacitors after dielectric breakdown.

ACKNOWLEDGMENT

The research is supported by Russian Science Foundation (project № 19-79-10075).

REFERENCES

[1] W. Hauschild, E. Lemke, *High-Voltage Test and Measuring Techniques*, 1st ed. Springer: Berlin, Germany, 2014.

[2] H.M Ryan, *High-Voltage Engineering and Testing*, 3rd ed. The Institution of Engineering and Technology, London, UK, 2013.

[3] M.V. Fazio and H.C. Kirbie, "Ultracompact pulsed power," Proceedings of the IEEE, vol. 92, no. 7, pp. 1197-1204, July 2004. DOI: 10.1109/JPROC.2004.829002.

[4] D.F. Welch, "A brief history of high-power semiconductor lasers," in IEEE Journal of Selected Topics in Quantum Electronics, vol. 6, no. 6, pp. 1470-1477, Nov.-Dec. 2000. DOI: 10.1109/2944.902203.

[5] W. J. Sarjeant, J. Zirnheld and F. W. MacDougall, "Capacitors," in IEEE Transactions on Plasma Science, vol. 26, no. 5, pp. 1368-1392, Oct. 1998. DOI: 10.1109/27.736020.

[6] O. G. Gnonhoue, A. Velazquez-Salazar, É. David and I. Preda, "Review of technologies and materials used in high-voltage film capacitors", in Polymers, vol. 13, no. 5, 766, Feb. 2021. DOI: 10.3390/polym13050766.

[7] D.Q. Tan, "Review of Polymer-Based Nanodielectric Exploration and Film Scale-Up for Advanced Capacitors", in Advanced Functional Materials, vol. 30, no. 18, 1808567, March 2020. DOI: 10.1002/adfm.201808567.

[8] J. Ho, T. R. Jow and S. Boggs, "Historical introduction to capacitor technology," in IEEE Electrical Insulation Magazine, vol. 26, no. 1, pp. 20-25, Jan.-Feb. 2010. DOI: 10.1109/MEI.2010.5383924.

[9] I. Fofana, "50 Years in the Development of Insulating Liquids," in IEEE Electrical Insulation Magazine, vol. 29, no. 5, pp. 13-25, 6585853, Aug. 2013. DOI: 10.1109/MEI.2013.6585853.

[10] J. H. Tortai, A. Denat, and N. Bonifaci, "Self-healing of capacitors with metallized film technology: Experimental observations and theoretical model," in Journal of Electrostatics, vol. 53, no. 2, pp. 159–169, Aug. 2001. DOI: 10.1016/S0304-3886(01)00138-3.

[11] V. O. Belko, O. A. Emelyanov, I. O. Ivanov and A. P. Plotnikov, "Self-Healing Processes of Metallized Film Capacitors in Overload Modes—Part 1: Experimental Observations," in IEEE Transactions on

Plasma Science, vol. 49, no. 5, pp. 1580-1587, May 2021. DOI: 10.1109/TPS.2021.3071187.

[12] I. Ivanov, K. Voloshin and K. Kulbako, "Modeling of metallized film capacitors segmented electrodes electrodynamic destruction," 2020 IEEE 3rd International Conference on Dielectrics (ICD), 2020, pp. 689-691. DOI: 10.1109/ICD46958.2020.9341865.

[13] V. Belko, I. Ivanov, A. Plotnikov, V. Belanov, "Energy characteristics of self-healing process in metallized film capacitors," 2019 International Scientific Conference on Energy, Environmental and Construction Engineering, 02006, 2019, pp. 1-4. DOI: 10.1051/e3sconf/201914002006.

The Phenomenon of Low Frequency Dielectric Losses Increasing in Metallized Film Capacitors

Ivan O. Ivanov
Higher School of High Voltage Engineering
Peter the Great St.Petersburg Polytechnic University
Saint Petersburg, Russia
ivanov.eicc@yandex.ru

Ahmet A. Hojamov
Higher School of High Voltage Engineering
Peter the Great St.Petersburg Polytechnic University
Saint Petersburg, Russia
akhodzhamov@mail.ru

Abstract — **Metallized film capacitors have the highest operating electric field (more than 200 kV/mm) compare another capacitor types. Such high operating electric field explains by self-healing ability which related with isolating of breakdown channel. Multiple self-healing acts lead to changing of capacitors parameters: capacitance, dielectric losses and equivalent parallel resistance. These parameters changing during capacitors aging was investigated in a wide frequency range (0.1 Hz – 1 MHz). It was found that as the capacitors aging, dielectric losses increase at low frequency of electric field (less than 10 Hz). This phenomenon of low frequency dielectric losses increasing can be explained by formation of graphite layers between polymer films near breakdown channels.**

Keywords — metallized film capacitors, dielectric breakdown, self-healing, dielectric losses

I. Introduction

Metallized film capacitors (MFCs) have the highest value of specific energy compare with another electrical capacitors' types. This fact explains by high value of operating electric field. It well known that specific energy and electric field are related by following expression

$$W_V = \frac{\varepsilon_r \varepsilon_0 E^2}{2} k, \qquad (1)$$

where E is electric field, ε_r is relative permittivity of dielectric material, ε_0 is the dielectric constant and k is a packaging factor.

The value of relative permittivity of used polymer films is not high. Non-polar films have the value of relative permittivity $\varepsilon_r = 2 - 2.8$ and polar films $\varepsilon_r = 3 - 12$ [1-3]. For comparison, ferroelectric ceramics have the value of relative permittivity $\varepsilon_r > 1000$ [1-6]. Nevertheless, the important advantage of polymer films is high value of electric strength – 500 kV/mm and more [7-11]. Using of thin nanometer metal layers as electrodes gives MFCs a unique feature. This feature is self-healing (SH), that is capacitor's ability to continue its operating in the case of dielectric breakdown [11-13]. High value of electric strength and SH ability cause high value of operating electric field (> 200 kV/mm), that gives the value of specific energy up to $2 - 3$ J/cm^3 [13-15].

SH acts definitely affect on electrical parameters of MFCs. First of all, there is a decreasing of capacitance due to a decreasing of electrodes area. In additional, SH process finishes with formation of a carbon layer in the breakdown channel and around it in the demetallized area (Fig. 1). This layer has increased electrical conductivity. Formation of a carbon layer leads to decreasing of capacitor's equivalent parallel resistance [16-19]. It was established that this MFCs parameter is the most sensitive when SH acts occur [20]. Decreasing of capacitance and equivalent parallel resistance

leads to increasing of dielectric losses. These capacitor's parameters are associated with the expression

$$tg\,\delta = \frac{1}{\omega R_p C}, \qquad (2)$$

where tg δ is dielectric losses, R_p is equivalent parallel resistance, C is capacitance and ω is angular frequency.

SH processes can be characterized by the value of dissipated energy. The part of capacitor's stored energy which dissipates during this process is SH energy that is determined by the expression

$$W_{SH} = \frac{C\left(U_{BD}^2 - U_{RES}^2\right)}{2}, \qquad (3)$$

where U_{BD} is a breakdown voltage and U_{RES} is a residual voltage. Depending on capacitance and breakdown voltage values SH energy can take value from a few units to several hundred of millijoules and range from 10 to 90 % of the total stored energy of the capacitor. Thus, it is advisable to use not the number of SH acts, but their cumulative energy to make a quantitative assessment of SH acts [20-21].

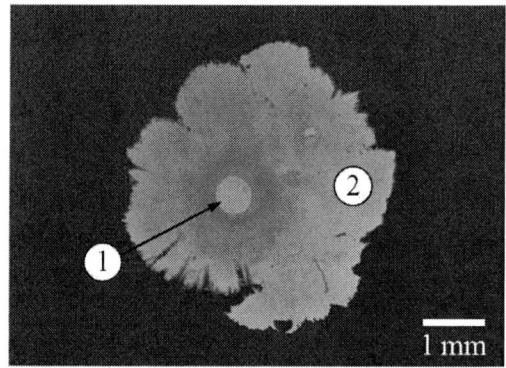

Fig. 1. The photograph of metallized film after self-healing: 1 – breakdown channel, 2 – demetallized area [13].

The values of capacitance and dielectric losses are usually used to evaluate MFCs' performance. As a rule, the parameters are measured at a frequency of 1 kHz. However, depending on capacitor's dielectric material, these parameters have a dependance on the voltage frequency. Therefore, the aim of present investigations was to determine changing of MFCs' parameters due to multiple impact of SH acts in a wide frequency range.

II. Materials and Methods

As the study objects different PP and PET metallized film capacitors were used. Nominal MFCs' parameters were capacitance $0.22 - 1$ µF and operating voltage $63 - 630$ V. Investigated capacitors had Al metallization with a thickness of $15 - 30$ nm. The parameters of investigated MFCs are presented in Table 1.

TABLE I.   PARAMETERS OF INVESTIGATED MFCS

Polymer film	Capacitance, µF	Nominal voltage, V	Metallization resistance, Ω/□
PP	0.22	250	3
PP	0.47	250	3
PP	1	250	3
PET	0.47	63	4
PET	0.47	250	3
PET	1	63	4
PET	0.33	630	2

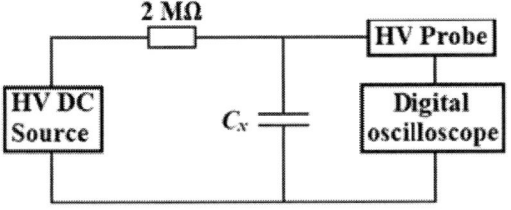

Fig. 2.   Block diagram of experimental setup.

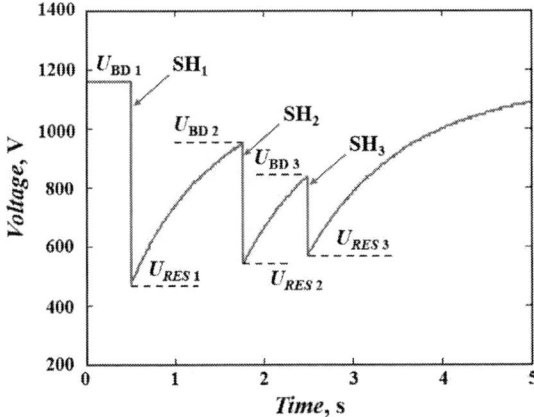

Fig. 3.   The example of voltage waveform consisting of 3 SH acts.

Block diagram of experimental setup is presented in Fig. 2. Experimental setup includes a high voltage DC source, current limiting resistance, high voltage probe and digital oscilloscope with segmented memory function. The example of voltage waveform is shown in Fig. 3. In the present waveform 3 SH acts were fixed.

The investigated capacitors were exposed to increased voltage that was $2.5 - 3$ of its nominal operating value. Every $10 - 15$ breakdown events capacitors' parameters were measured. Impedance analyzer was used to measure MFCs' parameters in the frequency range of $10^{-1} - 10^{6}$ Hz. The experiment finished when catastrophic failure had occurred. In this case investigated capacitor could not store a charge because of low value of its insulation resistance. After that, the SH acts were counted and their SH energy was determined by the equation (3).

III.   RESULTS AND DISCUSSION

More than 30 different MFCs were investigated according to described experimental method. As the capacitors' degradation under dielectric breakdowns, its parameters changing was observed. As an example, frequency dependencies of capacitance, dielectric losses and equivalent parallel resistance for initial and aged capacitors are shown in Fig. 4 – 6. In these figures W_Σ is cumulative energy of all SH acts, N is number of SH acts and ΔC is relative capacitance decreasing. These dependencies are related to following MFC: PP film, nominal capacitance 0.22 µF and operating voltage 250 V (see Table 1). The red lines correspond to the initial state, blue lines correspond to a slightly aged state and green lines correspond to a strongly aged state. Depending on MFC's type a strongly aged state may form at much higher values of cumulative energy and relative capacitance decreasing ($W_\Sigma = 5 - 7$ J, $\Delta C = 5 - 10\%$). The same character of parameters changing was observed for all types of investigated capacitors.

Capacitance decreasing was uniform over the entire frequency range as the MFCs degradation. The increasing of measured capacitance was observed at the frequency more than 10^{5} Hz. It should be noted that the impedance analyzer measures effective capacitance and not geometric capacitance. The Increasing of effective capacitance is related to capacitor's inductance and measuring near its resonant frequency. In the case of parallel equivalent scheme, effective and geometric capacitances are related by equation

$$C_{eff} = \frac{C_{geom}}{1 - \omega^2 L C_{geom}}, \qquad (4)$$

where L is capacitor's inductance. At the low frequency value $C_{eff} \approx C_{geom}$.

Frequency dependencies of dielectric losses in the initial MFCs increase in high frequency range (more than 10^{4} Hz). This fact explains by increasing of polarization losses. Following increasing of dielectric losses in the aged capacitors explains by decreasing of resonant frequency value. As the capacitors' degradation, increasing of dielectric losses in low frequency range (less than 10 Hz) was observed. At frequency 0.1 Hz dielectric losses of all investigated MFC took the value in the range of $0.3 - 0.8$.

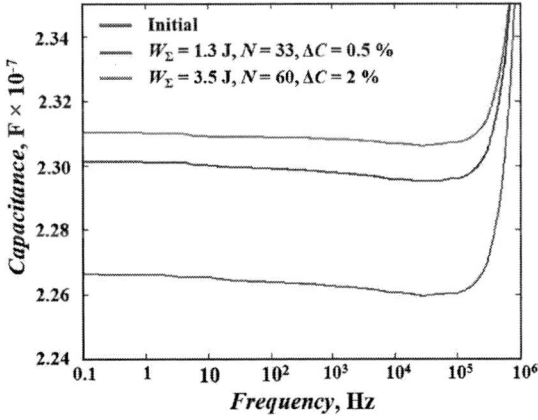

Fig. 4.   Typical dependencies of capacitance for initial and aged capacitors.

979-8-3503-8358-4/23 $31.00 © 2023 IEEE

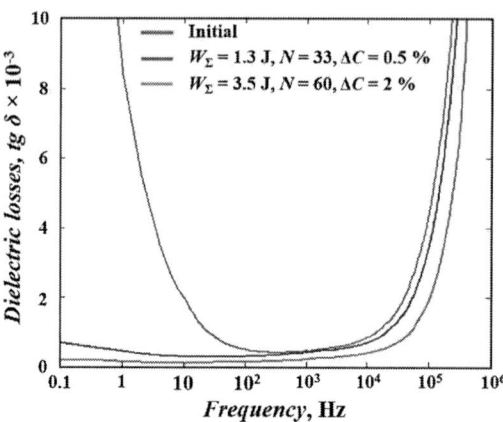

Fig. 5. Typical dependencies of dielectric losses for initial and aged capacitors.

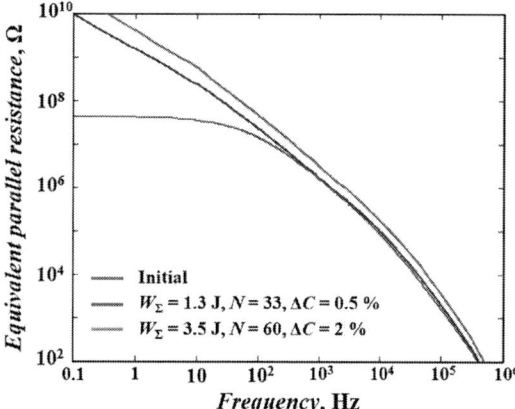

Fig. 6. Typical dependencies of equivalent parallel resistance for initial and aged capacitors.

It was established that the value of equivalent parallel resistance also changes non-uniform in investigated frequency range. Significant decreasing of this parameter was observed in low frequency range. Moreover, in the frequency range of 0.1 – 10 Hz equivalent parallel resistance took the constant value.

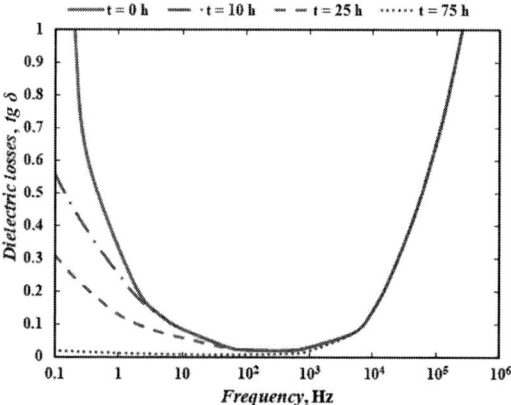

Fig. 7. The relaxation of low frequency dielectric losses.

According to equation (2) dielectric losses increasing may be explain by decreasing of capacitance and equivalent resistance at a constant frequency. Since the capacity value remains constant (Fig. 3), dielectric losses increasing in low

frequency range are not explained by slow kinds of polarization, such as interlayer or dipole polarizations and can be related to the value of equivalent parallel resistance only. This parameter is related to electrical conductivity of MFC's dielectric (polymer film). There is significant increasing of dielectrics electrical conductivity after multiple SH acts due to carbon layer formation in breakdown zones and appearance of free charge carriers. From the obtained experimental dependencies, it can be concluded that formed free charge carriers have strongly relationship of its mobility from electric field frequency. Moreover, the concentration of these free charge carriers changes over time. The last one manifests itself in the form of slow disappearance of low frequency dielectric losses after MFC's testing. The relaxation of low frequency dielectric losses is shown in Fig. 6. Decreasing of low frequency dielectric losses can be may indicate a decrease in free charge density in the breakdown zones. The average time relaxation of such free charge is 70 – 80 hours. After this time low frequency dielectric losses completely disappeared and only high frequency dielectric losses were observed. Therefore, the nature of free charge carriers which appear after SH acts needs additional experimental investigation.

IV. CONCLUSION

The phenomenon of low frequency dielectric losses increasing in metallized Film capacitors was found. Such increasing of dielectric losses is a consequence of multiple dielectric breakdowns and following SH acts. The final stage of each SH act is formation of a carbon layer in the breakdown channel and around it in the demetallized area. Carbon layer is a source of free charge carriers. The dependencies of dielectric losses and equivalent parallel resistance give us information that the density of charge carriers depends on electric field frequency. In addition, low frequency dielectric losses decrease in time after HV tests of MFCs that indicates a decrease in free charge density in the breakdown zones. The X-ray diffraction analysis of dielectric breakdown zone may give information about the nature of free charge carriers which appear after SH acts.

ACKNOWLEDGMENT

The research is supported by Russian Science Foundation (project № 19-79-10075).

REFERENCES

[1] G. G. Raju, "Dielectrics in Electric Fields," New-York, Marcel Dekker, Inc. Publ., 2003. DOI: 10.1201/9780203912270.

[2] O. G. Gnonhoue, A. Velazquez-Salazar, É. David, I. Preda, "Review of technologies and materials used in high-voltage film capacitors," Polymers, vol. 13, no. 5, 766, Feb. 2021, pp. 1-9. DOI: 10.3390/polym13050766.

[3] T.D. Huan, S. Boggs, G. Teyssedre, C. Laurent, M. Cakmak, S. Kumar, R. Ramprasad, "Advanced polymeric dielectrics for high energy density applications," Progress in Materials Science, vol. 83, 2016, pp. 236-269. DOI: 10.1016/j.pmatsci.2016.05.001.

[4] G. Wang, Z. Lu, Y. Li, L. Li, H. Ji, A. Feteira, D. Zhou, D. Wang, S. Zhang, I.M. Reaney, "Electroceramics for High-Energy Density Capacitors: Current Status and Future Perspectives," Chemical Reviews, vol. 21, no. 10, May 2021, pp. 6124-6172. DOI: 10.1021/acs.chemrev.0c01264.

[5] A. Zeb, S. J. Milne, "High temperature dielectric ceramics: a review of temperature-stable high-permittivity perovskites," Journal of Materials Science: Materials in Electronics, vol. 26, no. 12, Dec. 2015, pp. 9243-9255. DOI: 10.1007/s10854-015-3707-7.

[6] V.O. Belko, O.A. Emelyanov, I.O. Ivanov, A.P. Plotnikov, "Numerical Simulation of Electro-Thermal Processes in Multilayer Ceramic Capacitors," 2019 IEEE Conference of Russian Young

Researchers in Electrical and Electronic Engineering (ElConRus), 2019, pp. 78-80. DOI: 10.1109/EIConRus.2019.8656681.

[7] M. Ritamäki, I. Rytöluoto, K. Lahti, "Performance metrics for a modern BOPP capacitor film," in IEEE Transactions on Dielectrics and Electrical Insulation, vol. 26, no. 4, pp. 1229-1237, Aug. 2019. DOI: 10.1109/TDEI.2019.007970.

[8] H. Lia, H. Lia, Z. Lia, F. Lin, W. Wanga, B. Wanga, X. Huanga, X. Guo,"Temperature dependence of self-healing characteristics of metallized polypropylene film," Microelectronics Reliability, vol. 55, no. 12, Part B, Dec. 2015, pp. 2721-2726. DOI: 10.1016/j.microrel.2015.09.007.

[9] B. Yi, H. Li, H. Jiang, L. Li, Q. Chen, Z. Li, F. Lin, C. Zhang, "Breakdown Behavior of Metallized BOPP Film under the DC Superimposed Harmonic Condition," 2019 IEEE Conference on Electrical Insulation and Dielectric Phenomena (CEIDP), 2019, pp. 797-800. DOI: 10.1109/CEIDP47102.2019.9009686.

[10] V. Belko, O. Emelyanov, I. Ivanov, N. Fedotov, "Influence of Demetallization Processes on Capacitor Grade Polymer Films Breakdown Strength," 2020 IEEE 3rd International Conference on Dielectrics (ICD), 2020, pp. 166-168. DOI: 10.1109/ICD46958.2020.9341912.

[11] V. O. Belko, O. A. Emelyanov, I. O. Ivanov and A. P. Plotnikov, "Self-Healing Processes of Metallized Film Capacitors in Overload Modes—Part 1: Experimental Observations," IEEE Transactions on Plasma Science, vol. 49, no. 5, May 2021, pp. 1580-1587. DOI: 10.1109/TPS.2021.3071187.

[12] Y. Chen, H. Li, F. Lin, F. Lv, M. Zhang, Z. Li, D. Liu, "Study on Self-Healing and Lifetime Characteristics of Metallized-Film Capacitor Under High Electric Field," IEEE Transactions on Plasma Science, vol. 40, no. 8, Aug. 2012, pp. 2014-2019. DOI: 10.1109/TPS.2012.2200699.

[13] V. O. Belko, O. A. Emelyanov, I. O. Ivanov, A. P. Plotnikov, E. G. Feklistov, "Application of Numerical Simulation for Metallized Film Capacitors Electrodes Design," IEEE Access, vol. 9, 2021, pp. 80945-80952. DOI: 10.1109/ACCESS.2021.3085695.

[14] Y. Zhou, Q. Li, "High-temperature and high-energy-density polymer dielectrics for capacitive energy storage," 2018 IEEE 2nd International Conference on Dielectrics (ICD), 2018, pp. 1-4. DOI: 10.1109/ICD.2018.8514735.

[15] J. R. MacDonald, M. A. Schneider, J. B. Ennis, F. W. MacDougall, X. H. Yang, "High energy density capacitors," 2009 IEEE Electrical Insulation Conference, 2009, pp. 306-309. DOI: 10.1109/EIC.2009.5166362.

[16] C. W. Reed and S. W. Cichanowskil, "The fundamentals of aging in HV polymer-film capacitors," in IEEE Transactions on Dielectrics and Electrical Insulation, vol. 1, no. 5, Oct. 1994, pp. 904-922. DOI: 10.1109/94.326658.

[17] J. Kammermaier, G. Rittmayer, S. Birkle, "Modeling of plasmainduced self-healing in organic dielectrics," Journal of Applied Physics, vol. 66, no. 4, Apr. 1989, pp. 1594–1609. DOI: 10.1063/1.344373.

[18] V. Belko, O. Emelyanov, I. Ivanov, V. Starovoytenkov, "The Diagnostics of Metallized Film Capacitors under Soft Training Test," 2020 International Conference on Diagnostics in Electrical Engineering (Diagnostika), 2020, pp. 1-4. DOI: 10.1109/Diagnostika49114.2020.9214585.

[19] V. O. Belko, O. A. Emelyanov, "Self-Healing Processes of Metallized Film Capacitors in Overload Modes—Part II: Theoretical and Computer Modeling," IEEE Transactions on Plasma Science, vol. 49, no. 6, June 2021, pp. 1898-1905. DOI: 10.1109/TPS.2021.3076007.

[20] V. Belko, O. Emelyanov, I. Ivanov, "Critical Parameters of Metallized Film Capacitor's Failure," 2018 IEEE 2nd International Conference on Dielectrics (ICD), 2018, pp. 1-3. DOI: 10.1109/ICD.2018.8514586.

[21] V. Belko, I. Ivanov, A. Plotnikov, V. Belanov, "Energy characteristics of self-healing process in metallized film capacitors," 2019 International Scientific Conference on Energy, Environmental and Construction Engineering, 02006, 2019, pp. 1-4. DOI: 10.1051/e3sconf/201914002006.

979-8-3503-8358-4/23 $31.00 © 2023 IEEE

Investigation of Particles Obtained by Green Synthesis Using Plant Extract

Kamilya Khalugarova
Micro and Nanoelectronics Department
Saint-Petersburg Electrotechnical University "LETI"
Saint Petersburg, Russia
kamilya_kh@mail.ru

Valeriy M. Kondratev
Center for Nanotechnologies
Alferov University
Saint Petersburg, Russia
Center for Photonics and 2D Materials
Moscow Institute of Physics and Technology
Dolgoprudny, Russia
kvm_96@mail.ru

Alexey Kuznetsov
Center for Nanotechnologies
Alferov University
Saint Petersburg, Russia
Center for Photonics and 2D Materials
Moscow Institute of Physics and Technology
Dolgoprudny, Russia
leshiy2698@mail.ru

Alena Yu. Gagarina
Micro and Nanoelectronics Department
Saint-Petersburg Electrotechnical University "LETI"
Saint Petersburg, Russia
gagarina.au@gmail.com

Abstract—**The article studies the green synthesis method for obtaining nickel particles, which involves a more environmentally friendly method of obtaining through the use of raw materials of natural origin. In this work, we study the synthesis using an extract of the plant *Fumaria officinalis*. The obtained particles are studied by scanning electron microscopy, energy dispersive X-ray spectroscopy, and Raman spectroscopy.**

Keywords—nickel, nanoparticles, green synthesis, SEM, EDX, Raman spectroscopy

I. INTRODUCTION

Nanotechnology is one of the rapidly developing concepts in recent years. The resulting nanomaterials, which have different characteristic physical and chemical properties, have a wide potential for the development of new systems, structures and devices. Nanostructures have various applications due to their properties depending on the size, distribution and morphology: biomedicine, catalysis, chemical industry, targeted drug delivery, electronics, sensors, energy, etc. [1-4].

The production and study of metal nanoparticles using biological materials, accompanied by an environmentally friendly approach to production, is attracting increasing attention.

Currently, physical and chemical methods for obtaining nanoparticles are widely used, which require the use of toxic solvents and chemicals that pose a potential hazard to the environment and living organisms. In this connection, for the last 30 years, an alternative method for obtaining nanoparticles, green synthesis, has been studied and applied [5]. This method is carried out using extracts of plants and microorganisms (fungi, yeast, bacteria) [6].

In this work, we study the green synthesis method using an extract of the plant *Fumaria officinalis* to obtain nickel nanoparticles. *Fumaria officinalis* is an annual plant that is a source of fumaric acid. The aerial part also contains alkaloids (0.2-1.6%), tannins (2.9%), resins (4.7%), vitamins C and K.

II. THE THEORETICAL PART

Green synthesis of nanoparticles avoids many harmful effects by allowing particles to be synthesized at moderate pressure, temperature, pH and at a much lower cost.

Plant extracts are used as reducing and stabilizing agents to obtain nanoparticles of various morphologies from salts of the corresponding metals. There are a number of advantages of using plant extracts in relation to the use of microorganisms in the synthesis of nanoparticles [7,8]. When using plant extracts, the kinetics of this method is much higher. In the synthesis, almost all parts of the plant can be used: leaves, stems, roots, flowers. The phytochemicals found in the plant extract, such as polyols, terpenoids, polyphenols, are responsible for the reduction of metal ions by reaction with metal salts. [9].

Known methods for obtaining various nanoparticles using plant extracts: Ag, Ni, Au, Pd, Zn, ZnO, NiO, SnO_2, etc. [10]. Due to the non-use of toxic precursors, the obtained nanoparticles do not have toxic residues, due to which they are actively used in biomedicine as antimicrobial, anticancer agents, for bioimaging, targeted drug delivery, without the risk of causing toxic damage to a living organism [11]. Nickel and nickel oxide nanoparticles are also produced by scientists using green synthesis using various plant species. [12-14]. The article [13] studies nickel nanoparticles obtained using an extract of F. officinalis obtained by placing dried leaves in boiling water.

Extracts from plants can be obtained thick, dry and liquid. There are several ways to obtain a liquid extract from plants:

- maceration or infusion (the raw material is crushed and together with the extractant is placed in a closed vessel and infused at room temperature for 5-7 days);

- percolation or filtering (the raw material is moistened with an extractant in a closed vessel, insisted for 4 hours, the swollen raw material is transferred to the

---

This research was funded by the "Development program of ETU "LETI" within the framework of the program of strategic academic leadership" Priority-2030 No 075-15-2021-1318 on 29 September 2021.

979-8-3503-8358-4/23 $31.00 © 2023 IEEE

percolator, tightly packed, the outlet cock is opened and the extractant is added);

- repercolation (the required concentration is achieved using several percolators, when the liquid from one percolator is used to percolate in the next);

- circulation extraction (multiple extraction with the same portion of a highly volatile extractant).

Water, ethanol, methylene chloride, chloroform, etc. are used as extractants. Since the composition of the extractant can affect the result of obtaining nanoparticles, it is an urgent task to study the effect of types of extractants on the obtained particles.

III. THE EXPERIMENTAL PART

A. Obtaining plant extract

In this work, we studied a method for obtaining nickel particles using an extract of the F. Officinalis plant obtained using a Soxhlet extractor and an alcoholic solution as an extractant. The ratio of alcohol and water was taken 1:1. For 150 ml of distilled water, there were 25 g of dried leaves of fumes officinalis.

B. Synthesis of nickel nanoparticles

15 mmol of NiSO4 7-aqueous salt were mixed using a magnetic stirrer with 10 ml of the resulting plant extract and 1 ml of 2% NaOH solution. Then the solution was placed in an ultrasonic bath, washed, centrifuged, and dried at 50°C.

The morphology of the obtained particles was studied by scanning electron microscopy using a Zeiss Supra25 scanning electron microscope (Carl Zeiss, Germany), and the composition was studied by the EDX method. For these studies, the samples were deposited from an aqueous solution onto a single-crystal silicon substrate.

The characterization of the particles was also carried out by Raman spectroscopy; the particles for the study were dried on a glass substrate. To analyze the structural parameters of the nanostructures under study, a Jobin Yvon Horiba LabRAM HR 800 spectrometer combined with a confocal microscope was used.

IV. RESEARCH RESULTS

The results of studies by scanning electron microscopy are shown in Figures 1 and 2.

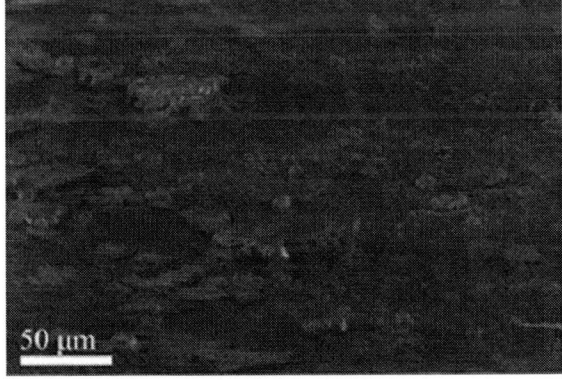

Fig. 1. SEM image of obtained nickel particles

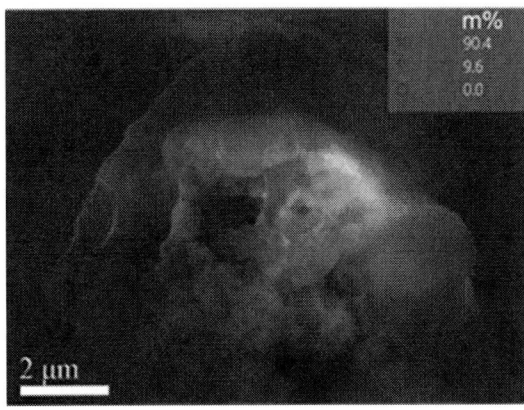

Fig. 2. SEM image of the obtained particles with the results of the EDX study of the elemental composition of the selected particle

Figure 3 shows the results of the study of the sample by Raman spectroscopy.

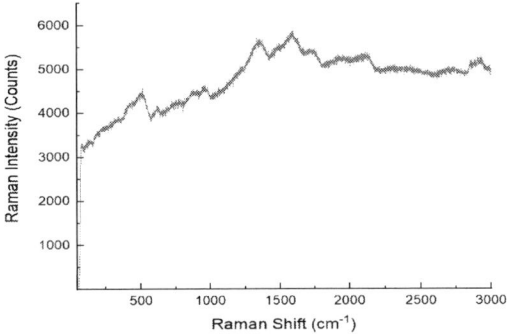

Fig. 3. The results of the study by the method of Raman spectroscopy

In Figure 1, you can see the presence of large agglomerates ranging in size from a few to tens of micrometers, while in Figure 2 it can be noted that these agglomerates consist of smaller particles with a size of tens of nanometers, while forming a porous structure (Fig. 2). According to the results of the study of these large agglomerates by the EDX method, it was determined that the main part is Ni, Si, O. In all cases, the percentage of nickel was more than 70%.

Based on the spectrum obtained by Raman spectroscopy (Fig. 3), key peaks can be identified: 500, 900, 1300, 1550, 2100 cm^{-1}.

V. CONCLUSIONS

In this work, a green synthesis of nickel nanoparticles was carried out using an alcoholic extract of F. officinalis. The obtained particles were characterized by SEM, EDX, Raman spectroscopy.

As a result of the work, it can be noted that the use of alcohol in the synthesis leads to particle agglomeration, since this was not observed when using only water. This effect may be related to the magnetic dipole interaction between neighboring particles, which is enhanced in the presence of a less coordinating solvent such as alcohol.

Analysis of SEM images made it possible to estimate the size of synthesized agglomerated particles from 10 μm to 50 μm.

ACKNOWLEDGMENT

Valeriy M. Kondratev and Alexey Kuznetsov thanks the Ministry of Science and Higher Education of the Russian Federation (Grant FSRM−2020−0011) and the strategic academic leadership program "Priority 2030" (Agreement 075-02-2021-1316 30.09.2021)".

REFERENCES

[1] Sharma P. K. et al. Nanotechnology and its application: a review. Nanotechnology in Cancer Management, 2021, pp. 1-33.

[2] Spivak Yu. M. et al. Porous silicon as a nanomaterial for dispersed transport systems of targeted drug delivery to the inner ear. *Zhurnal tekhnicheskoi fiiziki* (Journal of technical physics), 2018, V. 88, no. 9, pp. 1394. (in Russian)

[3] Maksimov A. I. et al. Fundamentals of sol-gel technology of nanocomposites, 2008. (In Russian)

[4] Moshnikov V. A. et al. Hierarchical nanostructured semiconductor porous materials for gas sensors. Journal of non-crystalline solids, 2010, V. 356, no. 37-40, pp. 2020-2025.

[5] Kharissova, O. V., Kharisov, B. I., Oliva González, C. M., Méndez, Y. P., & López, I. Greener synthesis of chemical compounds and materials. Royal Society Open Science, 2019, no. 6(11), pp. 191378.

[6] Singh J. et al. 'Green' synthesis of metals and their oxide nanoparticles: applications for environmental remediation. Journal of nanobiotechnology, 2018, V. 16, no. 1, pp. 1-24.

[7] Iravani, S. Green synthesis of metal nanoparticles using plants. Green Chemistry, 2011, no. 13(10), pp. 2638.

[8] Carolin C. F. et al. Efficient techniques for the removal of toxic heavy metals from aquatic environment: A review. Journal of environmental chemical engineering, 2017, V. 5, no. 3, pp. 2782-2799.

[9] Ovais M. et al. Role of plant phytochemicals and microbial enzymes in biosynthesis of metallic nanoparticles. Applied microbiology and biotechnology, 2018, V. 102, no. 16, pp. 6799-6814.

[10] Naikoo G. A. et al. Bioinspired and green synthesis of nanoparticles from plant extracts with antiviral and antimicrobial properties: A critical review. Journal of Saudi Chemical Society, 2021, V. 25, no. 9, pp. 101304.

[11] Selvakesavan R. K., Franklin G. Prospective application of nanoparticles green synthesized using medicinal plant extracts as novel nanomedicines. Nanotechnology, science and applications, 2021, V. 14, pp. 179.

[12] Pang H. et al. Porous nickel oxide microflowers synthesized by calcination of coordination microflowers and their applications as glutathione electrochemical sensor and supercapacitors. Electrochimica Acta, 2012, V. 85, pp. 256-262.

[13] Huang Y. et al. Green synthesis of nickel nanoparticles using Fumaria officinalis as a novel chemotherapeutic drug for the treatment of ovarian cancer. Journal of Experimental Nanoscience, 2021, V. 16, no. 1, pp. 368-381.

[14] Ahghari M. R., Soltaninejad V., Maleki A. Synthesis of nickel nanoparticles by a green and convenient method as a magnetic mirror with antibacterial activities. Scientific reports, 2020, V. 10, no. 1, pp. 1-10

Different Approaches to Increasing the Continuous Operation Time of IPMC Actuators

Ivan K. Khmelnitskiy
Saint Petersburg Electrotechnical University "LETI"
Saint Petersburg, Russia
khmelnitskiy@gmail.com

Dmitriy O. Testov
Saint Petersburg Electrotechnical University "LETI"
Saint Petersburg, Russia
dtestov@bk.ru

Vagarshak M. Aivazyan
Saint Petersburg Electrotechnical University "LETI"
Saint Petersburg, Russia
aivazyanvm@mail.ru

Victor V. Luchinin
Saint Petersburg Electrotechnical University "LETI"
Saint Petersburg, Russia
cmid_leti@mail.ru

Andrey V. Korlyakov
Saint Petersburg Electrotechnical University "LETI"
Saint Petersburg, Russia
akorl@yandex.ru

Alexandr M. Karelin
Saint Petersburg Electrotechnical University "LETI"
Saint Petersburg, Russia
karelin1999@tut.by

Abstract—**IPMC actuators have great potential in robotics due to low control voltages and high output displacements. A major limitation is the use of aqueous electrolytes, which reduces the time of their continuous operation in air due to water evaporation and electrolysis. To increase it, two approaches have been investigated: encapsulation and the use of non-aqueous electrolytes. It was experimentally found that the continuous operation time of the water-impregnated IPMC actuator after encapsulation with a 15 µm-thick parylene layer increased by 10 times at a control signal of 4 V 1 Hz. However, this approach does not solve the problem of water electrolysis in DC mode. The use of nonvolatile ionic liquids with a wide electrochemical stability window made it possible to solve this problem. By replacing water with EMIM-BF$_4$ ionic liquid, the continuous operation time of IPMC actuators increased to more than 24 h.**

Keywords—*IPMC actuators, encapsulation, polyurethane coating, parylene coating, ionic liquids*

I. INTRODUCTION

Due to the possibility of creating large displacements and forces, IPMC actuators have found application as propulsors for the implementation of walking miniature biosimilar robots [1, 2]. The most "vulnerable" parameter of actuators operating both in air and in water is the service life. This parameter represents the continuous operation time of the actuator from the voltage is applied and the motion starts until the motion is completely stopped.

IPMC actuator consists of a polymeric ion-exchange membrane (with polar sulfo groups) impregnated with a solvent (usually water or an aqueous solution) and metal electrodes deposited on both sides of it [3]. The principle of its operation is based on the redistribution of ions inside the membrane under the action of an external electric field, which leads to the appearance of an osmotic fluid flow from one electrode to another [4]. When the actuator operates in air, the membrane needs constant hydration. To increase the operation, the control voltages should be no more than 1 V, since when a voltage is applied above 1.23 V, water electrolysis can occur [5].

The service life of the actuator in air is approximately 15...20 min. In aqueous medium, the actuator service life turns out to be significantly longer, despite the fact that the processes of electrolysis and salt leaching through porous electrodes reduce it [6].

To study the possibility of increasing the service life of

The work has been funded by Russian Science Foundation (Grant No. 16-19-00107).

actuators in air, in this work, we consider IPMC actuators with dimensions of 20 × 5 mm, consisting of a 175 µm-thick Nafion 117 ion-exchange membrane and platinum electrodes deposited on its surface. The technique for depositing electrodes by chemical reduction from a Pt(NH$_3$)$_4$Cl$_2$ solution is presented in [7, 8].

II. ENCAPSULATION OF IPMC ACTUATORS WITH POLYMER LAYERS

One of the ways to increase the operation time of IPMC actuators in air is encapsulation, i.e., applying a protective film to the actuator that prevents the liquid from evaporating. The most commonly used materials for such a film are polyurethane applied from a solution and parylene applied from the gas phase.

To avoid moisture loss in the membrane, the possibility of encapsulating the IPMC actuator with a polyurethane polymer was investigated (Fig. 1). The coating was applied by dip-coating from a solution of polyurethane in butyl acetate using a brush.

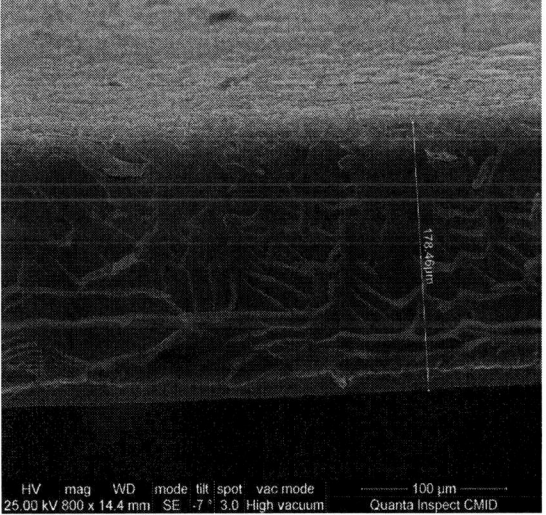

a

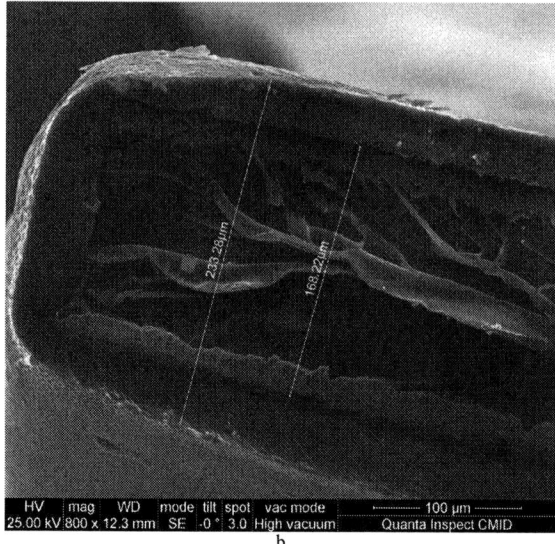

Fig. 1. SEM images of IPMC actuators: a – unencapsulated; b – encapsulated with polyurethane polymer.

To investigate the characteristics of IPMC actuators, we used a bench containing a 33210A waveform generator (Keysight Technologies, USA), an HG-C1100 laser distance sensor (Panasonic, Japan), a 34461A digital multimeter (Keysight Technologies, USA) and a personal computer with BenchVue licensed software (Keysight Technologies, USA).

The maximum displacement, blocking force and operation time were investigated for samples coated with a polymer layer. It is shown that the encapsulated actuators are characterized by greater blocking forces than the unencapsulated ones (Fig. 2), however, the resulting displacements become smaller (Fig. 3). The operation time of the actuators in air with complete encapsulation increases to 3…5 hours. Due to the fact that after complete encapsulation the actuator cannot be moistened, its reuse (after drying) is not possible. With partial encapsulation, the operation time increases to 30…40 minutes.

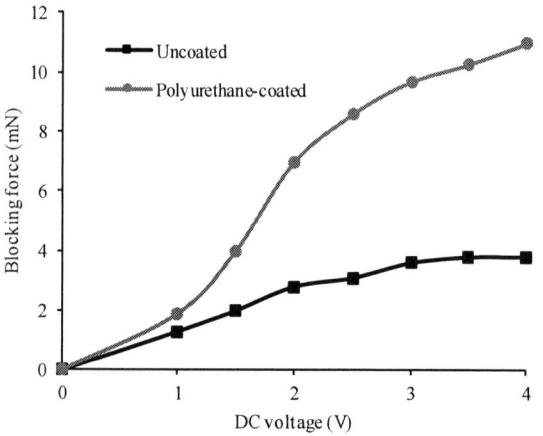

Fig. 2. Dependences of the IPMC actuator blocking force on the DC control voltage.

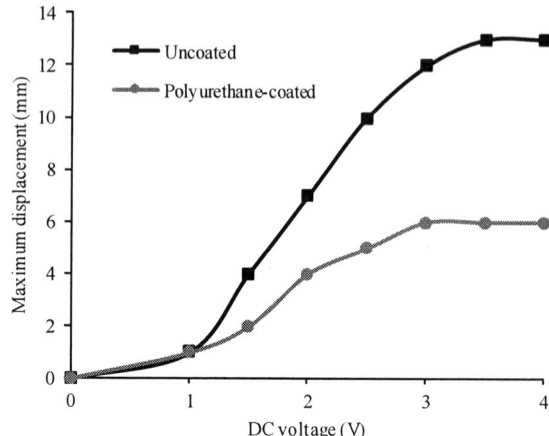

Fig. 3. Dependences of the IPMC actuator maximum displacement on the DC control voltage.

The use of parylene allows the most efficient preservation of liquid inside the ion-exchange membrane. For example, during 20 minutes of operation, the water content in the membrane decreases from 17% to only 14% in a parylene-coated actuator, and from 17% to 4% in a polyurethane-coated one. In this regard, in further studies, parylene was used to encapsulate IPMC actuators.

To apply parylene, a UNB-4 unit (LLC Bazalt, Russia) was used. The IPMC actuator preliminarily impregnated with liquid was placed in the working chamber, where parylene was deposited at room temperature. The coating thickness was 15 μm.

It was experimentally found that the continuous operation time of the IPMC actuator encapsulated with parylene at a control voltage of 4 V is correspondingly 10 times longer than that of the unencapsulated actuator (Fig. 4). The nature of the dependence of the continuous operation time of the actuator encapsulated with parylene on the control voltage is caused by electrolysis of water at voltages higher than 1.23 V: the higher the control voltage, the more intense the electrolysis.

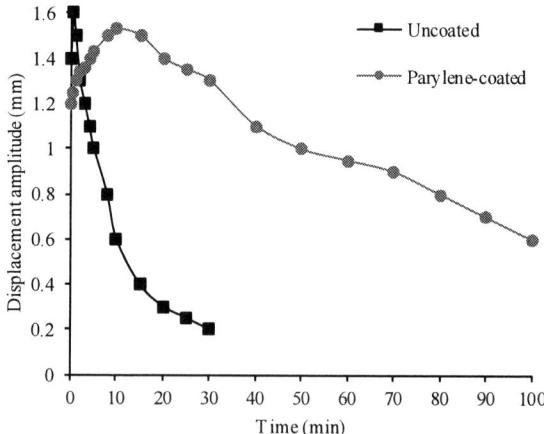

Fig. 4. Dependences of the IPMC actuator displacement amplitude on time at an amplitude of a sinusoidal control voltage of 4 V and at a frequency of 1 Hz.

III. IMPREGNATION OF IPMC ACTUATORS WITH IONIC LIQUIDS

Another way to increase the continuous operation time of IPMC actuators in air is to impregnate them with solvents with a higher boiling point than water, for example, dimethyl sulfoxide [5], ethylene glycol [9, 10] or ionic liquids [9-13]. This method reduces the effect of electrolysis and prevents evaporation of the electrolyte.

Ionic liquids (ILs) are salts whose melting point is below 100 °C. There are even ionic compounds that are room temperature ionic liquids. As a rule, their molecules consist of a bulky organic cation (imidazolium, pyridinium or pyrrolidinium) and an organic (trifluoroacetate, acetate) or inorganic (bromide, chloride, tetrafluoroborate, hexafluorophosphate) anion. ILs are electrochemically stable even at voltages above 4 V and thermally stable up to 400 °C. In addition, ILs have high ionic conductivity, which is very important for excitation of IPMC actuators [9-11].

The following ILs were used to impregnate the IPMC actuators: 1-allyl-3-methylimidazolium chloride (AMIM-Cl), 1-ethyl-3-methylimidazolium tetrafluoroborate (EMIM-BF$_4$) and 1-butyl-3-methylimidazolium tetrafluoroborate (BMIM-BF$_4$). Impregnation of the actuators with ILs was carried out by boiling them in a mixture of IL and ethyl alcohol (1:1). The displacement amplitudes of the IPMC actuators impregnated with three ILs were investigated: AMIM-Cl (AMIM$^+$ form), EMIM-BF$_4$ (EMIM$^+$ form) and BMIM-BF$_4$ (BMIM$^+$ form). The actuators impregnated with EMIM-BF$_4$ (EMIM$^+$ form) were characterized by the highest displacement amplitudes at a sinusoidal control voltage. Apparently, this is due to the lower melting point of EMIM-BF$_4$ than that of AMIM-Cl and lower viscosity than that of BMIM-BF$_4$.

At the same time, the displacement amplitudes of the actuators impregnated with EMIM-BF$_4$ ionic liquid (EMIM$^+$ form) turned out to be more than 3 times less than that of the actuators impregnated with deionized water (H$^+$ form) and an aqueous 1 M solution of copper sulfate (Cu^{2+} form) (Fig. 5, 6). This may be due to the weakening of the electroosmotic fluid flow in aprotic solvents due to the absence of hydrogen bonding system.

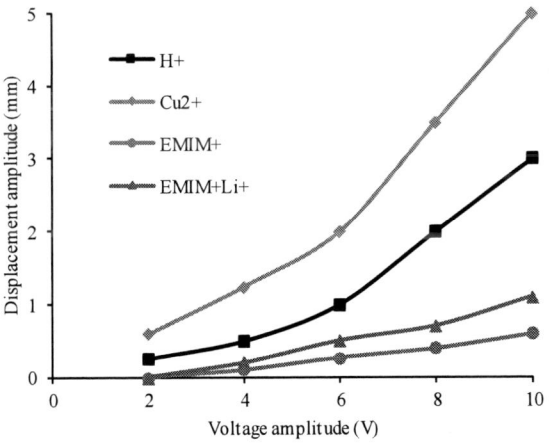

Fig. 5. Dependences of the IPMC actuator displacement amplitude on the amplitude of a sinusoidal control voltage at a frequency of 1 Hz.

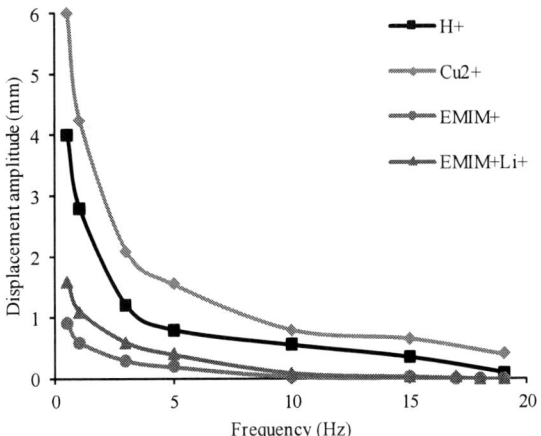

Fig. 6. Dependences of the IPMC actuator displacement amplitude on the frequency at an amplitude of a sinusoidal control voltage of 10 V.

The blocking force of the actuators impregnated with EMIM-BF$_4$ ionic liquid (EMIM$^+$ form) turned out to be more than 2 times lower than that of the actuators in H$^+$ form (Fig. 7).

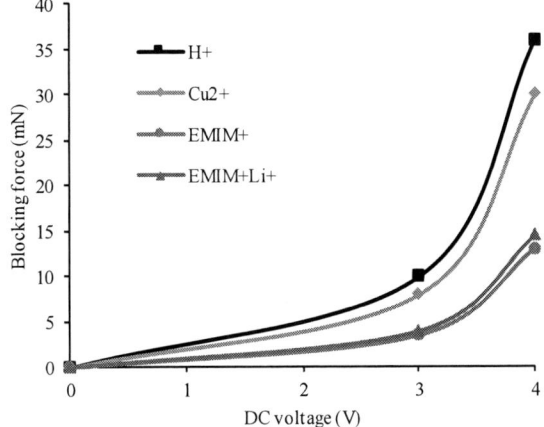

Fig. 7. Dependences of the IPMC actuator blocking force on the DC control voltage.

Due to the loss of the solvent, the continuous operation time in air of IPMC actuators in H$^+$ and Cu^{2+} forms is in the range of 15...20 minutes, and IPMC actuators impregnated with IL are capable of continuous operation for more than 24 hours (Fig. 8).

The introduction of LiBF$_4$ into the electrolyte composition allowed to increase the displacement amplitudes of IPMC actuators impregnated with IL by more than 50% (Fig. 5, 6).

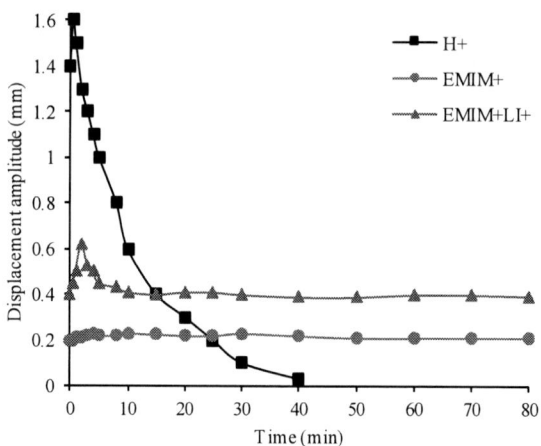

Fig. 8. Dependences of the IPMC actuator displacement amplitude on time at an amplitude of a sinusoidal control voltage of 4 V and at a frequency of 1 Hz.

IV. CONCLUSIONS

The possibility of encapsulating the IPMC actuators with polyurethane and parylene was investigated. It was shown that the blocking forces of the actuators encapsulated with polyurethane are superior to those of the unencapsulated actuators, however, the displacements created become smaller. It was found that encapsulation of the actuator with parylene leads to a 10-fold increase in the continuous operation time at a control voltage of 4 V compared to the unencapsulated actuator.

To reduce the effect of electrolysis on the continuous operation time of the IPMC actuators, it is reasonable to impregnate them with ionic liquids. The displacement amplitudes of the actuators impregnated with EMIM-BF$_4$ ionic liquid turned out to be more than 3 times smaller than those of the actuators impregnated with water. At the same time, the actuators impregnated with ionic liquid were able to operate continuously for more than 24 hours. By adding LiBF$_4$ to EMIM-BF$_4$, the displacement amplitudes of the actuators increased by more than 1.5 times compared to the actuators impregnated only with EMIM-BF$_4$.

REFERENCES

[1] Nguyen K.T., Ko S.Y., Park J.-O., Park S. Terrestrial Walking Robot With 2DoF Ionic Polymer–Metal Composite (IPMC) Legs. IEEE/ASME Transactions on Mechatronics, 2015, V. 20, pp. 2962-2972.

[2] Kalyonov V.E., Orekhov Y.D., Khmelnitskiy I.K., Alekseev N.I., Broyko A.P., Lagosh A.V., Testov D.O., Shpakovsky A.D. Walking Robot with Propulsors Based on IPMC Actuators. Proceedings of the 2019 IEEE International Conference on Electrical Engineering and Photonics (EExPolytech). St. Petersburg, Russia, 2019, pp. 169-172.

[3] Alekseev N.I., Bagrets V.S., Broyko A.P., Korlyakov A.V., Kalenov V.E., Luchinin V.V., Sevostyanov E.N., Testov D.O., Khmelnitsky I.K. Ionic Polymer Electroactive Actuators Based on the MF-4SK Ion-Exchange Membrane. Part 1. Ionic Polymer-Metal Composites. Journal of Structural Chemistry, 2020, V. 61, pp. 601-608.

[4] Broyko A.P., Khmelnitskiy I.K., Ryndin E.A., Korlyakov A.V., Alekseyev N.I., Aivazyan V.M. Multiphysics Simulator for the IPMC Actuator: Mathematical Model, Finite Difference Scheme, Fast Numerical Algorithm, and Verification. Micromachines, 2020, V. 11, 1119.

[5] Khmelnitskiy I., Vereschagina L., Kalyonov V., Krot A., Korlyakov A. Performance Improvement of an Ionic Polymer-Metal Composite Actuator by Using DMSO as Solvent. Proceedings of the 2016 IEEE NW Russia Young Researchers in Electrical and Electronic Engineering Conference (EIConRusNW). St. Petersburg, Russia, 2016, pp. 58-61.

[6] Kalyonov V.E., Testov D.O., Alekseev N.I., Broyko A.P., Luchinin V.V. Study of IPMC Actuators Based on MF-4SK Membrane in Aqueous Medium. Proceedings of the 2020 IEEE Conference of Russian Young Researchers in Electrical and Electronic Engineering (EIConRus). St. Petersburg, Russia, 2020, pp. 1027-1030.

[7] Khmelnitskiy I.K., Vereschagina L.O., Kalyonov V.E., Lagosh A.V., Broyko A.P. Improvement of Manufacture Technology and Investigation of IPMC Actuator Electrodes. Proceedings of the 2017 IEEE Conference of Russian Young Researchers in Electrical and Electronic Engineering (EIConRus). St. Petersburg, Russia, 2017, pp. 892-895.

[8] Khmelnitskiy I.K., Vereshagina L.O., Kalyonov V.E., Broyko A.P., Lagosh A.V., Luchinin V.V., Testov D.O. Improvement of manufacture technology and research of actuators based on ionic polymer-metal composites. Journal of Physics: Conference Series, 2017, V. 857, 012018.

[9] Khmelnitskiy I.K., Luchinin V.V., Kalyonov V.E., Lagosh A.V., Broyko A.P. IP^{2}C Electromechanical Actuators and Sensors. Proceedings of the 2018 IEEE Conference of Russian Young Researchers in Electrical and Electronic Engineering (EIConRus). St. Petersburg, Russia, 2018, pp. 415-418.

[10] De Luca V., Digiamberardino P., Di Pasquale G., Graziani S., Pollicino A., Umana E., Xibilia M.G. Ionic electroactive polymer metal composites: Fabricating, modeling, and applications of postsilicon smart devices. Journal of Polymer Science Part B: Polymer Physics, 2013, V. 51, pp. 699-734.

[11] Kikuchi K., Tsuchitani S. Nafion®-based polymer actuators with ionic liquids as solvent incorporated at room temperature. Journal of Applied Physics, 2009, V. 106, 053519.

[12] Safari M., Naji L., Baker R.T., Taromi F.A. The enhancement effect of lithium ions on actuation performance of ionic liquid-based IPMC soft actuators. Polymer, 2015, V. 76, pp. 140-149.

[13] Morozov O.S., Ivanchenko A.V., Nechausov S.S., Bulgakov B.A. Effect of Electrode Morphology on Performance of Ionic Actuators Based on Vat Photopolymerized Membranes. Membranes, 2022, V. 12, 1110.

Investigation of the Electro Physical Properties of the Components of Modern Paper-Impregnated Insulation

Dmitry V. Kiesewetter
Higher School of High Voltage
Engineering
Peter the Great St. Petersburg
Polytechnic University
Saint Petersburg, Russia
dmitrykiesewetter@gmail.com

Natalia M. Zhuravleva
Higher School of High Voltage
Engineering
Peter the Great St. Petersburg
Polytechnic University
Saint Petersburg, Russia
natalia_zhurav@mail.ru

Alexandr S. Reznik
Higher School of High Voltage
Engineering
Peter the Great St. Petersburg
Polytechnic University
Saint Petersburg, Russia
alexxxandr2803@mail.ru

Danila Litvinov
Higher School of High Voltage
Engineering
Peter the Great St. Petersburg
Polytechnic University
Saint Petersburg, Russia
0000-0003-1876-4335

Denis Trubin
Higher School of High Voltage
Engineering
Peter the Great St. Petersburg
Polytechnic University
Saint Petersburg, Russia
0000-0002-3494-9279

Le Sun
School of Electrical Engineering
Yanshan University
Qinhuangdao, China
0000-0002-7594-9371

Abstract—The results of measurements of the mechanical tensile strength, short-term electrical strength, as well as other electro physical characteristics of the components of modern paper-impregnated insulation: electrical insulation paper of various types and some used and promising dielectric liquids are presented. An assessment of the change in the tensile strength during artificial aging of both dry insulating paper and paper aged together with various dielectric liquids is given. Significant differences in the magnitude of the electric field strength at the breakdown, as well as different dynamics of the breakdown development in different types of electrical insulation paper, are noted. Special attention is paid to the measurement of electrical parameters – breakdown voltage, electrical resistance at direct current, the tangent of the dielectric loss angle of mixtures of silicone transformer fluid with petroleum oil. It is shown that colloidal particles formed as a result of the emulsification of silicone liquid in petroleum oil have a strong influence on the parameters of the mixture; and the parameters of the mixtures change over time or after heating, the analysis of the data obtained is carried out, on the basis of which a conclusion is made about the prospects of using modern materials in paper-impregnated insulation.

Keywords—*insulating paper, high-voltage insulation, liquid dielectrics, petroleum oils, polydimethylsiloxane fluids, dielectric mixtures*

I. INTRODUCTION

Paper-impregnated insulation (PII) is widely used in medium and high power transformers. The reliability of the transformers, and therefore of the entire power system as a whole, largely depends on the quality of the electrical insulation. Therefore, improving the insulation properties of power transformers – electrical insulating paper (EIP) and liquid dielectric is still an urgent technical task. In particular, promising areas are: the use of laminated electrical insulating paper [1-4]; paper from the cellulose which is modified by the biopolymers [5-8], as well as the use of synthetic esters and transformer silicone liquids (TSL) in power transformers. Despite the higher cost of such liquids compared to petroleum oils, their use may be appropriate, as this will reduce the cost of maintenance of power transformers. It is known that petroleum oils and silicone liquids are compatible substances [9] ("questionable compatibility"). This makes it possible to replace the oil transformer oil in the operated transformer with a synthetic liquid, in particular, with TSL in the form of polydimethylsiloxane (PDMS) liquid dielectric. However, the presence of some petroleum oil in the TSL, as well as some PDMS in the paper-impregnated insulation, has an impact on the electrical properties of the insulation.

An interest in the use of power electric cables with paper-impregnated insulation is returning with the improvement of materials and technologies. Taking into account the accumulated experience of successful operation of cables with paper-impregnated insulation of the first generation (in particular, cables with PII at the nominal voltage of 6 kV have been operated in St. Petersburg for more than hundred years [10]). It can be assumed that modern cables with PII will have better performance characteristics.

This work is devoted to the study of the main properties, in particular, mechanical and electrical strength, of electrical insulating paper of various types: commercially produced and experimental castings from nano-gel film of bacterial cellulose (NGF BC) [11-13], as well as the parameters of mixtures of petroleum oils with PDMS electrical insulating liquid.

II. MATERIALS AND METHODS

Measurements of the mechanical tensile strength and short-term electrical strength of various types of industrially produced electrical insulating paper – made of electrical insulation cellulose (EIC) of pine made using traditional technology, micro-reinforced (K120, KB130 et al.) and laminated three-layer paper, used in the production of modern power transformers and power cables with paper-impregnated insulation, as well as experimental castings of paper, manufactured in of the Russian Academy of Sciences on the basis of bacterial cellulose species *Komagataeibacter rhaeticus* (number B-13015 in the all-Russian collection of industrial microorganisms) with nano fillers in the form of carbon nano tubes [12] or nano silicon dioxide (nano SiO_2) [13] and without nano fillers were performed.

Transformer silicone fluids (TSL "Sofexil"), dielectric fluids PDMS-50, PDMS-12500 and their mixtures with petroleum oils T1500C and GK were studied also. The most

979-8-3503-8358-4/23 $31.00 © 2023 IEEE

detailed study was performed for the mixtures of T1500C oil and TSL (PDMS-50 with the kinematic viscosity of 50 mm²/s).

Preparation and conditioning of paper samples was carried out in accordance with state standards [15] and [16].

The mechanical strength and strain curves were measured using MARK-10 ESM301/ESM301L device [17]. The breakdown voltage of the insulating paper was measured in accordance with existing standards [16-17] at the frequency of 50 Hz. The experimental setup was used (the description of which is described in detail in [18-19]) to study the dynamics of the development of electrical breakdown in paper using: the standard electrode design, the adjustable high voltage source with the frequency of 50 Hz, voltage dividers and an analog-to-digital converter connected to the personal computer. The rate of voltage rise was approximately 200 V/s.

PDMS dielectric liquids and their mixtures with petroleum oils were also studied. The breakdown voltage of dielectric fluids was measured in a cell witch dimensions and parameters of the electrodes corresponded to the standard [20] (analogous of IEC 60156: 1995), in particular, the distance between the electrodes was 2.5 mm. The tangent of the dielectric loss angle was measured in the metal two-electrode cell with interelectrode gap of 0.78 mm using the automatic bridge MEP-5C in accordance with standard [21]. The electrical resistance to direct current-using the METREL MI3200 T-Ohmmeter.

Artificial aging of dielectric liquids and insulating paper together with dielectric liquids was performed at a temperature of 140⁰ C in closed glass vessels, similar to the method described in [22-24]. Aging was performed in the presence of the catalyst in the form of a copper wire.

III. TENSILE STRENGTH OF ELECTRICAL INSULATING PAPER

The results of measurements of the tensile strength of the insulating paper across the fibers in the initial state (0 h) and after artificial aging (with petroleum oil GK or TSL and catalyst; K120, BK130, #2 – cellulose paper of various manufacturers, #1 – laminated electrical insulation paper) are shown in Table 1.

From the obtained data, it follows that the artificial aging of cellulose paper in TSL occurs at approximately the same rate as in GK oil. The above-mentioned artificial aging technique was not used for three-layer laminated paper since at the temperature of 140⁰ C the laminate layer could dissolve in dielectric liquid.

TABLE I.   TENSILE STRENGTH OF INSULATING PAPER IN MPA ($10^6 \cdot KG \cdot M^{-1} \cdot C^{-2}$)

Type of paper	Conditions of artificial aging					
	0 h	20 h GK	60 h GK	154 h GK	60 h TSL	154 h TSL
K120	85±8	78±3	67±5	61±5	80±6	53±8
BK130	68±1	60±2		-		-
#1 (lam.)	48±1	-	-	-	-	-
#2 (BC)	94±2	-	-	-	89±2	-

TABLE II.   INSULATING PAPER'S BREAKDOWN PARAMETERS

Type of paper	Parameters		
	Thickness (µm)	Average (kV)	Variance (kV)
K120	120±5	11	2
BK130	128±5	11	2
NGF BC	29±2	38	5

The resulting deformation curves corresponded to the technical existing specifications for electrical insulating paper. Theoretically, an important advantage of laminated paper is the greater elongation before breaking than that of standard cellulose electrical insulating paper.

It can be assumed that laminated paper is used as the component of paper-impregnated insulation, is more resistant to mechanical influences, in particular, bends and vibrations of the cable, compared to standard cellulose paper. However, such insulation may be inferior to standard insulation in terms of the long-term permissible temperature.

IV. SHORT-TERM ELECTRICAL STRENGTH OF INSULATING PAPER AND BREAKDOWN DEVELOPMENT DYNAMICS

Probabilistic distributions of the breakdown stress value for K120, BK130 paper types, as well as for castings made of nano-gel film of bacterial cellulose (NGF BC) were measured. The statistical data obtained are summarized in Table 2.

The waveforms illustrating the dynamics of the breakdown of electrical insulating paper of various types were also obtained.

From the obtained waveforms, it follows that the typical time interval τ_p between the moment of occurrence of the first streamer and the moment of breakdown for the studied paper types is significantly different. For K120 paper, the typical value is 0.2..2 s, for paper BK130 – 2..6 s, for castings NGP BC – 5..15 s. It can be assumed that an increase in the density of the cellulose material leads to an increase in the value of τ_p due to decrease in the distance between the fibers.

It also follows from the results of a preliminary study carried out by analogy with the works [9, 15] that the presence of nano fillers in the form of carbon nanotubes or nano silicon dioxide in castings from NGP BC did not have a noticeable effect on the breakdown voltage and the dynamics of the development of electric breakdown.

The breakdown voltage of the studied samples of laminated EIP at alternating voltage, as expected, was significantly higher than that of non-laminated paper. However, at an electric field strength of approximately 10 kV/mm, the cellulose paper layer began to break down under the action of the resulting barrier discharge.

V. SHORT-TERM ELECTRICAL STRENGTH OF MIXTURES OF PETROLEUM OILS AND SILICONE LIQUIDS

A feature of the properties of mixtures of petroleum oils and PDMS liquids is the significant influence of the resulting colloidal particles formed during the mixing of liquids (which is noted, among other things, in [25]) on the electrical parameters of the mixture, in particular, on the breakdown voltage. As an example, this paper considers mixtures of T1500C oil and silicone transformer fluid (TSL) with different relative concentrations of components.

TABLE III. DIELECTRIC LIQUID'S BREAKDOWN PARAMETERS

Type of paper	Percentage of TSL in GK (%)							
	0	*0.2*	*0.2 after 20h*	*0.7*	*1.7*	*1.7 after 92h*	*6.6*	*6.6 after heating*
Average (kV)	39	37	41	34	36	38	34	45
Variance (kV)	4	6	4	3	6	5	5	3

The probabilistic characteristics of the breakdown voltage of mixtures T1500C and TSL were obtained. The statistical characteristics of the obtained distributions (arithmetic mean and variance of the breakdown voltage) are summarized in Table 3.

From the obtained data, it follows that the short-term electrical strength of the mixture can be either less than the electrical strength of the initial components, or more. The U_b (peak) value of the mixtures immediately after mixing the components is less than that of the starting liquids, and as the silicone liquid is emulsified in the T1500C oil, the breakdown voltage increases.

In particular, at a concentration of 0.2% TSL, the breakdown voltage at room temperature is restored after a few hours, and at 7% – only after a few weeks or even months. The increase in the U_b of the mixture from the time immediately after mixing the components at a low TSL concentration can be observed during the breakdown voltage measurements: there is a clear tendency to increase the breakdown voltage after each breakdown.

At the initial stage of emulsification at the concentration of TSL 2% dependence of U_b on time can be approximated by linear dependence with the initial value of 29 V and slope coefficient of 0.43 V/s. It should also be noted that the breakdown voltage of the specified mixture after 20 hours became higher than the U_b value of the original T1500C oil by approximately 10% (an increase from 37 kV to 41 kV), which is presumably due to the partial dissolution of the moisture in the oil in the emulsified state by the silicone liquid. However, this effect is less pronounced for TSL than for synthetic ether MIDEL 7131.

At TSL concentration of 6.6% or more, the change in $U_b(t)$ is significantly slower. Heating the mixture of these dielectric liquids significantly accelerates the emulsification process and, as a result, significantly reduces the recovery time of the electrical strength of the mixture after mixing the liquids, which is confirmed by the results of measurements of the breakdown voltage (Table 3). Thus, it follows from the obtained data that the arithmetic mean value of U_b after heating for 15 minutes at the temperature of 90^0 C at atmospheric pressure increased by more than 30% (increase from 34 kV to 45 kV). It should be noted that after heating the mixture, the liquid becomes transparent, that is, colloidal particles that scatter optical radiation disappear.

Similar patterns also occur for mixtures of PDMS liquids of different viscosities with other types of petroleum oils.

VI. ELECTRICAL RESISTANCE AND TANGENT OF THE DIELECTRIC LOSS ANGLE OF MIXTURES OF PETROLEUM OILS AND SILICONE LIQUIDS

The results of measuring the electrical resistance and currents flowing through a mixture of petroleum oils (T1500C, GK, TKP) and PDMS-50 dielectric liquids when applying the constant voltage are presented in [26]. In particular, the effect of the appearance of colloidal particles in mixtures of petroleum oils and PDMS liquids was noted in [25]. In this work, the parameters of mixtures of dielectric liquids were measured both at an alternating voltage of industrial frequency and for direct current, which made it possible to compare all the measured electrical parameters of the mixtures.

The following values of the tangent of the dielectric loss angle (tgδ) and the resistivity of the mixtures at the effective voltage U=500 V were obtained: at TSL content of 6.6% at room temperature, the value of tgδ was approximately 0.00025 and at 90^0 C – approximately 0.001. The electrical resistance of the mixture in the measuring cell was approximately 1 TΩ volume at room temperature and 2 GΩ at 900 C.

The following identified rules should also be noted. The dependence of the measured value of tgδ on the voltage applied to the measuring cell (tgδ(U)) is not a constant: for mixtures of TSL and petroleum oils at room temperature an increase in the value of tgδ on U is observed, but for pure petroleum oils, as well as for mixture subjected to heating (in this example, 15 minutes at the temperature of 90^0 C), the dependence of tgδ(U) was slightly decreasing. For example, for the mixture with TSL concentration of 6.6% the value of tgδ was 0.00025 at U=500 V and 0.0006 at U=1000 V, and at 90^0 C at the same voltages – 0.0013 and 0.001. For the studied sample of the mixture of TSL and T1500C, the dependences of tgδ(U) in the specified voltage range are well approximated by the linear function.

For a constant voltage, the obtained dependence of the current passing through the test mixture in the measuring cell $I(U)$ was well approximated by the linear function with the initial value (intercept) of 0.09±0.03 nA and the slope of $8.4 \cdot 10^{-4} \pm 0.4 \cdot 10^{-4}$ nA/V. At U=500 V, the typical time to reach the steady-state current value (estimated at 0.5 nA) was approximately 5 minutes. When changing the polarity of the voltage applied to the measuring cell by mutual replacement of the electrodes, the current jump of approximately 1.25 nA (from 0.5 nA to 0.75 nA with change in polarity) occurred, followed by the achievement of steady current value also in approximately 5 minutes. It can be assumed that this current jump is due to the polarization of the liquid dielectric.

VII. DISCUSSION

Approximately the same values of the electric field strength of the breakdown of electrical insulation transformer paper K120 and cable paper BK130 from different manufacturers is the expected result, since these types of paper are produced from the same raw materials and using the same technology. The significantly higher electrical strength of castings made of nano-gel films of bacterial cellulose is a well-known fact [5, 8]. However, the use of such material for the production of insulation of transformers and power cables is currently economically and technologically impractical. The use of laminated electrical insulation paper as the component of cable insulation theoretically allows us to reduce the thickness of the insulation, respectively, the outer diameter of the cable. However: firstly, the insulation thickness of cables with paper-impregnated insulation is determined by the standard [26] and cannot be reduced without adjusting the existing standard, and secondly, a significant reduction in the insulation thickness can lead to a

979-8-3503-8358-4/23 $31.00 © 2023 IEEE

decrease in the long-term electrical strength of the insulation, due to its destruction by barrier or partial discharges.

The study shows that the rate of artificial aging of cellulose paper in both petroleum oil and transformer silicone liquid is approximately the same.

The performed measurements of the electrical parameters of mixtures of petroleum oil and transformer silicone liquid suggest that when replacing petroleum oil in a power transformer with PDMS, the resulting colloidal formations may affect the operational characteristics of the transformer. To avoid such effect, it is advisable to turn on the power transformer after the restoration of the dielectric properties of the insulation: either after a short-term heating of the dielectric liquid and paper-impregnated insulation, or after a long time interval.

VIII. CONCLUSION

The use of modern materials makes it possible to improve the performance characteristics of the components of paper-impregnated insulation, as well as the properties of paper-impregnated insulation of power transformers and power cables in general. However, their widespread implementation requires further development of existing state standards, as well as additional research.

REFERENCES

[1] H. Kubo, B. Noda, I. Nishino, R. Hata, T. Miyazaki, "Development of 275 kV Oil-Filled Cable Insulated with Polypropylene Laminated Paper (PPLP)", IEEE Trans. Power Appar. Syst., 1982, 101, pp. 4472–4483.

[2] C.S. Esendal, R.D. Findlay, "A review of the composite dielectric insulated underground transmission cable." In Proceedings of the Second International Conference on Properties and Applications of Dielectric Materials, Beijing, China, 1988; pp. 754–757.

[3] D. Arsenyev, S. Dubitsky, D. Kiesewetter, V. Malyuin, "Numerical simulation and calculation of resistance of laminated paper-impregnated insulation of power cables", Energies, 2022, 15(19), 7403.

[4] R. Hata, Y. Yoshino, T. Shimizu, "Application of PPLP to EHV and UHV class underground and submarine cables." In Proceedings of the IEEE Power Engineering Society Winter Meeting, Cat. No.00CH37077, Singapore, 2000; pp. 676–680.

[5] N.M. Zhuravleva, A.S. Reznik, D.V. Kiesewetter, A.M. Stolpner, E.G. Smirnova, A.K. Khripunov, "Improving the efficiency of power transformers insulation by modifying the dielectric paper with bacterial cellulose," IOP Journal of Physics: Conf. Series, 2019, 1236(1), 012002, pp. 1-4.

[6] N.M. Zhuravleva, A.S. Reznik, D.V. Kiesewetter, A.M. Stolpner, E.G. Smirnova, "Study of the Electrophysical Properties of the Composite of Plant and Bacterial Cellulose". In: Proc. 2019 IEEE conf. of Russian young researches in electrical and electronics engineering, St. Petersburg, 2019, pp. 838-842.

[7] N. Zhuravleva, A. Reznik, D. Kiesewetter, A. Stolpner, E. Smirnova, V. Budaeva, "Improvement of properties of cellulose dielectrics by their structure modification with nanocel-lulose produced as wastes of agricultural crops," IOP Journal of Physics: Conf. Series, 2019, 1410(1), 012068, pp. 1-4.

[8] N. Zhuravleva, A. Reznik, D. Kiesewetter, A. Stolpner, A. Khripunov, "Possible applica-tions of bacterial cellulose in the manufacture of electrical insulating paper," IOP Journal of Physics: Conf. Series 2018, 1124(1), 031008, pp. 1-4.

[9] I. Fernandez, A. Ortiz, F. Delgado, C. Renedo, S. Perez, "Comparative evaluation of alter-native fluids for power transformers. Review," Electric Power Systems Research 98, 2013, pp. 58–69.

[10] V.M. Barinov, D.V. Kiesewetter, "Experience in Enhancing the Reliability of Operation of Power Cable Lines in St. Petersburg." In Proc. 10th Electric Power Quality and Supply Reli-ability Conference, Tallinn, 2016, pp. 55-58

[11] D.V. Kiesewetter, A.S. Reznik, N.M. Zhuravleva, A.K. Khripunov, "Investigation of elec-trophysical properties of composite from nano-gel film of bacterial cellulose with addition of carbon nanotubes." In Proc. 2020 IEEE Conference of Russian Young Researchers in Electrical and Electronic Engineering (EIConRus), St. Petersburg, 2020, pp. 1039-1041.

[12] D.V. Kiesewetter, N.M. Zhuravleva, A.S. Reznik, A.K. Khripunov, A.V Migunova, "Investigation of Dielectric Properties of Composite Films of Bacterial Cellulose with Carbon Nanotubes." In Proc. IEEE 3rd International Conference on Dielectrics (ICD), Valencia, Spain, 2020, pp. 245-248.

[13] D.V. Kiesewetter, A.S. Reznik, N.M. Zhuravleva, D.A. Trubin, A.K. Khripunov, "Study of the dynamics of electric breakdown in cellulose composite materials with nanofillers." In 2021 IEEE Conference of Russian Young Researchers in Electrical and Electronic Engineering (EIConRus), St. Petersburg, 2021, pp. 1220-1223.

[14] State Standard 13523-78 RU. Fiber semi-finished products, paper and board. Method for conditioning of samples. Standard Publ., 1978, rev. 1989 (in Russian).

[15] State Standard 26130-84 RU. Electrical insulating paper. Methods for determination of elec-trical strength at alternating (frequency of 50 Hz) and constant voltage. Standards Publ., 1984 (in Russian).

[16] State Standard 30180.2-99 RU. Paper of electrically insulating cellulose. Specification. Part 2. Test methods. Standards Publ., 2000 (in Russian).

[17] N. Zhuravleva, A. Reznik, D. Kiesewetter, D. Tashlanov, "The impact of the degree of polymerisation of the cellulose molecules on the electrical and mechanical properties of insulating paper." In Proceedings of the 2017 IEEE Russia Section Young Researchers in Elec-trical and Electronic Engineering Conference, St. Petersburg, 2017, pp. 1220–1223.

[18] D.V. Kiesewetter, A.S. Reznik, N.M. Zhuravleva, D.A. Trubin, "Study of the dynamics of insulating paper electrical breakdown." In Proc. 2020 IEEE Conference of Russian Young Researchers in Electrical and Electronic Engineering (EIConRus), St. Petersburg, 2020, pp. 1036-1038.

[19] D. Kiesewetter, V. Malyugin, A. Reznik, A. Yudin, N. Zhuravleva, "Application of the spectral-correlation method for diagnostics of cellulose paper," IOP Journal of Physics: Conf. Series, 2017, 917(4), 042020, pp. 1-4.

[20] State Standard 60156 – 2013 RU (IEC 60156:1995). Insulating liquids – Determination of the breakdown voltage at power frequency – Test methods (IDT), Standards Publ., 2014 (in Russian).

[21] State Standard 6581-75 RU. Liquid electrical insulating materials. Electric test methods. Standards Publ., 2008 (in Russian).

[22] N. Zhuravleva, A. Reznik, D. Kiesewetter, A. Tukacheva, E. Smirnova, "About the Possibilities of Increasing the Reliability of Paper-Impregnated Insulation of Power Transformers." In Proc. 57th Int. Scientific Conference on Power and Electrical Engineering of Riga Technical University (RTUCON), Riga, 2016, pp. 1-4.

[23] N. Zhuravleva, A. Reznik, A. Tukacheva, D. Kiesewetter, E. Smirnova, "The study of thermal aging components paper-impregnated insulation of power transformers," In Proc. 2016 IEEE North West Russia Young Researchers in Electrical and Electronic Engineering Conference, (EIConRus) St. Petersburg 2016, pp. 747 – 751

[24] N. Zhuravleva, A. Reznik, D. Kiesewetter, A. Tukacheva, E. Smirnova, "On the increasing of the sorption capacity and temperature resistance of cellulosic insulation dielectrics," In Proc. ELEKTRO-2016, High Tatras, Slovac Rep., 2016, pp. 649-653.

[25] D.V. Kiesewetter, A.S. Reznik, N.M. Zhuravleva, D.V Litvinov, "Research of dielectric properties of the mixtures of petroleum transformer oils and silicone liquids." In 2021 IEEE Conference of Russian Young Researchers in Electrical and Electronic Engineering (ElConRus), St. Petersburg, 2021, pp. 1224-1227.

State Standard 8410-83 RU. Power paper insulated cables. Specifications. Standards Publ., 1998 (in Russian).

Research of Low-Cost Technological Process for Manufacturing of One Side Silicon Interposer

Mikhail D. Kochergin
Institute of Nano and Mycrosystem Technology
National Research University of Electronic Technology
Moscow, Russia
misha.kochergin1999@yandex.ru

Denis V. Vertyanov
Institute of Nano and Mycrosystem Technology
National Research University of Electronic Technology
Moscow, Russia
vdv.vertyanov@gmail.com

Igor A. Belyakov
Institute of Nano and Mycrosystem Technology
National Research University of Electronic Technology
Moscow, Russia
igor-terra@yandex.ru

Raiymbek N. Zhumagali
Institute of Nano and Mycrosystem Technology
National Research University of Electronic Technology
Moscow, Russia
zhumagali_991@mail.ru

Abstract — to manufacture silicon interposers, back-end (BEOL) semiconductor production processes are commonly used. This processes providing minimal topological norms, but at the same time leading to a more complex and expensive production. However, minimum norms are not always a priority for interposers, especially if we are talking about small-scale production. In this work we studied the possibility of obtaining single-sided silicon interposers with 10 μm norms using low-cost standard electronics manufacturing processes, such as vacuum metal deposition, liquid etching of the copper pattern and plasma chemical etching of dielectric in oxygen plasma through a photoresist mask.

Keywords—interposer, copper etching, plasma etching, polyimide passivation

I. INTRODUCTION

In this work, single-sided silicon interposers with topological norms of 10 μm were fabricated to mount crystals with a pad size of 50x50 μm and a pad spacing of 20 μm. Advanced BEOL (back-end of line) processes are not necessary for such norms. Therefore, the possibility of making interposers with a simple vapor deposition and plasma chemical etching equipment's was investigated. For this reason, after titanium layer with thickness of 50 nm and copper layer with thickness of 2 μm were deposited on oxidized silicon wafers, etching process was performed by wet method. Several different compositions of etchant solutions were investigated. After obtaining the topology, a polyimide protective coating was formed on the plate from a polyamide acid solution based on aromatic diamine and aromatic tetracarboxylic acid dianhydride in N-methyl-2-pyrrolidone. The opening of contact pad in polyimide for the interposer assembly was performed using plasma-chemical etching. However, the process was performed directly through the thick layer of the photoresist mask, rather than through the metal mask previously formed on the dielectric by PVD (Physical Vapor Deposition) sputtering. The last operation was the formation of a finish coating on the interposer areas uncovered from the polyimide.

II. RESEARCH AREA

Silicon commutations boards are now increasingly being used to miniaturize electronics. There are many approaches to the design and manufacturing technology of such boards. For example Intel's EMIB solution is a combination of both interposer and substrate. Rather than taking a large interposer, Intel uses a small slither of silicon that embedded directly into the substrate (Intel calls this a bridge) [1]. Pads of two chips are connected via the bridge using hundreds or thousands of connections. This allows data to be transmitted through high density traces on silicon without restrictions that a large interposer might bring.

We can see a more classical approach to the interposer as a big silicone substrate in the 3D integration technology from Intel's named die-to-die stacking or Feveros. In this technology we have one or more units of dies (chips ore interposers) with through silicon vias on top of another. There are several types of Foveros options. One of them is logic-on-logic technology, where is the interposer, or base die, has active circuitry relevant for the full operation of the main compute processors found in the top piece of silicon 3D assembly [2].

Similar to the logic on logic technology from Intel is a POSSUM chip-on-chip technology from the Amkor technologies OSAT company. POSSUM technology allows you to mount a smaller chip under a large chip that acts as a carrier. In this case, there is no need for through silicon vias [3].

Silicon interposers have many advantages over substrates based on printed circuit boards. Due to the small topological norms interposers allow almost any existing die to be mounted on them, no matter how small the pitch of their contact pads may be. In addition, interposers do not require the CTE (coefficient of thermal expansion) of the die and the substrate to be matched, since both the die and the substrate are made of silicon. The disadvantage of such structures is expensive manufacturing technology, especially if we are talking about interposers with metallized holes in the silicon [4-6].

The traditional approach to obtaining small topological norms for aluminum metallization on silicon was plasma chemical etching. However, when copper replaced aluminum as commutation material, there were difficulties in plasma chemical etching of copper due to the absence of volatile reaction products. The substrate was quickly contaminated with products that blocked etching. This forces the use of expensive back-end-of-line damascene processes in the manufacture of interposers.

Dual damascene process includes many operations, such as dielectric chemical vapor deposition (CVD), two stages of dielectric plasma chemical etching, physical vapor deposition (PVD) of metal layers, electroless filling of trenches and vias with copper and chemical-mechanical polishing, which require special reagents and equipment [7].

979-8-3503-8358-4/23 $31.00 © 2023 IEEE

First stage of dual damascene process after low-k dielectric deposition is the plasma etching process forms vias to underlying metal and trenches for layer metals line in low-k dielectric. Then, a thin barrier layer of tantalum (Ta) or titanium nitride (TiN) materials are deposited on trench and vias edges using physical vapor deposition. The barrier layer is coated over by a copper seed layer using PVD. Finally, the structure is electroplated with special solution of copper metallization and ground flat using chemical mechanical polishing [8-9].

However for most crystals, the pad spacing is still greater than 20 µm. Interposers with this pin spacing can be fabricated without using of damascene process (with the help of only wet etching for copper with underlayer and using traditional oxygen plasma etching for opening pads in dielectric). Multilayer interposers with nanometer norms make sense for chips with 20 µm pitches only if we are talking about complex on-chip systems, FPGAs or processors with thousands of pads. Low expensive variant of interposers is especially relevant for research labs and companies that are solely involved in packaging.

III. WET ETCHING

The etchants involved in the study, as well as the etching times, are shown in Table 1.

TABLE I. ETCHANTS STUDY

Etching solution	Etching time, s	Undercut, µm
$FeCl_3$	10	6
$H_2SO_4 + H_2O_2$	8	5-6
$CuCl_2 \cdot 2H_2O + K_4P_3O_7$ + Monoethanolamine	10	4-5
$HCl + H_2O_2$	8	4-5
$CrO_3 + HCl + HNO_3$	5	3-4
Citric acid + H_2O_2	10	3
$HCl + H_2O_2$	8	2,5-3
$CrO_3 + H_2SO_4$	4	1,5-2

The classical anisotropic copper etching solution based on ferric chloride, which showed good results with copper thicknesses of 20-30 µm and topological limits of 50 µm, turned out to be unsuitable for obtaining standards of 10 microns. The reason for this was too heterogeneous etching of copper over the substrate area - when in some places the copper was completely removed, in other places the process was just beginning. Also, ferric chloride strongly contaminated the substrate. As a result, the solution showed the largest undercut under the resist Fig.1.

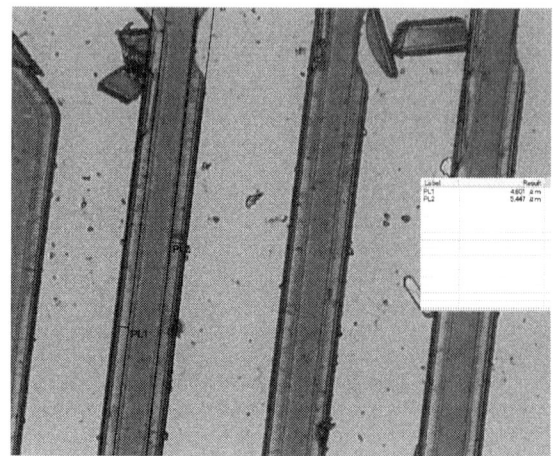

Fig. 1. Topology fragment after etching of commutation pattern with FeCl3 solution (60%) - undercut more than 5 µm, after etching of titanium - delamination even on large polygons.

Among alkaline solutions, the alkaline copper etching solution based on monoethanolamine was chosen - it was assumed that the smothered photoresist would last long enough to remove the 2 µm copper layer. However, due to the low plasticity of the positive resist, the destruction process of the protective coating was observed during etching under the influence of alkali. The peroxide and sulfuric acid solution performed slightly better than in the cases of ferric chloride and alkaline solution on monoethanolamine. This is most likely due to the uniform etching rate over the entire substrate. In the case of hydrochloric acid and peroxide, minimal undercut of 2.5 to 3 µm was observed, but only if the concentration of peroxide in the solution was 2 or 3 times higher than that of acid. The best etching results were obtained for solutions based on a mixture of sulfuric acid and chromium oxide as an oxidizer - 1.5-2 µm undercut. Fig.2 shows fragment of the topology after etching.

Etching of a thin adhesion sublayer of titanium was performed in a solution of peroxide, nitric acid and ammonium fluoride salt. The beginning of etching could be determined by the beginning of gas emission from the surface, and the end by the cessation of gas emission and a change in the color of the surface from yellowish to dark violet.

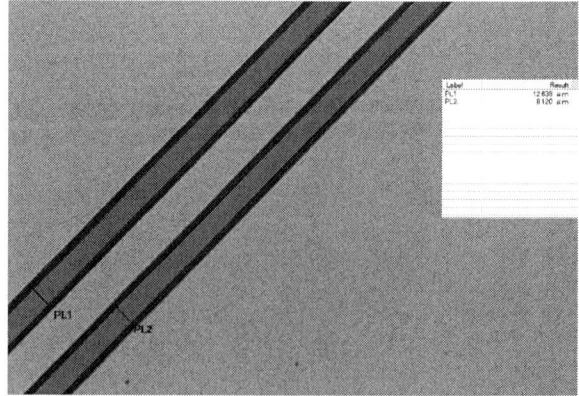

Fig. 2. Fig.2. Topology fragment with 8-10 µm norms and 2 µm undercut.

IV. PLASMA-CHEMICAL ETCHING

After obtaining the commutations with 10 μm norms, it was necessary to form the finish coating on the copper and the protective mask. The final coating in the form of a layer of chemical nickel and immersion gold was planned to cover the entire topology in order to avoid the negative effects of metallization reagents on the protective mask. However, experiments showed that the chemical nickel in the growth process closes the tracks located at a distance of 10 μm due to deposition of the chemical nickel in the gaps between the topology elements and expansion of the tracks to the sides. Therefore, it was decided to first form a protective mask of polyimide or photoresist on the plate, and only then to cover the areas opened in the mask with a layer of finishing metallization.

The photoresist coating as a mask showed low temperature and mechanical resistance, so it was decided to abandon it. Instead, polyimide coating was formed on the plates. By centrifugation a layer of polyimide varnish with a thickness of about 7 μm was applied to the plate with commutation. After the three-step process of varnish imidization, a thick layer (about 4-7 μm) of photoresist was also applied to the plate, in which, after curing, a pattern of pads was formed for opening in polyimide. The polyimide over the pads was further etched to copper through the photoresist by plasma-chemical etching in argon-oxygen plasma. The etching depth was measured with a profilometer. Underetching was determined both by the depth of groove etching and by the structure of the polyimide inside the groove by examining the samples on an optical profilometer and SEM (scanning electron microscope). The most suitable etching mode was selected by varying the etching time and the gases in which the process took place. Some modes and their results are shown in Table 2.

TABLE II. MODES OF PLASMA-CHEMICAL ETCHING

Mode №	Power, W	Gas	Etching time, min	Etching depth, μm
1	200	O_2, Ar, SF_6	10	1.0
2	200	O_2, Ar	10	2.0
3	200	O_2, Ar, SF_6	5+5	1.5
4	200	O_2, Ar	10+2	8.0
5	200	O_2, Ar, SF_6 +Ar	4+4+0.5	6.0
6	200	O_2, Ar, SF_6	4+4	6.0
7	200	O_2, Ar + Ar	10+2	6.5
8	200	O_2, Ar + Ar	10+1	7.0
9	150	O_2, Ar + Ar	11+1	7.0
10	150	O_2, Ar + Ar	11+1.5	7.5

Initially, the plates were etched in oxygen, argon and sulfur hexafluoride plasma. However, experiments showed that using sulfur hexafluoride makes the surface of contact pads too rough, and its use was abandoned. Argon treatment after the main etching process was performed to remove contaminants from the previous step. Fig.3 shows a graph of the change in the etch velocity as a function of the etch time.

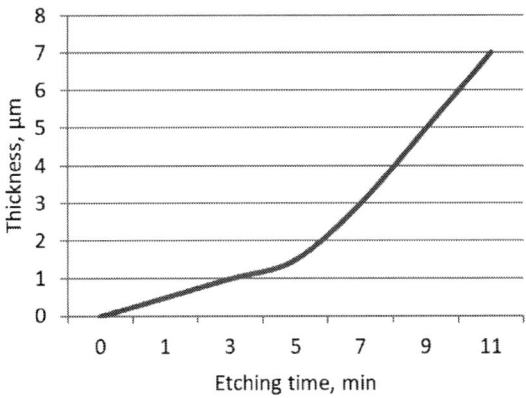

Fig. 3. Dependence of the polyimide thickness change on the etching time (mode №9 from Table II).

Fig.4 shows an image of an under-etched hole in polyimide with a diameter of 500 μm, obtained using an optical profilometer.

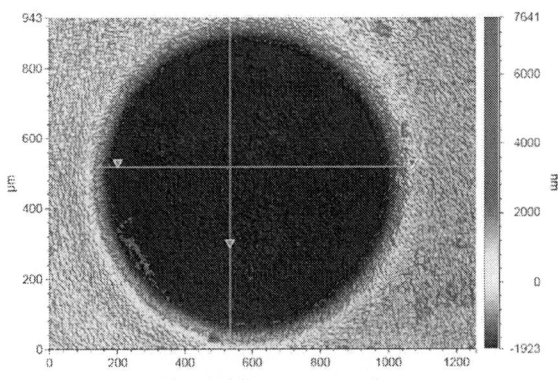

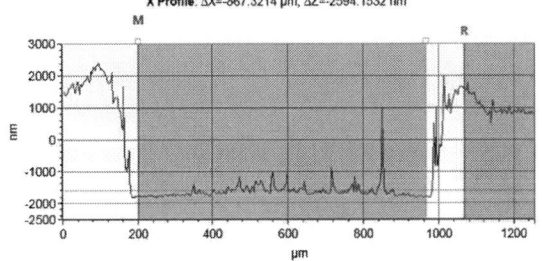

Fig. 4. A polyimide film about 3.5 μm thick, remaining after etching.

Fig.5 shows an image of the same site, but with the polyimide completely etched over it.

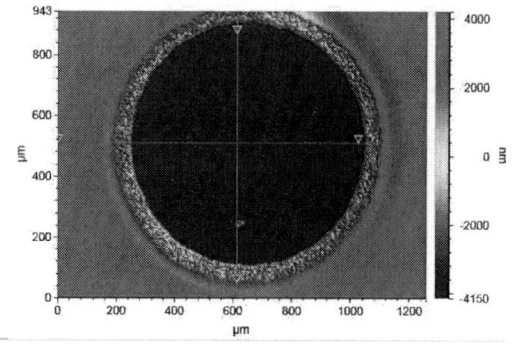

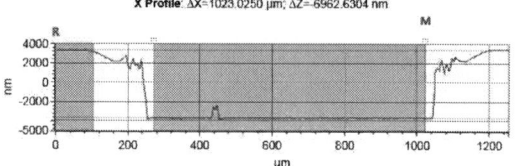

Fig. 5. Polyimide completely etched by plasma chemical etching to copper over the pad.

Fig.6 shows SEM image of 55 μm diameter pads after plasma-chemical etching of polyimide through a photoresist mask after mode №6 from Table II.

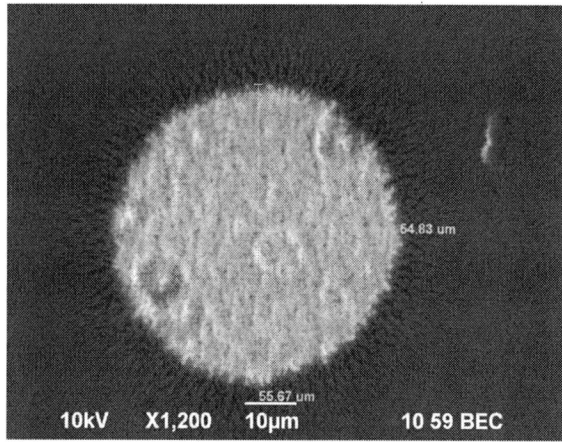

Fig. 6. SEM image of the opened contact pad in the polyimide protective coating for flip chip mounting (mode №6 from Table II).

After opening the areas in the polyimide, the photoresist mask through which the etching was performed was removed from the plate and the finish coating was formed on the opened areas. As part of this work, the formation of two types of finish coating was investigated - on the basis of NiP/Au layers and on the basis of immersion tin. During the formation of NiP/Au layers on copper pads, polyimide delamination was observed in areas where the dielectric lay not on copper but on silicon under the effect of high temperature of metallization solutions, most likely due to the difference in CTE between the substrate and polyimide. Significantly better results were achieved with the formation of immersion tin on the sites - Fig.7. After the finish coating was formed, the interposer plate was separated into individual substrates on which ball pins were formed to mount the interposer to the board.

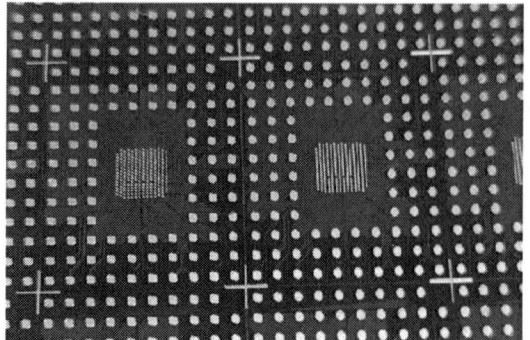

Fig. 7. Finish coating (immersion tin) on interposer contact pads opened in the protective mask before plate separation.

V. CONCLUSIONS

This article describes the process of creating an interposer with topological norms of 10 μm using processes such as PVD sputtering, wet etching, and plasma chemical etching. Topological norms of 10 μm with the smallest undercut of 2 μm were achieved using a solution based on a mixture of sulfuric acid and chromium oxide as an oxidizer for copper etching and a solution of peroxide, nitric acid and ammonium fluoride salt for titanium etching. The required results of plasma chemical etching were achieved in a argon-oxygen atmosphere in an etching mode lasting 11 minutes at a power of 150 watts followed by purification in the argon atmosphere for 1.5 minutes. Immersion tin showed the best results as the final coating.

ACKNOWLEDGMENT

This work was carried out with the financial assistance of the Ministry of Education and Science in the framework of state task FSMR-2022-0002.

REFERENCES

[1] Mahajan, R., Sankman, R., Patel, N., Kim, D. W., Aygun, K., Qian, Z., Mallik, D. Embedded multi-die interconnect bridge (EMIB)--a high density, high bandwidth packaging interconnect. *2016 IEEE 66th Electronic Components and Technology Conference (ECTC)* (pp. 557-565). IEEE.

[2] Prasad, C., Chugh, S., Greve, H., Ho, I. C., Kabir, E., Lin, C., Pantuso, D. Silicon reliability characterization of Intel's Foveros 3D integration technology for logic-on-logic die stacking. *2020 IEEE International Reliability Physics Symposium (IRPS)* (pp. 1-5). IEEE.

[3] Sutanto, J. POSSUM die design as a low cost 3D packaging alternative. *3D Packaging*, 2012, №25, pp.17-18

[4] Vertyanov, D., Sidorenko, V., Belyakov, I., Timoshenkov, S. Forming interlayer holes in the dielectric of redistribution layers of microcircuits with embedded die in plasma-forming atmospheres. Nanoindustry, vol. 13, no. 5. Russia, 2020. pp. 560–566. DOI: 10.22184/1993-8578.2020.13.5s.560.566.

[5] Burakov, M., Vertyanov D., Boyko A., Sosnovsky A. Investigation of TSV metallization for MEMS encapsulation technology. *2018 ElConRus Conference* (pp. 1599-1603). IEEE.

[6] John H. Lau. *Fan-Out Wafer-Level Packaging.* SPRINGER Verlag, Singapor, 2018. 303 p.

[7] Carpio R., Jaworski A. Review—Management of Copper Damascene Plating. *Journal of The Electrochemical Society.* 2019. 166. D3072-D3096. DOI:10.1149/2.0101901jes.

[8] Korobova N. *Piezoelectric MEMS technologies.* The book Advanced Piezoelectric Materials. Science and Technologies. Second ed. Kenji Uchino. Elsevier. 2017. pp.533-574

[9] Pratt A. Overview of the Use of Copper Interconnects in the Semiconductor Industry. Advanced Energy Industries, Inc. 2004.

LNA IC Development for LTE Bands Based on Bulk CMOS 180nm Digital Library

Evgeny Yu. Kotlyarov
JSC MERI
National Research University of
Electronic Technology
Moscow, Zelenograd, Russia
orcid.org/0000-0002-6314-3178

Mikhail Putrya
National Research University of
Electronic Technology
Moscow, Zelenograd, Russia
mishapmg@gmail.com

Viktor Mikhailov
JSC MERI
Moscow, Zelenograd, Russia
vmikhaylov@niime.ru

Igor Zubov
JSC MERI
Moscow, Zelenograd, Russia
izubov@niime.ru

Sergey Timoshin
JSC MERI
Moscow, Zelenograd, Russia
stimoshin@niime.ru

Abstract— **The article discusses the process of developing an analog cell of a low-noise amplifier for operation as part of an LTE band transceiver for an Internet of Things system. The CMOS 180nm digital process, which requires adaptation to work with RF blocks, has been selected as the basic technology. The main aspects of the adaptation and optimization of the library are described in this publication. The developed A-class low-noise amplifier is built in three stages with linear gain up to 25 dB at 1.7 dB of its own noise and a VSWR level of no more than approximately 1.22. The 1 dB compression point is 6.7 dBm.**

Keywords—CMOS, LNA, LTE, RF, 180nm.

I. INTRODUCTION

The current level of development of computer-aided design (CAD) allows us to achieve reliable results even at the early stages of the device's development. This has contributed to the acceleration of the development process and the emergence and introduction of new technologies in the microelectronic industry [1]. The Internet of Things (IoT) ecosystem in the telecommunications industry is constantly developing new standards and protocols that are intended to complement or improve the user experience with mobile devices, make it easier to interact with smart home systems, and enable sensors in industrial facilities [2]. Along with this, semiconductor industry technologies are progressing rapidly. Technological standards for digital processes in leading factories have reached 7 and 5 nm. These values are currently at a level of 40–28 nm for analog high-frequency processes, depending on the boundary frequencies and the power limitations of active devices. However, with the steady escalation of limitations on the distribution of high-tech even for non-military use, it is currently becoming difficult to plan the development of modern mixed signal SoCs at huge factories like TSMC, GlobalFoundries, and Samsung.

The implementation of mixed-signal nodes is required by the development of a transceiver for LTE communications networks. Despite having the capability to achieve the best parameters, using conventional analog high-frequency technological processes based on GaAs and GaN is impracticable. Except if we are talking about 3D-SoC [3] assembly, where it is possible to integrate various technological processes within a single IC package to maximize their advantages. At this moment, silicon-based processes (CMOS, SoI, and SiGe) are the only alternative technological bases for the development of mixed-signal systems. The transceiver node prototypes projects were made using a 180 nm CMOS process, taking into account the technology's accessibility and relative affordability.

This article discusses the process of developing one of the nodes for the LTE frequency band transceiver to work with the narrowband Internet of Things standard[4].

II. SPECS AND THEORY

Sensitivity is one of the key parameters of the receiving device, which must be paid attention to at the prototyping stage. To increase sensitivity, it is necessary to use a low-noise input stage, the main node of which is a low-noise amplifier (LNA). When designing the input stage, it is necessary to estimate the minimum level of signal power expected at the input of the receiving device. By simulating how radio waves move in different situations, the minimum signal power level can be found. Taking into account the results of system modeling [5], the requirements for a receiving and transmitting device were formulated. This device is being made based on a prototype. It is supposed to use one or more LNAs, which are multi-band receivers with the following features:

- noise factor ≤4.5 dB;

- gain ≥10 dB;

- compression point P1dB ≥ -1 dBm;

- Operating frequency range 400 - 5000 MHz.

The low-noise amplifier is the first active node in the system's receive channel. It is right behind the antenna and the preselector filter, which determines the operational frequency band. The functional purpose of the LNA is to amplify small levels of the received signal with minimal signal distortion (IP2, IP3) and the lowest possible noise level (Noise Figure) to maintain the signal-to-noise ratio (SNR) set within the system. For successful detection and interpretation of SNR, the signal strength must significantly exceed the noise floor. The noise level of the receiver can be kept as low as possible if the input circuits of the low-noise amplifier are matched and the noise level of the first (input) gain stage is kept as low as possible.

When standard CMOS technical processes are used to make integrated circuits, there are a few things that make it hard to meet the required parameters in the radio frequency ranges:

- Low values of the dielectric permittivity of oxides and the thickness of the conductors do not allow to

979-8-3503-8358-4/23 $31.00 © 2023 IEEE

achieve a high Q–factor of resonant circuits of conductors, especially critical for noise degeneration elements such as inductors, which generally affect the growth of noise characteristics and deterioration of nonlinear parameters of devices and, as a consequence, a decrease in efficiency (% PAE). Also, this factor is associated with a decrease in resonant frequencies (characterizing resonators operating bands) in inductors;

- To get high inductance values while keeping the Q-factor of the coils at a fair level, you need a large layout area. As the layout area goes up, the frequency of self-resonance goes down, and the resistance goes up.

- Due to the effect of the skin layer of metallization (skin depth), the small thicknesses of the conductors stop the signal from spreading at frequencies in the decimeter and lower part of the centimeter wave range. They also cause the resistance to go up, which causes the signal to weaken and the noise to get higher.

- Due to the need for a high number of conductors per unit area, parasitic capacitive bonds are formed.

III. LNA IC DESIGN FLOW

As part of the development of the receiver-transmitter, projects for several nodes were prepared for prototyping an NB-IoT system based on a CMOS process with a design norm of 180 nm. The choice of the basic technology is determined by several criteria. Firstly, it is a fundamental possibility to build a "system on a chip" of mixed type, including analog RF and LF blocks, as well as digital logic. Secondly, it is the possibility of localizing production and launching prototypes in domestic factories. Thirdly, the price of the final product is an important factor. Thus, the development of prototype nodes was carried out on the basis of HCMOS8D technology. This library is mostly for making digital devices. It is not a tool for making high-frequency projects, but you can make complex high-frequency devices, like transceivers, by adding the necessary elements of the planar path.

A methodical approach to developing this chip includes several additions to the traditional analog integrated circuit flow diagram. The proposed design flow in the form of an algorithm is shown in Figure 1.

Like any development, this path begins with obtaining technological requirements, in this case as a result of preliminary system modeling. The selection of technology is frequently associated with factors unrelated to technology. The selection of a specific technology, library, and available sets of options is more frequently justified by pragmatic arguments: availability of technology, price, and production time. In this case, it is a low-frequency set of libraries based on the CMOS 180 nm process.

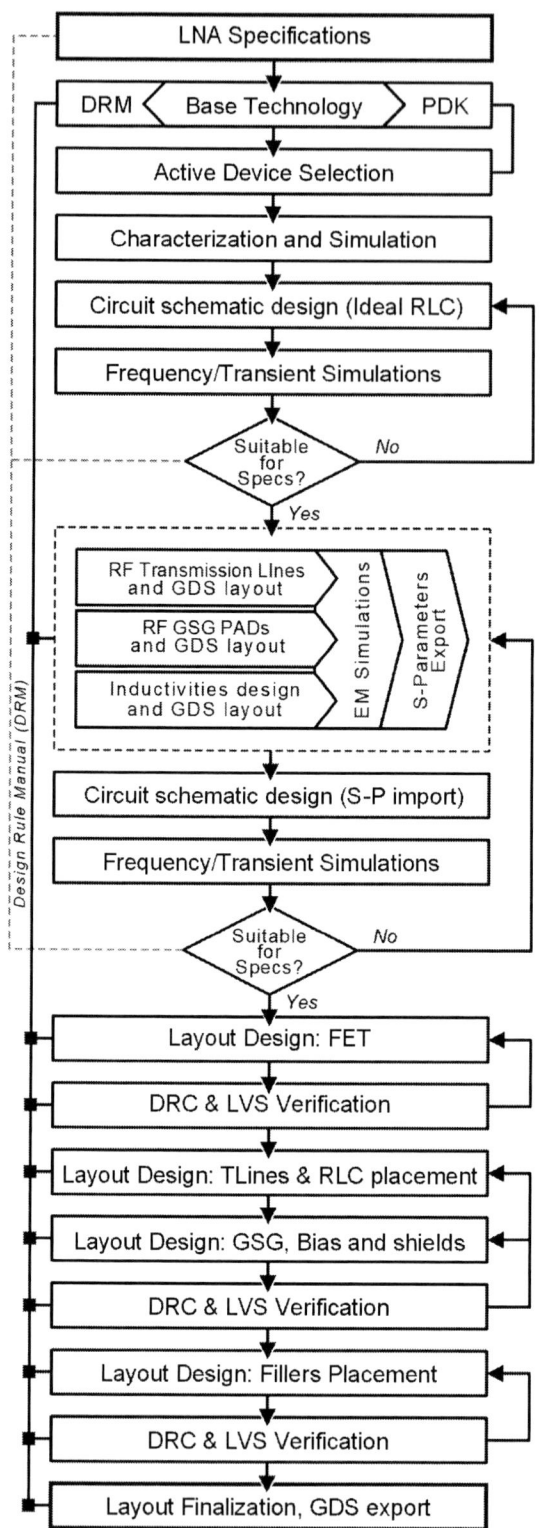

Fig. 1. Project design flow algorithm.

Modern automatic design systems provide many opportunities for implementing a flexible approach to the design of analog ICs. Traditional calculations can be used to choose and set up devices, parametric sweeps can be done by gradually reducing the sampling range, parameter optimization can be set up, and active devices can be tuned with smart tools and wizards. Among the active devices in the selected library are metal-oxide-semiconductor field-effect transistors (FET, MOSFET), and based on a low-voltage NMOS 2V transistor, we will try to build a class A amplifier chip. At this point, the initial parameters must be evaluated in order to determine the primary geometry of the device and the operating point transistor.

After the tests and parameterization of the active device are completed, it is necessary to select the type of transistors to include in the amplifying stage. These can be options with common source, common drain, or cascode scheme options for switching on transistors each of them has its own advantages and disadvantages. For the prototype low-noise amplifier, a variant with three common-source switching stages was used. Initially, the circuit is run using the idealized RLC values of the elements to evaluate their influence without taking into account their frequency parameters or other parasitic factors. It is necessary to evaluate the possibility of operation and get primary data on electrical parameters in the frequency domain to compare with the specification.

In view of the fact that the library is not intended for the development of RF devices, minimal adaptation is required. At this stage, additional development of planar tract elements is required. We are talking about the calculation of the transmission line, the development of inductor sets, and the contact pads (GSG) required for measuring the microwave parameters of devices on the plate. Here you will have to resort to the help of 3rd party CADs, which are multi-physics packages that can help in the calculation, synthesis, and EM analysis of planar structures. The development process uses information from the design rule manual (DRM) of the chosen technology so that the results can be integrated into the overall design flow. Electromagnetic simulation results (usually using finite element methods - FEM) are exported as blocks of S-parameters for secondary analysis in the amplifier circuit. At the same time, in order to parameterize the additional blocks developed, a topology (Fig. 2) was created with test cells for on-wafer measurements. Each block has RF GSG IO pads for calibration and de-embedding, set of test measures (through, open, short, and load) located in the same layout. The obtained data will allow us to correct the calculated parameters of passive planar elements as well as update the values of transistor models, MIM capacitors, and resistors in the frequency range, obtained using simulations through SCPICE models.

Further along the design flow, the schematic design is updated with extracted S-parameters blocks that were represented by idealized cells in the previous iteration. There is inevitably a change in high-frequency characteristics (most likely for the worse), which required significant changes in the original concept of the circuit. Upon reaching acceptable results in the specification, the development of the project layout begins. The design of the low-noise amplifier circuit topology begins with the topology of active elements (nmos). It is necessary to bring the drain source and gate into the upper layers of metallization, for which the topology tracing technique with noise reduction described in the article [6] was used.

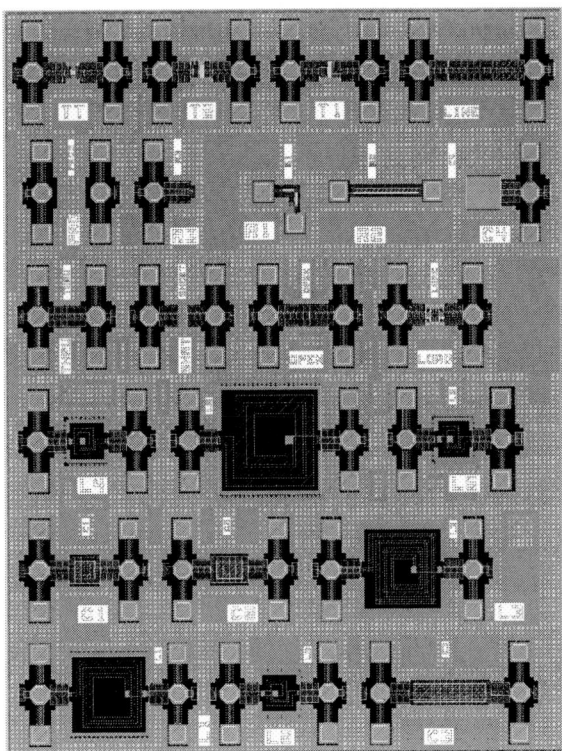

Fig. 2. Testsuit layout.

Each stage in the design of the LNA prototype topology is associated with constant cycles of design rule check (DRC) and layout versus schematic (LVS) verifications, which are integral parts of any IC design route. After placing the transistor cells, the drawing of high-frequency signal transmission lines, RLC elements, and contact pads with passivation openings (RF-GSG & DC bias) is performed, followed by verification cycles. Once the topology core is ready, the fillers are placed in accordance with the DRM requirements to ensure the density of conductors per unit area of the crystal. Fillers can be generated by means of special tools that provide filling according to the embedded algorithms, or they can be placed manually in accordance with the requirements of DRM (defined gaps, angles, slotting, etc.). Each of these approaches has its pros and cons, as manual placement of the dummy filler allows for more flexible area-filling work, while automatic placement saves a lot of time, which is important when developing multiple nodes in a large project. After the final checks, the project's *.gds file is exported and placed in the multi-project wafer (MPW) frame's topology so that it can be used to make photomasks.

IV. LNA IC RESEARCH AND DESIGN

To build a low-noise amplifier block, an N-channel MOSFET with low leakage currents with an oxide thickness of 3.2 nm and a voltage of 1.8 V. The IV chart of a transistor with a gate width of 670 microns and a length of 200 nm is shown in Fig. 3. The frequency properties of field-effect transistors are characterized by their speed. Quantitative measures for assessing the frequency properties of the transistor are the transition (Ft) and maximum (Fmax) frequencies of the transistor. The transistor transition frequency (Ft) is the frequency at which the transistor gain becomes equal to one (0 dB). Ft manifests itself at high frequencies and is explained by the influence of parasitic reactivities (capacitances) of the transistor as well as features of the semiconductor structure (mobility of charge carriers, barrier capacitance, etc.). For a low-signal transistor model, the boundary frequency can be calculated from the pass function of the scattering matrix H-parameters – H21. The point where the Mason gain tends to unity, K = 1 (Max = 0 dB), defines the maximum frequency (Fmax). This point is found by the pass-through expression of the scattering matrix S21. The potential possibility of constructing high-frequency circuits based on this transistor can be judged by the boundary and maximum frequencies of the transistor. For example, the operating frequency range of the proposed device should be on a segment twice or three times less than the boundary frequency (0.5–5 GHz for systems of the 5G FR1 family), with a single point gain (0 dB). The boundary (Ft) and maximum frequency for the selected transistor when the gate is powered (Vg = 0.9 V) and the drain (Vd = 0.9 V) are shown as functions from the array of the scattering matrix on the graph (Figure 4).

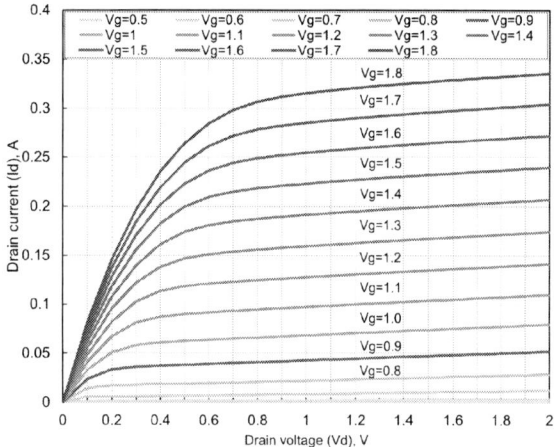

Fig. 3. nmos2 IV chart.

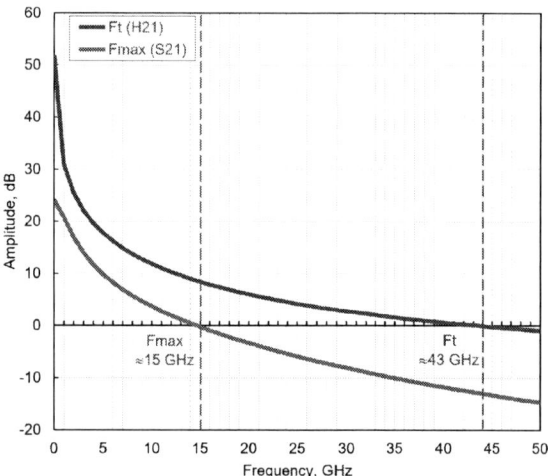

Fig. 4. nmos2 Ft and Fmax frequencies.

Based on these data, a scheme with a minimum noise level and high linearity for the n7 (Russia: 2560-2570 / 2680-2690 MHz) range was developed with the highest possible level of output compression, which will be in demand in a number of cases.

According to the calculations made in [4], the optimal gain level of the LNA circuit should be 17 dB. The transistors selected as part of the HCMOS8D process, in the selected configuration, can achieve an average of ~8.5 dB of gain per cascade. Therefore, in order to obtain the required gain level, it is required to build a circuit with at least two gain stages. To compensate for the attenuation in the matching circuits, it is not possible to reach the required gain (G = 17 dB) on a two-stage circuit therefore, it is necessary to increase the number of amplifier stages to 3x. This paper presents a scheme for a linear, low-noise A-class amplifier based on three cascades. The amplifier circuit is shown in Figure 5.

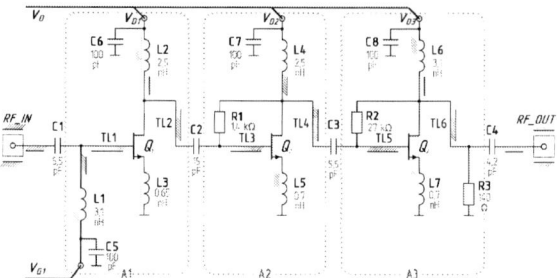

Fig. 5. Scheme of a three-stage low-noise S-band amplifier.

The circuit (Fig. 5) consists of 3 stages. The design of the first stage is made according to the switching scheme with a common source (common source) to obtain a large gain, and noise degeneration is performed using the L3 coil. Through the inductor L1, the external power supply to the gate of the first stage lets you change the working point for matching the input circuits. This lets you change the gain and noise of the input stage and the whole device. The second and third stages of the circuit have feedback circuits by means of large resistances R1 and R2, which allow you to save the crystal area, eliminating the need for discrete gate power, as well as improve matching, which will positively affect the growth of the stability coefficient despite the input of the feedback circuit. The R3 output resistor sets the output impedance at the expense of a small loss in signal gain. Capacitors C1–C4 perform the function of galvanic isolation and matching, and capacitance C5–C6 filters low-frequency interference. It was mentioned above that, due to the fact that this process is not adapted for the development of high–frequency devices, the library lacks parameterized cells of inductors and elements of the planar path—cells of primitives of microstrip lines and RF GSG contact pads. These elements were performed in a third-party CAD using electromagnetic simulations using the finite element method [7]. Integral inductors are used in the power supply to reduce noise, match signals, and keep RF out. Inductors with nominal values of 2.5 nH (in two modifications: with a small width and a large width to meet the requirements of the technology for electromigration), 3.1 nH, 0.7 nH, and 0.65 nH were used to develop the microcircuit. These nominal values are chosen so that the geometry of the cells matches the maximum values of the Q-factor with the highest value of inductance and meets the design rules, especially the density of metals and alloyed layers per unit area of the plate. These factors also limit the achievement of the highest gain parameters and minimization of the noise coefficient and indirectly cause the presence of a third gain stage. All this is connected with the features of the basic technological process, the parameters of transistors, and the absence of "thick" top metal conductor layers, which affect the quality of resonant structures and are a limiting factor for the use of coils with large inductance ratings. Because there aren't "thick" layers of metal, we have to deal with problems like the substrate capacity, the "skin layer" effect, and so on when making inductors, which aren't in the library of elements and have to be made from scratch.

The linear parameters of the scattering matrices are shown in the graphs - Figure 6.

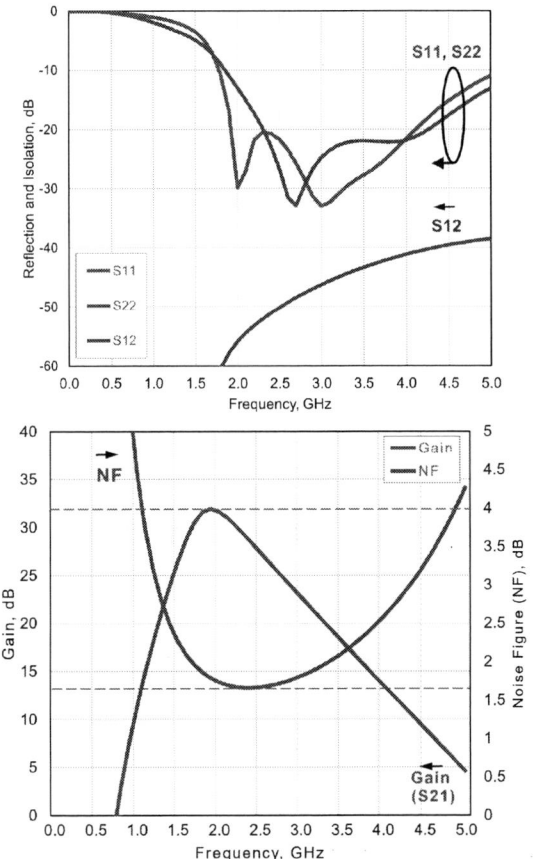

Fig. 6. Scattering parameters and noise figure graph.

The signal gain in the operating range n7 is more than 25 dB, with reverse isolation (S12) exceeding the level of 40 dB. The reflection coefficients of the input (S11) and output (S22) reach 33 dB and exceed 20 dB values throughout the operating range and cover most of the 5G frequencies. The noise coefficient is 1.7 dB in the n7 range and does not exceed 2.5 dB over the frequency band from 1.4 to 3.9 GHz. The values obtained are a consequence of the low Q-factor of resonant structures, which results from the design features of the technological process.

It is possible to adjust the offset of the input stage in the range of ± 0.1 V from the nominal power supply of the gate. By changing the voltage, the matching of the input circuits can be improved by up to 28 dB of the additional reflection coefficient S11 in a narrow frequency band. This will change the noise coefficient, which will be between 1.62 dB and 1.81 dB.

The operating point of the first stage is selected for the operation of the amplifier in the most linear region with low and minimal consumption. In addition, each of the amplification stages is optimized for an unconditionally stable operating mode, and as a result, the stability coefficient

K >> 1, which potentially eliminates the possibility of generating side harmonic components (Fig. 7).

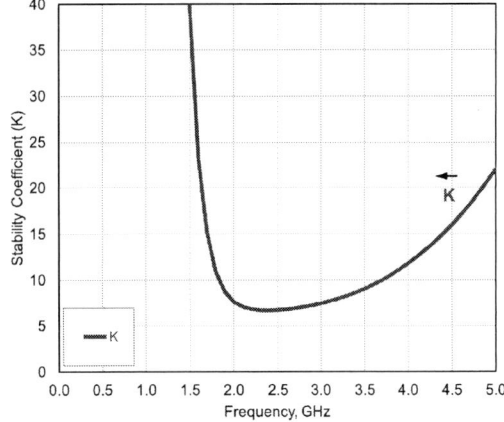

Fig. 7. LNA stability factor.

The nonlinear parameters of the LNA obtained during computer simulations are presented in the form of input/output power dependencies on graphs (Figures 8 and 9). The 1dB compression level (P1dB = 6.69 dBm), which characterizes the boundary of the linear region, is achieved at an input power value of -19 dB, i.e., the power gain is approximately 27 dB at the 1dB compression point. The input level of intermodulation distortion of the second order, relative to the 5.4 GHz harmonic (IP2), was 13.91 dBm, and the output level (OIP2) is approximately 39.0 dBm. The level of third-order intermodulation distortion at the input (IP3) with respect to the 8.1 GHz harmonic was 1.03 dBm, and the level at the output (OIP3) was about 27.63 dBm.

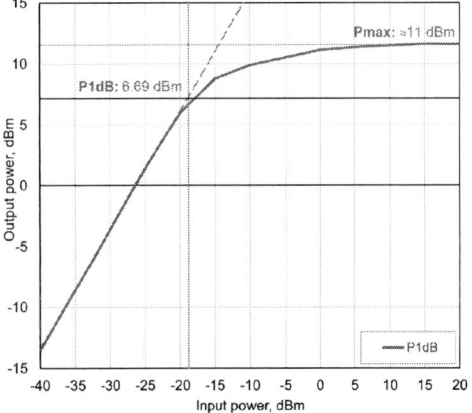

Fig. 8. P1dB compression point.

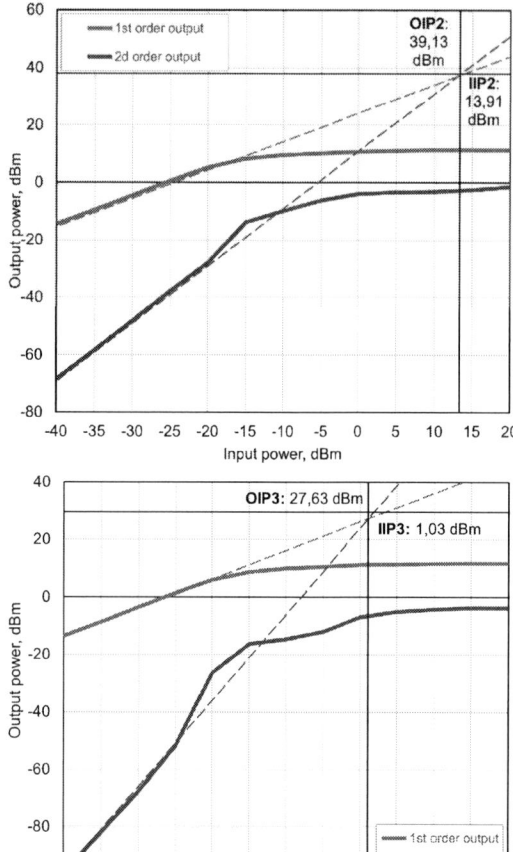

Fig. 9. Nonlinear LNA parameters.

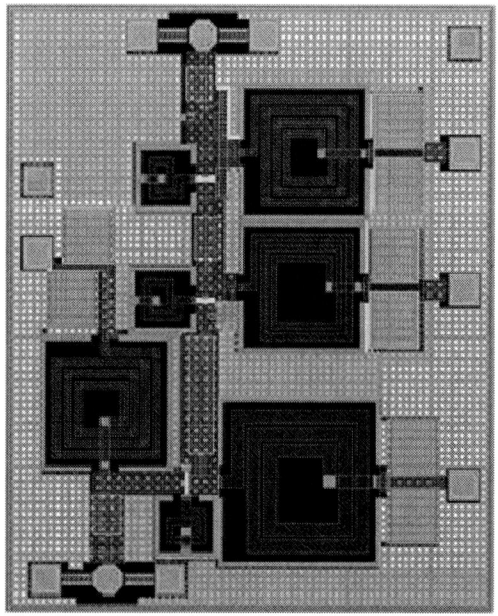

Fig. 10. LNA IC topology.

The topology of the low-noise amplifier is shown in Figure 10, and the dimensions of the crystal are 1.19 ×

1.43 mm. The topology of the chip is made taking into account all the requirements of the chosen technology, including reliability requirements that determine the design features of planar elements. The circuit has G-S-G inputs and outputs for measurements using microwave probes on a plate or for mounting in a housing for further research using modern measuring systems.

TABLE I. TABLE 1. LNA PARAMETERS.

Parameter	Value
Operating frequency range (-10 dB S11 and S22)	1.8-5.0 GHz
Gain in the n7 band	25.5 dB
Noise figure (n7 band)	1.7 dB
Input reflection coefficient (n7 band)	≥ 25 dB
Output reflection coefficient (n7 band)	≥ 30 dB
Output compression point P1dB (n7 band)	6.69 dBm
Intermodulation distortions of the 2nd and 3rd order for the n7 band range	
By output (OIP3):	39.13 dBm
By input (IIP3):	13.91 dBm
By output (OIP3):	27.63 dBm
By input (IIP3):	1.03 dBm
Supply voltage, V	
By the Vg gate:	0.9 ± 0.1
By Vd drains:	0.9 ± 0.1
IC Current consumption	90 mA
Chip dimensions	1.19×1.43 mm

V. CONCLUSION

Based on the calculated data and the parameters in the standard for working with narrow-band signals in the Internet of Things (NB-IoT) ecosystem, the primary requirements for a low-noise amplifier of the path of the transceiver device of the LTE subscriber terminal's frequencies were made. The design flow for industrial radio frequency device development based on a digital library was proposed.

Based on the primary requirements, a 3-stage class A amplifier chip optimized for operation in the n7 frequency range has been developed. Within this sub-band, the amplifier has a high gain level of +25 dB. With an input noise level of 1.7 dB, the reflection coefficients at the device's inputs and outputs are higher than +25 dB. The high linearity of the device is confirmed by its compression of 6.7 dB at levels of OIP3 = 27.6 dBm and IIP3 = 1.03 dBm. The amplifier is based on a digital library of CMOS processes with a design norm of 180 nm, which demonstrates the possibility of building a low-noise S-band amplifier based on the HCMOS8D library. As part of the same technological process, it is possible to build an NB-IoT system that can both receive and send data and work in LTE bands on a single chip.

REFERENCES

[1] Krasnikov G.Ya., Gornev E.S., Matushkin I.V. General theory of technology andmicroelectronics: Part 1. Levels of technology // / Electronic Engineering Series 3 Microelectronics, Scientific and Technical Journal, Volume 1 (165), 2017. pp. 51-69 (accessed 8 November 2022).

[2] Nuykin A.V., Kravtsov A.S., Mytnik K.Ya. Developing ultra-low power universal protected software and hardware platform for iot devices // Nanoindustry, Issue S (89), 2019. pp. 328-329 (accessed 8 November 2022).

[3] Heterointegrated multichip micromodules based on Si for microwave applications / G. Ya. Krasnikov, P. V. Panasenko, A. V. Volosov // Nanoindustry. - 2017. - No. S (74). - S. 529-530. – EDN ZDRTEB.

[4] Design aspects of the "Internet of Things" transceiver / V. YU. Mikhaylov, Ye. YU. Kotlyarov, I. A. Zubov et. Al. // Elektronnaya tekhnika. Seriya 3: Mikroelektronika. – 2020. – № 4(180). – pp. 43-57. – EDN QZECQV (in Russian).

[5] V. Mikhailov, A. V. Nuykin, I. Zubov and E. Kotlyarov, "NB-IoT Transceiver Research and Development," 2021 IEEE Conference of Russian Young Researchers in Electrical and Electronic Engineering (ElConRus), 2021, pp. 2707-2713, doi: 10.1109/ElConRus51938.2021.9396474 (accessed 8 November 2022).

[6] J. Jeon and M. Kang, "Circuit Level Layout Optimization of MOS Transistor for RF and Noise Performance Improvements," in IEEE Transactions on Electron Devices, vol. 63, no. 12, pp. 4674-4677,

Dec. 2016, doi: 10.1109/TED.2016.2614275 (accessed 8 November 2022).

[7] Kotlyarov E.Yu., Mikhailov V.Yu., Zubov I.A., Nuykin A.V., Iljin A.F., Putrya M.G. CMOS inductor design features for LTE devices. Computing, Telecommunications and Control, 2021, Vol. 14, No. 1, Pp. 22–32. DOI: 10.18721/JCSTCS.14102 (accessed 8 November 2022).

Complex Dielectric Permittivity Measurement of 3D Printing Resin FTD Nano Clear in the 1-10 GHz Band

Mikhail M. Migalin
Southern Federal University
Taganrog, Russia
migalin@sfedu.ru
0000-0003-1600-7349

Andrey V. Kovalev
Southern Federal University
Taganrog, Russia
avkovalev@sfedu.ru

Samir R. Gadzhiev
*Platov South-Russian
State Polytechnic University*
Novocherkassk, Russia

Vladislav S. Kuzmin
Southern Federal University
Taganrog, Russia
kuzmin@sfedu.ru

Lev N. Libin
JSC «Vega»
Moscow, Russia

Vladimir A. Fleyteng
Southern Federal University
Taganrog, Russia
flyayteng@sfedu.ru

Abstract—**Additive manufacturing became a valuable part of the antennas and RF devices prototyping process. However, the electromagnetic properties of 3D printing resins are usually not provided. This paper presents a complex permittivity determination of a cured high-resolution 3D printing resin, «Fun To Do Nano Clear», in the frequency range from 1 to 10 GHz. Microstrip differential phase, ring resonators and filled rectangular waveguide methods were used to obtain tangent loss and relative dielectric permittivity. Both experimental and electromagnetic simulation results are presented. MATLAB scripts were developed for automated measurement data processing.**

Keywords—additive manufacturing, complex permittivity, ring resonators, material characterization, microstrip lines

I. INTRODUCTION

Additive manufacturing is versatile in terms of its applications: ceramics, cured resins and melted powders are used in biomedical applications [1-3], aerospace industry [4-5], automotive industry [6], robotics [7], microfluidics [8], also electronics and RF applications [9 -12]. For RF devices development, obtaining accurate electrical properties of the material used for 3D printing is vital to ensure the simulations are correct before manufacturing. Usually, physical properties like viscosity or density are provided, while cured or melted material's permittivity and permeability stays unknown.

There are resonant and non-resonant methods of material characterization that can be used for 3D printing materials' electrical properties extraction [13-14]. The resonant methods are more accurate but usually narrowband, while non-resonant ones provide a wideband material characterization with lower accuracy than the first ones. For cured polymers used in 3D printing, resonant material characterization can be done with numerous printed resonant structures [15-16], including ring resonators [17-19] substrate integrated waveguides (SIWs) [20-22]. Also, resonant cavities [23-25] and dielectric rod resonators [26] can be used. Non-resonant methods include different transmission methods, including filled transmission lines [27-28], microstrip lines [29-30] and free space measurements [31].

This work discusses a cured 3D printing resin FTD "Nano Clear" material characterization using multiline differential phase, ring resonators and waveguide transmission methods. Several copper plating attempts were made on the cured resin, ending with copper tape. Planar structures etching is briefly described, and complex permittivity results are shown.

II. PLASTIC PLATING

Multiple copper plating attempts on the polymer were made to produce printed structures for material characterization. In the first batch, 300 nm of chrome with a subsequent 3-um copper layer was applied in the deposition chamber. Due to low adhesion, the SMA connectors could not be soldered, and end-launch connectors destroyed the microstrip lines after placement.

In the second batch, electroplating was used. A conductive graphite layer was deposited on the pre-degreased plastic. Then, the sample underwent electrochemical treatment in copper-plating sulfuric acid electrolyte. A low current density of 0.5 A/dm2 was used until the entire surface was visually covered with copper, and then the current density was doubled. The thickness of the applied copper coating varied from 118 to 300 microns, and it had low adhesion.

Finally, a 60 – um copper scotch tape was applied for prototyping in lab conditions – it can be easily put on the plastic and has a constant thickness and good adhesion. Planar structures were made using a dry photoresist film: the photoresist was exposed to UV light through a negative mask, then it was developed in Na2CO3 diluted solution five g/L, and finally, copper was etched in citric acid 300 g/L, 3% hydrogen peroxide solution and five g/L sodium chloride.

III. POLYMER CHARACTERIZATION

A. Differential phase method

This simple method was described initially by Das [32]. At least two microstrip lines on the same substrate with thickness h, copper cladding t, width w and different lengths l_1 and l_2 should be produced. The phase angle difference between two microstrip lines at wavelength λ_0 can be calculated as $\Delta\Phi = 2\pi\sqrt{\varepsilon_{eff}}(l_1 - l_2)/\lambda_0$; hence the effective dielectric permittivity can be found as:

$$\varepsilon_{eff} = \left(\frac{\Delta\Phi\lambda_0}{2\pi(l_1 - l_2)}\right)^2 \qquad (1)$$

This research was carried out as a part of the strategic academic leadership program "Priority 2030", SFedU № SP-111-22-02.

The relative dielectric permittivity ε and the effective one are linked through the following equations [33]:

$$\varepsilon_{eff}(f) = \varepsilon(f) - \frac{\varepsilon(f) - \varepsilon_{eff}(0)}{1 + P(f)}, \qquad (2)$$

where $P(f) = P_1 P_2 [(0.1844 + P_3 P_4) f 10^{-6} h]^{1.5763}$,

$$P_1 = 0.27488 + [0.6315 + \frac{0.525}{(1 + 0.0157 f 10^{-7} h)^{20}}] \frac{w}{h}$$

$$-0.065683 e^{-8.7513 \frac{w}{h}}, \; P_2 = 0.33622[1 - e^{-0.03442 \varepsilon(f)}],$$

$$P_3 = 0.0363 e^{-4.6 \frac{w}{h}} [1 - e^{-\left(\frac{f 10^{-7} h}{38.7}\right)^{4.97}}], \quad P_4 = 1 + 2.751[1 - e^{-\left(\frac{\varepsilon(f)}{15.916}\right)^8}],$$

$$\eta_0 = 120\pi, u = W/h, u_0 = u + \frac{t}{\pi h} \ln(1 + (\frac{4 exp(1)}{\frac{t}{h} \coth \sqrt{6.517 u}}),$$

$$u' = u + (u_0 - u)/2(1 + \frac{1}{\cosh \sqrt{\varepsilon - 1}}),$$

$$\varepsilon_{eff}(0) = \frac{(\varepsilon_{eff} + 1)}{2} + \frac{(\varepsilon_{eff} - 1)}{2}(1 + \frac{10}{u'})^{(-ab)},$$

$$a = 1 + \frac{1}{49} \ln(\frac{u^4 + (u/52)^2}{u^4 + 0.432}) + \frac{\ln(1 + (u/18.1)^2)}{18.7},$$

$b = 0.564(\frac{\varepsilon - 0.9}{\varepsilon + 0.3})^{0.053}$, h - thickness of the substrate, w – width of the line, t – copper cladding thickness.

The accuracy of the method improves with the growth of the electrical length difference. A set of microstrip lines was produced with lengths $l_1 = 58.9$ mm, $l_2 = 40.3$ mm, $l_3 = 60.2$ mm, $l_4 = 98.6$ mm, $w = 2$ mm, $t = 60$ um and $h = 1.5$ mm. The measurements were performed using a calibrated VNA Agilent N5222A with ECal N4691B.

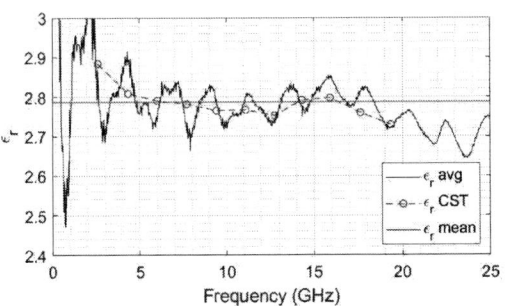

Fig. 1. Produced microstrip lines

The relative permittivity results derived from (1) by the secant method were averaged among pairs of microstrip line samples and presented below:

Fig. 2. Relative permittivity derived by phase method

The calculated value of the average relative permittivity oscillates from 2.65 to 2.9 in the frequency range from 2.5 GHz to 25 GHz. The mean value $\varepsilon_r = 2.78$ is shown as a red line in Fig. 2. Permittivity oscillations can be caused by the measurement setup: the studied polymer is flexible, and it could be bent during measurements. The resulting relative permittivity was confirmed using the CST macros described by Sokol [34]. The extracted tangent loss varied from 0.01 to 0.03 in the band from 1 to 20 GHz.

B. Ring resonator method

Ring resonators are planar structures allowing material characterization at frequencies up to 110 GHz [17]. Their length should be at least five wavelengths long to apply straight microstrip line dispersion equations and minimize the effects of the coupling gaps on the resonant frequencies [35]. The n-th resonant frequency of a ring resonator can be calculated as follows:

$$f_n = \frac{nc}{2\pi r_m \sqrt{\varepsilon_{eff}(f)}}, \qquad (3)$$

where n – resonant mode number, c – speed of light, r_m – mean ring radius.

A ring resonator with $r_m = 23.9$ mm was produced on a $h = 1.5$ mm thick substrate with width a coplanar line of width $w = 3.1$ mm and a gap of 1 mm.

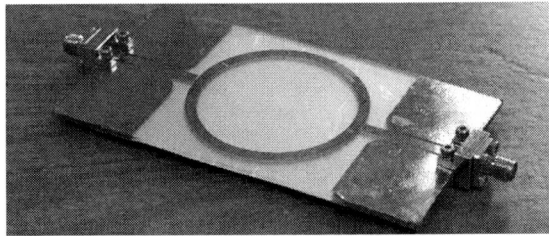

Fig. 3. Produced ring resonator

It is transmission coefficient is presented below:

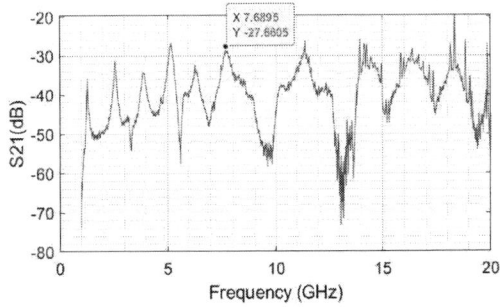

Fig. 4. Ring resonator's transmission coefficient

It is clear from Fig. 4 that measurements results at frequencies higher than 8 GHz become noisy. We assume the setup causes it: fixed plastics, coplanar line with vias, improved matching with the end-launch connectors, and multiple samples might improve the transmission coefficient results. The resulting relative permittivity with an average value of $\varepsilon_r = 2.99$ is presented in Fig.5:

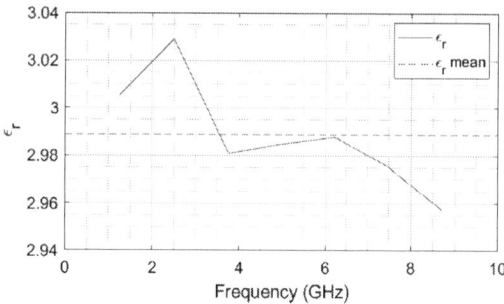

Fig. 5. Relative permittivity derived by ring resonator method

The surface roughness value of the copper tape was measured using an atomic-force microscope. The maximum measured RMS value of surface roughness was 116 nm. The tangent loss was calculated using the same resonance peaks [18]:

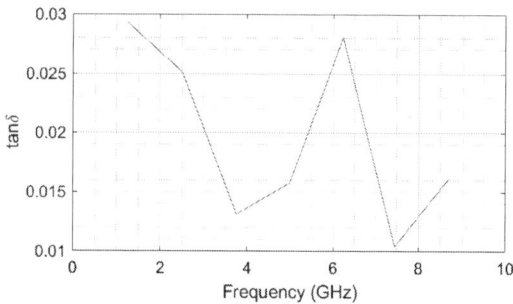

Fig. 6. Tangent loss derived by ring resonator method

C. Filled rectangular waveguide method

Nicolson-Ross-Weir (NRW) is an algorithm widely used for material characterization using waveguides [15]. A 23x10 mm waveguide was used for material characterization in the 9 – 12 GHz band. The VNA was calibrated using Through-Reflect-Line (TRL) technique. The line section of 10 mm length was manufactured using electrical discharge machining to provide a 20° to 160° phase shift in the desired frequency band.

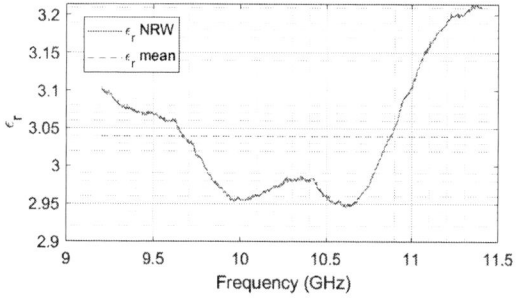

Fig. 1. Relative permittivity derived by filled rectangular waveguide method

A 1.5 mm thick sample was placed inside the waveguide holder, and the NRW algorithm was applied to measure S - the parameters results as shown in Fig.7. Relative permittivity with a mean value of $\varepsilon_r = 3.03$ varies from 2.95 to 3.2. The tangent loss data is not shown due to the large oscillations from 0.1 to 0.01. The polymer sample was

milled, and as a result, the corners were rounded, so the air gaps appeared.

CONCLUSION

This work presents an experimental determination of the complex permittivity of the 3D printing resin FTD Nano Clear in the 1-10 GHz band. The resulting relative permittivity $\varepsilon_r = 2.9 \pm 0.1$ and tangent loss $\tan\delta = 0.02 \pm 0.01$ were determined using three different methods implemented in a MATLAB script for automated measurement data processing. As a part of the further polymer study, the following steps would be considered:

- Metal fixture development for planar structures measurements.

- Thinner waveguide samples will be produced using DLP printing instead of CNC milling.

- The studied frequency range will be broadened to a 0-40 GHz band.

- Launching microstrip lines will be shortened and de-embedded to estimate the tangent loss using a ring resonator accurately.

- Resonant cavity usage method for cross-validation of the results.

ACKNOWLEDGMENT

We want to thank Ivan Bobkov and Sergey Chapek for the polymer samples preparation and the «Communication systems» lab for providing VNA access.

REFERENCES

[1] Tahayeri, Anthony, et al. "3D printed versus conventionally cured provisional crown and bridge dental materials," in *Dental Materials*, 2018, vol. 34. no.2, pp. 192-200.

[2] S. Sivashankar, S. Agambayev, U. Buttner and K. N. Salama, "Characterization of solid UV curable 3D printer resins for biological applications," *2016 IEEE 11th Annual International Conference on Nano/Micro Engineered and Molecular Systems (NEMS)*, 2016, pp. 305-309, doi: 10.1109/NEMS.2016.7758255.

[3] N. A. K. Rosli, M. A. H. M. Adib, N. N. M. Sukri, I. M. Sahat and N. H. M. Hasni, "The CardioVASS Heart Model: Comparison of Biomaterials between TPU Flex and Soft Epoxy Resin for Biomedical Engineering Application," *2020 IEEE-EMBS Conference on Biomedical Engineering and Sciences (IECBES)*, 2021, pp. 224-229, doi: 10.1109/IECBES48179.2021.9398731.

[4] M. Kalender, S. E. Kılıç, S. Ersoy, Y. Bozkurt and S. Salman, "Additive Manufacturing and 3D Printer Technology in Aerospace Industry," *2019 9th International Conference on Recent Advances in Space Technologies (RAST)*, 2019, pp. 689-694, doi: 10.1109/RAST.2019.8767881.

[5] G. J. Schiller, "Additive manufacturing for Aerospace," *2015 IEEE Aerospace Conference*, 2015, pp. 1-8, doi: 10.1109/AERO.2015.7118958.

[6] D. -N. Pagonis, G. Kaltsas, T. Koutsis and A. Pagonis, "A Novel Engine Air Intake Sensor based on 3D Printing and PCB technology," *2021 IEEE Sensors*, 2021, pp. 1-4, doi: 10.1109/SENSORS47087.2021.9639681.

[7] M. Ntagios, S. Dervin and R. Dahiya, "3D Printed Capacitive Pressure Sensing Sole for Anthropomorphic Robots," *2021 IEEE International Conference on Flexible and Printable Sensors and Systems (FLEPS)*, 2021, pp. 1-4, doi: 10.1109/FLEPS51544.2021.9469839.

[8] K. Adamski, W. Kubicki and R. Walczak, "Inkjet 3D printed microfluidic devices," *2016 MIXDES - 23rd International Conference Mixed Design of Integrated Circuits and Systems*, 2016, pp. 504-506, doi: 10.1109/MIXDES.2016.7529795.

[9] Y. Cao, S. Yan, S. Wu and J. Li, "3D Printed Multi-Beam OAM Antenna Based On Quasi-Optical Beamforming Network," in *Journal of Lightwave Technology*, 2022, doi: 10.1109/JLT.2022.3196643.

[10] M. Cuevas, F. Pizarro, A. Leiva, G. Hermosilla and D. Yunge, "Parametric Study of a Fully 3D-Printed Dielectric Resonator Antenna Loaded With a Metallic Cap," in IEEE Access, vol. 9, pp. 73771-73779, 2021, doi: 10.1109/ACCESS.2021.3081068.

[11] R. J. Beneck, G. Mackertich-Sengerdy, S. Soltani, S. D. Campbell and D. H. Werner, "A Shape Generation Method for 3D Printed Antennas With Unintuitive Geometries," in IEEE Access, vol. 10, pp. 91294-91305, 2022, doi: 10.1109/ACCESS.2022.3202536.

[12] F. Sun et al., "A Millimeter-Wave Wideband Dual-Polarized Antenna Array With 3-D-Printed Air-Filled Differential Feeding Cavities," in IEEE Transactions on Antennas and Propagation, vol. 70, no. 2, pp. 1020-1032, Feb. 2022, doi: 10.1109/TAP.2021.3111502.

[13] L. Chen, et al., Microwave electronics: measurement and materials characterization. NJ: John Wiley & Sons, 2004.

[14] P. I. Deffenbaugh, R. C. Rumpf and K. H. Church, "Broadband Microwave Frequency Characterization of 3-D Printed Materials," in IEEE Transactions on Components, Packaging and Manufacturing Technology, vol. 3, no. 12, pp. 2147-2155, Dec. 2013, doi: 10.1109/TCPMT.2013.2273306.

[15] D. Markovic, B. Jokanovic, M. Marjanovic and M. Djordjevic, "Improved Method for Measurement of the Dielectric Properties of Microwave Substrates Using Microstrip T-resonator," *2007 IEEE Instrumentation & Measurement Technology Conference IMTC 2007*, 2007, pp. 1-3, doi: 10.1109/IMTC.2007.379138.

[16] C. -S. Lee and C. -L. Yang, "Complementary Split-Ring Resonators for Measuring Dielectric Constants and Loss Tangents," in *IEEE Microwave and Wireless Components Letters*, vol. 24, no. 8, pp. 563-565, Aug. 2014, doi: 10.1109/LMWC.2014.2318900.

[17] D. C. Thompson, O. Tantot, H. Jallageas, G. E. Ponchak, M. M. Tentzeris and J. Papapolymerou, "Characterization of liquid crystal polymer (LCP) material and transmission lines on LCP substrates from 30 to 110 GHz," in IEEE Transactions on Microwave Theory and Techniques, vol. 52, no. 4, pp. 1343-1352, April 2004, doi: 10.1109/TMTT.2004.825738.

[18] C. D. Morcillo, S. K. Bhattacharya, A. Horn and J. Papapolymerou, "Conductor surface-roughness effect in the loss tangent measurement of low-loss organic substrates from 30 GHz to 70 GHz," *2010 Proceedings 60th Electronic Components and Technology Conference (ECTC)*, 2010, pp. 727-734, doi: 10.1109/ECTC.2010.5490769.

[19] A. Rashidian, M. T. Aligodarz and D. M. Klymyshyn, "Dielectric characterization of materials using a modified microstrip ring resonator technique," in IEEE Transactions on Dielectrics and Electrical Insulation, vol. 19, no. 4, pp. 1392-1399, August 2012, doi: 10.1109/TDEI.2012.6260016.

[20] D. Zelenchuk, V. Fusco, J. Breslin, and M. Keaveney, "Millimetre wave dielectric chartacterisation of multilayer LTCC substrate," 2015 European Microwave Conference, pp. 1033-1036, 2015.

[21] Y.W. Wu, et al., "A Simple and Accurate Method for Extracting Super Wideband Electrical Properties of the Printed Circuit Board," IEEE access, vol. 7, pp. 57321-57331, 2019.

[22] H. B. Wang and Y. J. Cheng, "Broadband printed-circuit-board characterization using multimode substrate-integrated-waveguide resonator," IEEE Transactions on Microwave Theory and Techniques, vol. 65, no. 6, pp. 2145-2152, 2017

[23] K. Heinikoski, L. Hynynen and T. Tarvainen, "Wideband Complex Permittivity Tester for 5G Materials," 2018 48th European Microwave Conference (EuMC), 2018, pp. 198-201, doi: 10.23919/EuMC.2018.8541534.

[24] J. Krupka, "Measurements of the Complex Permittivity of Low Loss Polymers at Frequency Range From 5 GHz to 50 GHz," in *IEEE Microwave and Wireless Components Letters*, vol. 26, no. 6, pp. 464-466, June 2016, doi: 10.1109/LMWC.2016.2562640.

[25] P. Korpas, W. Wojtasiak, J. Krupka and W. Gwarek, "Inexpensive approach to dielectric measurements," *2012 19th International Conference on Microwaves, Radar & Wireless Communications*, 2012, pp. 154-157, doi: 10.1109/MIKON.2012.6233486.

[26] Y. Kobayashi and M. Katoh, "Microwave Measurement of Dielectric Properties of Low-Loss Materials by the Dielectric Rod Resonator Method," in IEEE Transactions on Microwave Theory and Techniques, vol. 33, no. 7, pp. 586-592, Jul. 1985, doi: 10.1109/TMTT.1985.1133033.

[27] A. Bogle, M. Havrilla, D. Nyquis, L. Kempel and E. Rothwell "Electromagnetic Material Characterization using a Partially-Filled Rectangular Waveguide," in *Journal of Electromagnetic Waves and Applications*, 2005, 19:10, 1291-1306, DOI: 10.1163/156939305775525909

[28] D. Misra, M. Chabbra, B. R. Epstein, M. Microtznik and K. R. Foster, "Noninvasive electrical characterization of materials at microwave frequencies using an open-ended coaxial line: test of an improved calibration technique," in IEEE Transactions on Microwave Theory and Techniques, vol. 38, no. 1, pp. 8-14, Jan. 1990, doi: 10.1109/22.44150.

[29] W. Liu and W. Song, "Broadband Dielectric Measurement Of LCP Substrate Materials By Differential Phase Length Method," 2020 IEEE MTT-S International Microwave Workshop Series on Advanced Materials and Processes for RF and THz Applications (IMWS-AMP), 2020, pp. 1-3, doi: 10.1109/IMWS-AMP49156.2020.9199742.

[30] N. K. Das, S. M. Voda and D. M. Pozar, "Two Methods for the Measurement of Substrate Dielectric Constant," in *IEEE Transactions on Microwave Theory and Techniques*, vol. 35, no. 7, pp. 636-642, Jul 1987, doi: 10.1109/TMTT.1987.1133722.

[31] D. K. Ghodgaonkar, V. V. Varadan and V. K. Varadan, "Free-space measurement of complex permittivity and complex permeability of magnetic materials at microwave frequencies," in IEEE Transactions on Instrumentation and Measurement, vol. 39, no. 2, pp. 387-394, April 1990, doi: 10.1109/19.52520.

[32] N. K. Das, S. M. Voda, and D. M. Pozar, "Two methods for the measurement of substrate dielectric constant," IEEE Transactions on Microwave Theory and Techniques, vol. 35, no. 7, pp. 636-642, 1987.

[33] R. Garg, et al., "Microstrip antenna design handbook," London, U.K.: Artech house, 2001.

[34] Sokol, J. Eichler and M. Riitschlin, "Calibration of EM simulator on substrate complex permittivity", 83rd ARFTG Microwave Measurement Conference, pp. 1-5, Jun. 6, 2014.

[35] K. Chang and L.-H. Hsieh, Microwave Ring Circuits and Related Structures, New York, NY, USA:Wiley, 2004.

Investigation of the Power Source Parameters Influence for the Plasma Jet of DC Plasma Torch

Iurii Murashov
Higher school of Electric Power Systems, Institute of Energy
Peter the Great St.Petersburg Polytechnic University
Saint-Petersburg, Russia
0000-0001-7839-6007

Ruslan Zhiligotov
Higher school of Electric Power Systems, Institute of Energy
Peter the Great St.Petersburg Polytechnic University
Saint-Petersburg, Russia
0000-0002-0244-8998

Nikita Obraztsov
Higher school of Electric Power Systems, Institute of Energy
Peter the Great St.Petersburg Polytechnic University
Saint-Petersburg, Russia
0000-0003-1107-3259

Natalia I. Kurakina
Higher school of Electric Power Systems, Institute of Energy
Peter the Great St.Petersburg Polytechnic University
Saint-Petersburg, Russia
0000-0002-4234-9425

Abstract— The parameters of the power source are one of the most important factors determining the behavior of the plasma jet. Thus, simulation the operation of the DC plasma torch is impossible without the electrical parameters of the power source. Mathematical model of the DC plasma torch operation taking into account the influence of the parameters of the power source is presented in the article. The model is based on the fundamental physics laws. Parametric simulation is carried out at different flow rates of the plasma-forming gas to assess the influence of the plasma jet parameters on the current-voltage characteristics of the power source. The optimal modes of operation of the plasma torch have been determined to ensure high efficiency of the plasma spraying technology.

Keywords—plasma torch, power source, simulation, local thermodynamic equilibrium

I. INTRODUCTION

The development of coating technologies is due to the need to increase the durability of devices, parts and assemblies of various mechanisms and surfaces exposed to the working or natural environment, as well as operating conditions [1-3].

There are different coating methods: electrochemical (chemical) coatings, electrophysical coatings, thermal spraying, surfacing, cladding and others [4-6].

The plasma spraying technology has become widespread in industry, because allows you to solve a large number of problems (the operation of structural elements of machines in harsh conditions: high-temperature gas flows, aggressive gas, abrasive materials and etc.) with minimal costs. The quality of the applied coatings and the efficiency of the technology are determined by the characteristics of the applied material and the parameters of the equipment used. The main effect on the melting of the sprayed powder is exerted by the behavior of the plasma flow (laminar, turbulent or intermittency) [2, 3, 7-12]. In addition to classical gas-dynamic instabilities, the arc burning mode has a significant effect on the plasma flow (the plasma flow is disturbed by the oscillations of the electric arc). Thus, the development of equipment for the plasma spraying technology cannot be regarded as the design of an exclusively plasma torch, it must be done in conjunction with a power source.

This research is financially supported by The Russian Science Foundation, Agreement No. 22-29-20223, and St-Petersburg Fund for Support of Scientific, Scientific and Technical, Innovative Activities Project, Agreement No. 64/2022

The idealized structure of the mathematical model for the spraying process is shown in Fig. 1.

Description of the symbols shown in Figure 1: *EN* – electrical network; *G* – gas consumption; *I(U)* – power source current (voltage); *PP* - plasma properties; *T* – plasma flow temperature distribution; v - plasma flow velocity distribution; T_p – particle temperature; v_p – particle velocity; *P* – porosity; *A* – adhesion; *MUR* – material utilization ratio.

The important links in the construction of a mathematical model of the spraying process are the description of the plasma torch model, the power source simulation and their interaction. Simulation of the plasma torch operation is described in the works of a lot of different authors [2, 3, 11-18]. The proposed models are based on solving a multiphysical task, which includes solutions to differential equations: heat balance equation, momentum equation and Maxwell's equations.

The most popular software products for simulating plasma torch operation are Ansys and Comsol Multiphysics [15, 17, 19-22]. Using the latter one allows you to connect a power source to the plasma torch using the "Electrical circuit" interface or create a hybrid model with Simulink.

II. DESCRIPTION OF PLASMA TORCH AND POWER SOURCE MODEL

The main goal of the investigation is to assess the influence of the characteristics of the power source on the plasma flow stability. As a result, studies of identical operating modes of the plasma torch were carried out for different power sources (APR-403 and APR-404). Electrical scheme of power sources APR-403 and APR-404 are shown in Fig. 2 and Fig. 3, respectively. Where RH is a load (plasma torch). The implementation of the power supply model for the plasma torch is most convenient in MATLAB/Simulink. Thus, in this article, a hybrid model was implemented based on the conjugation of Comsol Multiphysics and Simulink. The ripple factor of the load voltage (see Fig. 4) is 0.16 for the APR-403 power source, and 0.49 for the APR-404. The above values of the ripple factors demonstrate the plasma flow is more stable when operating from an APR-403 power source. According to the results of the study it is advisable to use the APR-403 as a power source for the plasma spraying technology. These results confirmed by the Fourier analysis (see Fig. 5) too.

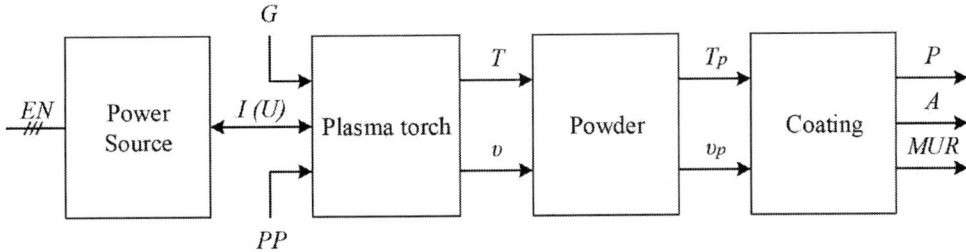

Fig.1. Block scheme of the mathematical model for plasma spraying technology.

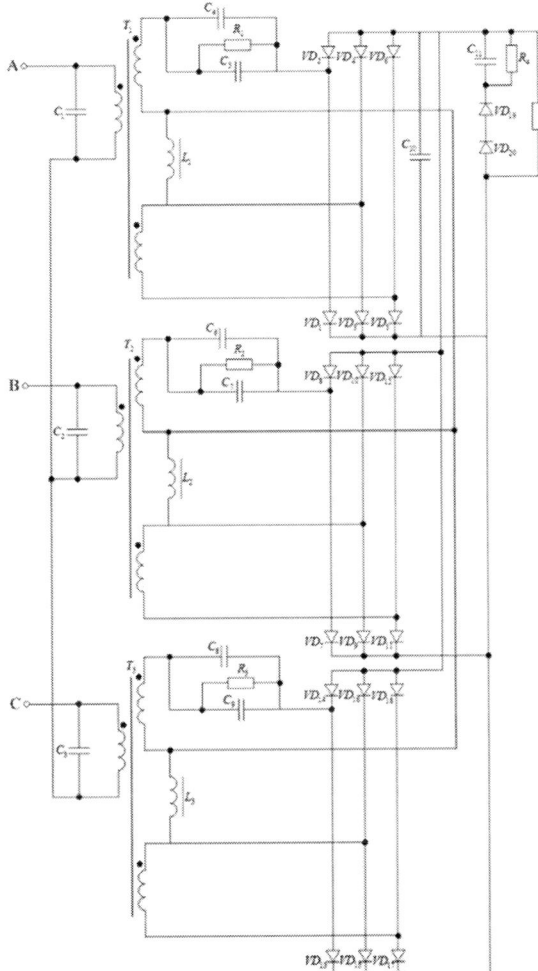

Fig. 2. Electrical scheme of power source APR-403.

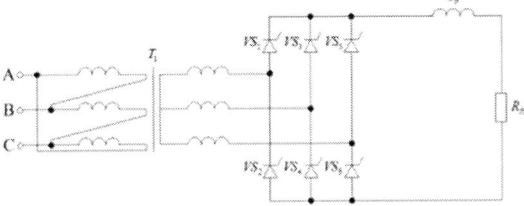

Fig.3. Electrical scheme of power source APR-404.

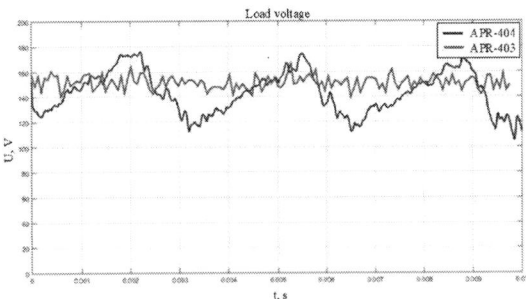

Fig. 4. Time dependences of the arc voltage for APR-403 and APR-

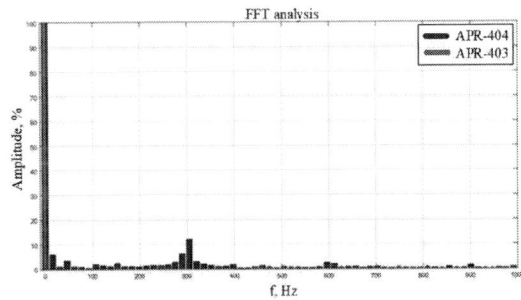

Fig.5. Electrical scheme of power source APR-403.

However, the size and weight of the APR-404 power source are more attractive and make it possible to create a mobile installation for plasma spraying technology. In addition, the use of controlled semiconductor elements makes it possible to implement feedback in the control system to stabilize a given parameter(s) of the technological process.

Therefore, the focus is on the results of plasma torch simulation with the APR-404 power source and the analysis of the influence of the power source parameters on the stability of the plasma flow.

One of the main ways to reduce the Joule heating pulsations, which has a significant effect on the stability of the plasma flow, is the use of a damping inductance (power low-

979-8-3503-8358-4/23 $31.00 © 2023 IEEE   87

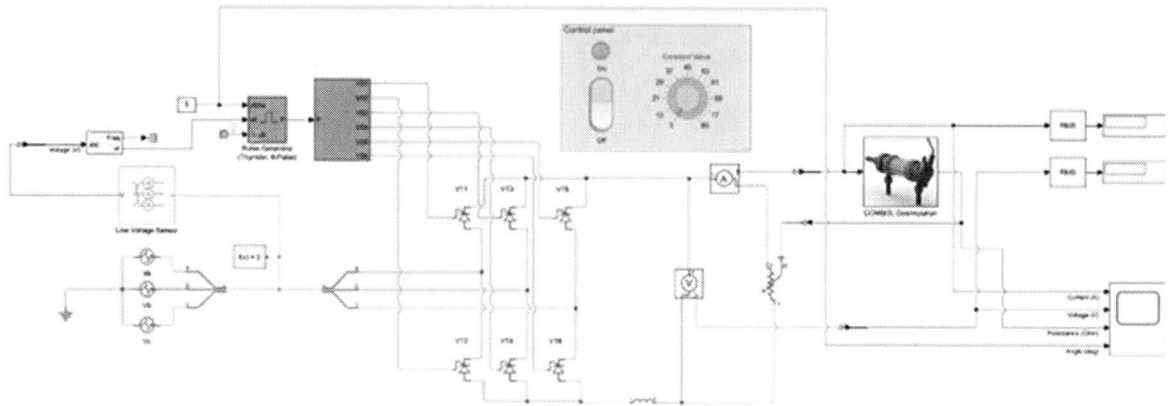

Fig. 6. The APR-403 with plasma torch model in MATLAB/Simulink.

pass filter) to smooth out the pulsations of the load (plasma torch) current. However, numerical and qualitative analysis has not been performed previously.

Understanding the need for smoothing inductance and the economic feasibility is critical. The inductance of the low pass filter is 1mH (0.9…1,2mH from datasheet) in an existing installation, but whether this inductance is sufficient or excessive is un-known.

The APR-403 implementation with plasma torch in MATLAB/Simulink is shown in Fig. 6.

The detailed description of the mathematical model of the DC plasma torch operation is presented in the articles [23, 24].

III. SIMULATION RESULTS

Numerical simulation was carried out for typical operating modes of the PN-V1 plasma torch for the lowest plasma-forming gas flow rate (0.5 g/s) and the highest flow rate (1 g/s) corresponding to the laminar flow, which was identified at the early stages of the study [23, 24].

One of the key results of simulation is the analysis of fluctuations in the value of the maximum temperature of the plasma flow depending on the value of the inductance of the smoothing filter (see Fig. 7-8).

The results obtained are expected and do not represent scientific novelty according to preliminary estimates. But everything is not so unambiguous, when analyzing the distribution of the temperature field in the case with the largest pulsations of the maximum temperature value (L=0.5 mH and the flow rate is 0.55 g/s), there are no visually observed differences in the temperature distribution (see Fig. 9) in the working region of the melting of the sprayed material, which is one of the main indicators for the plasma spraying technology.

In accordance with the obtained simulation results, the analysis of the maximum temperature fluctuation along the axis of the plasma torch is carried out (see Fig. 10).

The temperature fluctuations significantly decrease in the plasma torch anode region, point with coordinates (0, 50), and are completely compensated due to the chosen [23, 24] angle of the diffuser.

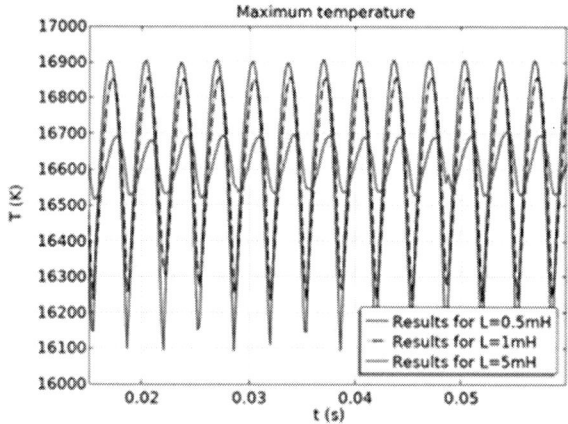

Fig. 7. The maximum temperature fluctuation for the flow rate 0.55g/s.

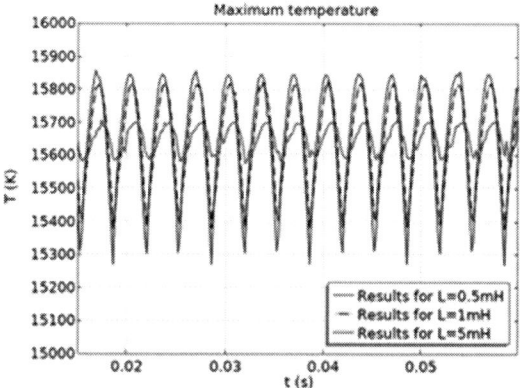

Fig. 8. The maximum temperature fluctuation for the flow rate 1.0g/s.

Feedback is introduced when it is necessary to compensate for pulsations. Power sources design uses current, voltage, or power feedbacks. The feedback type is determined based on the reaction of the controlled parameter to a change in the input signal. So, in order to analyze and establish the response of the controlled parameter (temperature), the dependences of the change in current and power on time were built and combined with temperature fluctuations (see Fig. 11-12). The dependences are presented for the most critical operating

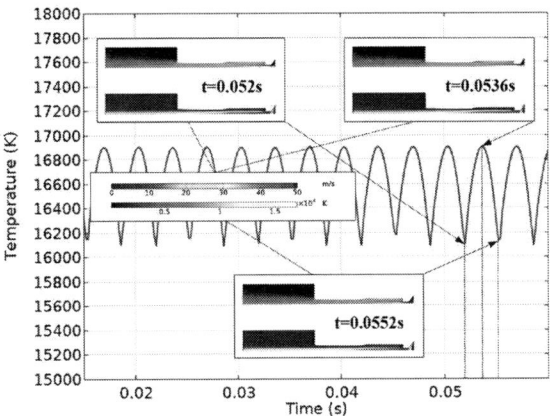

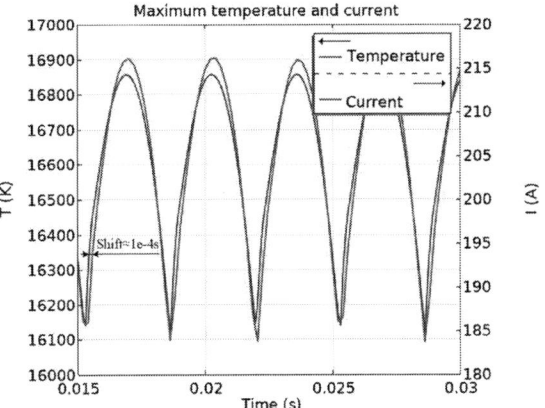

Fig. 9. Analysis of the temperature distribution with the maximum temperature fluctuations.

Fig. 11. Maximum temperature and current fluctuations.

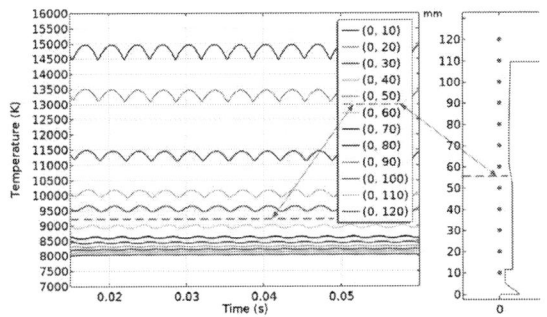

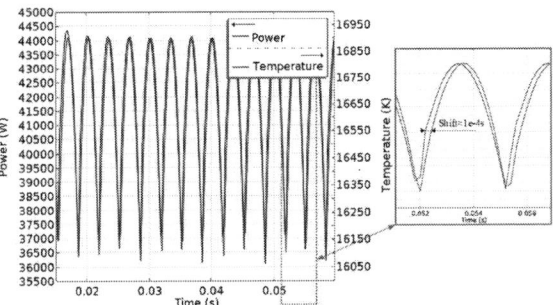

Fig. 10. Analysis of the temperature distribution with the maximum temperature fluctuations.

Fig. 12. Maximum temperature and power fluctuations

mode (L=0.5 mH and the flow rate is 0.55 g/s) of the plasma torch.

In accordance with the simulation results, a response delay of the output parameter (temperature) with rising is observed, which is 0.1 ms, and there is no delay with falling.

IV. DISCUSSIONS

The simulation of the operation of DC plasma torch using the power source to take into account the parameters of it for the stability of the plasma flow is presented. Based on the simulation results, it was found that the developed design of the plasma torch PN-V1 makes it possible to compensate for the pulsations of the temperature field arising from fluctuations in the Joule heating.

The effect of the smoothing filter on the stability of the plasma flow is described, the results obtained will allow us to assess the validity of the use of inductors with high inductance.

In addition, the developed model makes it possible to obtain a transfer characteristic for the introduction of feedback in order to increase the plasma flow stability.

ACKNOWLEDGMENT

This article was prepared with the support of the Russian Science Foundation, Project. No. 22-29-20223 and St-Petersburg Fund for Support of Scientific, Scientific and Technical, Innovative Activities Project. No. 64/2022: "Foundations of the heat transfer theory in thermal plasma flows. Interaction of the "power source - plasma torch" system. Simulation and experimental study."

The computation resources were provided by the Supercomputer Center of Peter the Great St. Petersburg Polytechnic University.

REFERENCES

[1] M. Baeva et all. Study of plasma torches used for air plasma spraying of protective and decorative coatings. Paper presented at the 19th Symposium on Physics of Switching Arc 2011, FSO 2011, 105-108. (2011)

[2] Z. Duan and J. Heberlein. Arc instabilities in a plasma spray torch. Journal of Thermal Spray Technology, 11(1), 44-51. (2002)

[3] J. F. Coudert and P. Fauchais. Arc instabilities in A D.C. plasma torch. High Temperature Material Processes, 1(2), 49-166. (1997)

[4] V. Frolov et all. Air–plasma technologies of spraying of coatings VII International Conference on Plasma Physics and Plasma Technology, 608–611. (2012)

[5] B. Ushin. Development of air–plasma technology of spraying of protective and decorative coatings (2010)

[6] Yu. Tsvetkov. Thermal plasma and new materials technology, 2, 291–322. (1995)

[7] R. Zhukovskii et all. Control of the arc motion in DC plasma spray torch with a cascaded anode. Journal of Thermal Spray Technology, 29(1-2), 3-12. (2020)

[8] R. Zhukovskii et all. Effect of electromagnetic boundary conditions on reliability of plasma torch models. Journal of Thermal Spray Technology, 29(5), 894-907. (2020)

[9] I. Murashov et all. Numerical simulation of DC air plasma torch modes and plasma jet instability for spraying technology. Paper presented at the Proceedings of the 2016 IEEE North West Russia Section Young

Researchers in Electrical and Electronic Engineering Conference, ElConRusNW 2016, 625-628. (2016)

[10] D.I. Ivanchenko and E.V. Iakovleva. Simulation of switching overvoltage in mine production unit. Proceedings of the 2017 IEEE Russia Section Young Researchers in Electrical and Electronic Engineering Conference, ElConRus 2017, 873-876. (2017)

[11] J. P. Trelles. Advances and challenges in computational fluid dynamics of atmospheric pressure plasmas. Plasma Sources Science and Technology, 27(9). (2018)

[12] X. Xu et all. A 2-D axisymmetric magneto-hydrodynamic model of a DC arc plasma torch and its solution methodology. IEEE Transactions on Magnetics, 56(1). (2020)

[13] N. V. Obraztsov et all. A non-stationary model of the AC plasma torch. Paper presented at the IOP Conference Series: Materials Science and Engineering, 643(1). (2019)

[14] C. Rehmet et all. A comparison between MHD modeling and experimental results in a 3-phase AC arc plasma torch: Influence of the electrode tip geometry. Plasma Chemistry and Plasma Processing, 34(4), 975-996. (2014)

[15] M. Baeva. Non-equilibrium modeling of tungsten-inert gas arcs. Plasma Chemistry and Plasma Processing, 37(2), 341-370. (2017)

[16] J. P. Trelles et all. Multiscale finite element modeling of arc dynamics in a DC plasma torch. Plasma Chemistry and Plasma Processing, 26(6), 557-575. (2006)

[17] N. V. Obraztsov et all. Time dependent 2-dimensional model of an alternating current arc. Paper presented at the Journal of Physics: Conference Series, 1135(1). (2018)

[18] A.A. Kadyrov et all. Development of plasma technology for metal powders used in additive technologies. Paper presented at the Journal of Physics: Conference Series, 1753(1). (2021)

[19] S. Nguyen-Kuok. Theory of Low-Temperature Plasma Physics, Springer Series on Atomic, Optical, and Plasma Physics (Springer, 2017).

[20] N.Y. Bykov et all. Modeling of an AC plasma torch - part I: Electrical parameters and flow temperature. IEEE Transactions on Plasma Science, 49(3), 1017-1022. (2021)

[21] N.Y.Bykov. Modeling of an AC plasma torch - part II: Gasdynamic pattern and effect of flow rate. IEEE Transactions on Plasma Science, 49(3), 1023-1027. (2021)

[22] N.V. Obraztsov et all. The usage of low-voltage AC plasma torch for polystyrene gasification. Paper presented at the IOP Conference Series: Materials Science and Engineering, 643(1). (2019)

[23] I. Murashov et all. Numerical simulation of DC air plasma torch modes and plasma jet instability for thermal spraying technology. Paper presented at the MATEC Web of Conferences, 245. (2018)

[24] I. Murashov. The development of dc plasma torch for the spraying technology taking into account the phenomena of plasma flow instability (2016)

[25] Transl. J. Magn. Japan, vol. 2, pp. 740–741, August 1987 [Digests 9th Annual Conf. Magnetics Japan, p. 301, 1982].

[26] M. Young, The Technical Writer's Handbook. Mill Valley, CA: University Science, 1989.

Hardware Realization Version of a Neuron for a Self-routing Analog-to-digital Converter

Anton A. Naborshikov
Automation and Telemechanics Department)
Perm National Research Polytechnic University (PNRPU)
Perm, Russia
anton.naborshikov@gmail.com

Ilya A. Artemyev
Automation and Telemechanics Department)
Perm National Research Polytechnic University (PNRPU)
Perm, Russia
ilya.artemjev2017@yandex.ru

Anton I. Posyagin
Automation and Telemechanics Department
Perm National Research Polytechnic University (PNRPU)
Perm, Russia
line 5: posyagin.anton@gmail.com

Vyacheslav A. Tsyganstsev
Automation and Telemechanics Department)
Perm National Research Polytechnic University (PNRPU)
Perm, Russia
slava.tsygantsev@yandex.ru

Abstract — **This article presents an implementation of the basic measuring neuron for a self-routing analog-to-digital converter that uses an additional microcontroller to realize operation. The use of a microcontroller allows changing the operation and architecture of the hardware network of neurons flexibly, preserving the circuit diagram of the neuron itself. The article also provides a method and model of self-diagnostics, which allows taking into account changes in the calibration characteristic of neurons and is implemented as a physical model by changing the value of the digital potentiometer. The results of the simulation showed the effectiveness of self-diagnostics at the stage of network initialization, however additional studies are required to work in the background in parallel with the measurement to improve the proposed method**

Keywords — *analog-to-digital converter, neural network, self-routing, self-diagnostics.*

I. INTRODUCTION

Automatic control systems (ACS) find their application in a wide variety of fields of technology from high-tech production to household appliances. The use of ACS allows, having obtained data on the state of the automation object (AO), to form control actions on the actuators, ensuring control and stable operation of the system, significantly increasing its characteristics in terms of speed and efficiency in general. The continuous development and modernization of the AO leads to the complication of the ACS, which need to receive more information about the status of the AO. The AO status information can be obtained by using measuring sensors, which typically convert the various measured parameters into a voltage or current representing an analog electrical signal. Analog-to-digital converter (ADC) converts analog signal coming from measuring sensor into digital signal understandable for ACS. As the number of sensors increases, the number of ADCs increases, since the capabilities of the equipment in terms of the number of processed signals at a given speed are limited. There is also an operational need to increase the reliability of the ADC and the degree of trust in the information received from it, since the quality and reliability of regulation depends on it [1].

An increase in the degree of trust in the ADC is achieved due to the implementation of the built-in control and/or self-diagnostics system (hereinafter referred to as the CDS) [1, 2]. Classical CDS are centralized; the hardware of such systems can be taken out into a separate module, on the basis of which the remaining units of the device are checked. As a rule, the hypothesis of negligible probability of failure

situations is applied to the module itself [2]. This allows performing the check procedure while the module is healthy, but in case of failure, the test results will be incorrect. This principle is the main drawback of classical CDS and leads to the problem of «controller over controller».

II. SELF-ROUTING ANALOG-TO-DIGITAL CONVERTER BASED ON A NEURAL NETWORK

One way to solve this problem is to abandon the principle of centralization in the development of CDS. The proposed decentralized approach using local CDS is being developed for devices implemented on the basis of a neural network (NN) architecture [3, 4]. The proposed device is a hardware NN, which consists of a finite number of elements of the same type with different types of connections between them. It is based on a self-routing analog-to-digital converter based on a neural network (SADC NN) with a fragmentary local control unit (FLCU), principles of which are described in [5]. The structure of the SADC NN is a single-layer neural network, which can be represented as a multi-ring (Fig. 1) [5, 6].

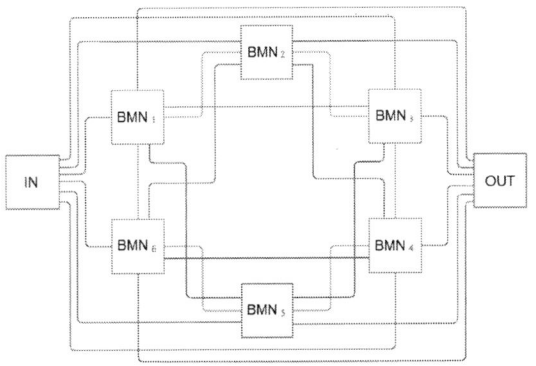

Fig. 1. Example of a single-layer NN with one input and output neurons and 6 BMN in a hidden layer and one additional link between them.

Each basic measuring neuron (BMN) in a single-layer NN is a combination of an analog measuring component (AMC) with a digital measurement control system and the formation of individual ADCs (IADCs) for each incoming input signal (Fig. 2) [6].

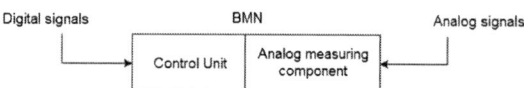

Digital signals   BMN   Analog signals

Fig. 2. Structure of BMN.

To perform full control in a single-layer network, it is necessary to combine several units and form an FLCU (Fig. 3), which will allow processing information and generating control impulses [5]. The measurement is controlled on the basis of the generated IADC using the same FLCU implemented in the FPGA. One of the drawbacks of the SADC NN presented in [7] is the limited scaling of the network, which does not allow the full implementation of a multi-pass and highly reliable SADC NN due to the limited number of FPGA outputs required to control analog keys in AMC, which are implemented as a separate board.

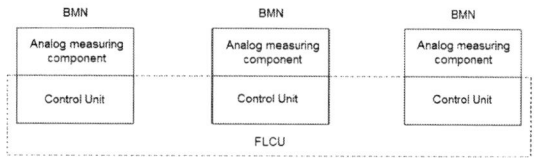

Fig. 3. Structure of FLCU.

To solve this problem, a hardware implementation of the measuring part of the BMN has been developed. A ready-made KR590KN4 solution (four-channel analog keys with a control circuit) was chosen as the keys. It is proposed to use microcontroller STM8s003F3P6 (8-bit, 16 MHz) to control keys and digital resistance. Following the program, the controller receives 16 bits of information including the BMN number and command on the UART serial port, then decodes them and sets the corresponding values at the key outputs.

III. SELF-DIAGNOSTICS SYSTEM

In order to create a self-diagnostics system, at this stage, the problem of changing the calibration characteristic («resistance aging») of resistances was highlighted, when during the service life their nominal value can change significantly, while in each AMC these changes will be independent, which will lead to an imbalance in the matrix. Modeling the occurrence of such disturbances in the matrix revealed the following patterns:

- when increasing (decreasing) the resistance of the element $2R$ corresponding to the i-th category ($2R_i$), the voltage when setting the i-th bit will be lower (higher), and for all others it will be higher (lower) than the nominal value;

- by increasing (decreasing) the resistance of the element R corresponding to the i-th bit (R_i), the voltage at the setting of the i-th and higher bits will be higher (lower), and for all lower bits lower (higher) nominal.

When creating a self-test method, it is necessary to determine at what point in time and under what conditions to check [8, 9]. There are several options for initializing the self-diagnosis process:

- The process is initialized when the power is turned on to the SADC NN, it allows you to provide the necessary time for a complete check of the network before the start of processing the signals coming for measurement;

- In the course of operation, the process is initialized periodically with a period of time between checks not exceeding the specified t, at the same time, during the check, the SADC NN ceases to process all incoming signals, but at the same time the entire network is checked;

- The process is initialized in the background, in the presence of the required number of free BMNs located in close proximity to each other, this allows maintaining the ability to measure signals, however, there is no guarantee of checking the entire network.

In order that the self-diagnostics system does not depend on any reference signals that may be subject to distortion or accumulate error during operation, it is proposed to use one BMN group to generate a «reference» voltage and another to measure it during the test. At the same time, in order to exclude errors that may be in the first group, the check is carried out in several iterations, changing the order of the BMN connection for each. The process of self-diagnosis is as follows:

1) A request for *2n* free BMN is generated;

2) A group of *2n* connected free BMNs generates a signal to start the self-diagnostics process;

3) Neurons *1* to *n* form DACs, and neurons *n+1* to *2n* form ADCs;

4) The DAC alternately generates voltages corresponding to the values of 2^{i-1} numbers in the binary coordinate system, where $i = 1...n$ (the duration of holding each voltage corresponds to the time spent by the ADC for digitization);

5) The ADC digitizes the received signal n times, with each clock cycle playing the role of the most significant bit of the next BMN;

6) Based on the results of each measurement of the ADC, an error is calculated between the specified and calculated digital number, which is summed up. After n^2 measurements, the checksum is estimated, based on the value of which the faulty element is determined;

7) If the failure is critical, when it is not possible to compensate for the error (for example, circuit break or short circuit), then the shutdown and isolation of the BMN occurs;

8) If the failure is not critical, then the algorithm is started, which must correct the operation of the faulty element in such a way that the measurement accuracy of the SADC NN does not change.

IV. MODELING

The next step was to model the behavior of the system taking into account the obtained dependencies and the method of self-diagnostics based on the existing implementation of the FLCU in the single-layer network of the SADC NN. The MATLAB/Simulink environment was chosen as the design environment, which made it possible to implement the SADC NN, while the digital part at this stage was implemented in the form of program code (using MATLAB Function blocks), which made it possible to debug self-diagnostics algorithms and obtain large samples of data when automatically modeling several network parameters at the same time.

Analysis of the data obtained by the model allows us to distinguish patterns corresponding to certain types of failures in certain neurons, which were introduced into the software part of the built model and allowed the implementation of a search algorithm that accurately and unambiguously determines the place and type of failure. When conducting a large number of experiments using the obtained fault search method, measurements were made of the time required for a complete network check (Fig. 4).

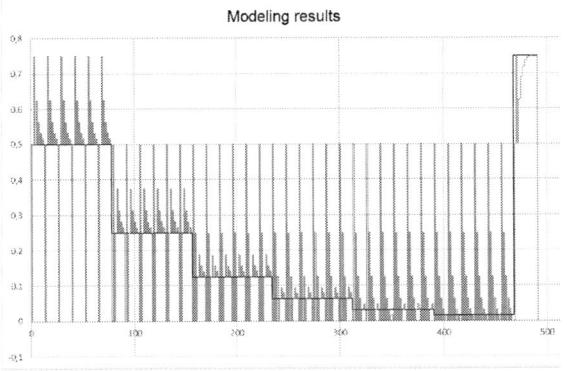

Fig. 4. Schedule of the Processes of Self-Diagnostics of 6 BMN and Measurement by 12-Bit IADC.

The obtained time for checking 6 BMN in the example shows that the self-diagnostics process takes an order of magnitude longer than the measurement itself using the generated IADC converter, therefore, this method can be used only when powering up the SADC NN until the input signals arrive and the measurement begins or when organizing cyclic network checks with the service stop of incoming measurement requests. It is also important that an increase in the number of tested BMNs increases the required time exponentially, since each additional BMN increases the size of the matrix of possible switches between tested neurons n^2.

In order to solve this problem, diagnostics can be provided in several sections of the network at once in order to reduce the number of switches within each diagnosed fragment. However, for this approach it will be necessary to provide for the behavior of the self-diagnostics system at the boundaries of the sections, when the calibration in one section may not coincide with the calibration in another section.

The method of self-diagnostics was implemented in the hardware implementation of the BMN by using a digital potentiometer DS1801, which has 65 positions for setting different resistance ratings. In order to simulate the change in resistance as it «ages», the microcontroller code adds a feature that allows you to change the position of the digital potentiometer, which disrupts the matrix balance. The obtained self-diagnostics algorithms during modeling at the current stage are also implemented on the basis of a microcontroller, as part of the FLCU of each BMN.

For a more convenient and visual health check, a system was developed containing a ESP8266 controller, on the basis of which a REST server was created, which stores data on all BMNs and a client application for Android OS, with which one can control the resistance in the BMN separately and also receive up-to-date information on the state of the BMN. The interaction scheme is shown in Fig. 5.

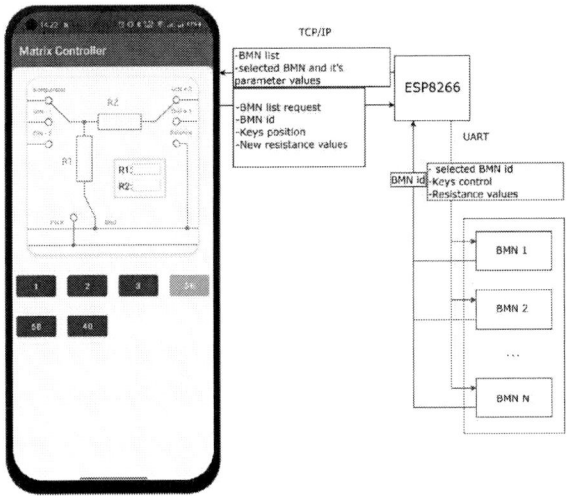

Fig. 5. Diagram of interactions of the matrix of BMN elements, ESP8266 and client application.

When selecting the appropriate BMN, the displayed positions of the keys and resistors resistance are updated. When entering values in the R_1 and R_2 fields, a query is sent containing the new resistance values. After processing the request, a UART message is sent to the server for the BMN, as well as a response to the request containing the set key positions and resistor resistance.

Advantages of the proposed BMN scheme using a microcontroller:

• a single communication bus for each BMN between the FPGA and the BMN measuring part;

• possibility to change the NN architecture without changing the measuring hardware.

The disadvantage of this connection is the reduced data transfer rate due to the limitations of the UART interface built into the microcontroller. However, this drawback can be eliminated by changing to faster counterparts.

In the future, it is planned to implement a multi-pass SADC NN, to assess the impact of the FLCU circuit with the microcontroller on the measurement speed, as well as to investigate the possibilities of using the proposed method of self-diagnostics on a larger number of BMNs both during initialization and during the operation of the SADC NN.

REFERENCES

[1] E. A. Emelianov and M. P. Artemov, "Primary program processing of results of measurements after analog-digital transformation", 2012 International Conference on Actual Problems of Electron Devices Engineering. 2012. pp. 282-285.

[2] V.I. Freyman, A.I. Posyagin "The soft decoding of control systems elements test diagnostics results", 2017 XX IEEE international conference on soft computing and measurements (SCM). 2017. pp. 329-332.

[3] S. V. Chelebaev, O. V. Melnik, Y. A. Chelebaeva "Application of simulation modeling for the analysis of neurons of converters of time-and-frequency parameters of signals in a digital code", 2018 7th Mediterranean Conference on Embedded Computing (MECO). 2018. pp. 1-4.

[4] D.N. Smirnov. "Analysis of neuron AD conversion in golden ratio codes", Telecommunications and Radio Engineering. 2007. № 8. pp. 20-22. (in Russian).

[5] N.G. Makagonov, A.I. Posyagin, A.A. Yuzhakov. "Principles of signals self-routing in analog-digital converter based on one-layer

neural network", Russian electrical engineering. 2016. № 11. pp. 3-6. (in Russian).

[6] Vasbieva A.F., Oniskiva L.M., Posyagin A.I., Yuzhakov A.A. "Basic measuring neuron structure of self-routing analog-to-digital converter", Journal Information-measuring and Control Systems. 2015. vol. 13. № 9. pp. 3-8. (in Russian).

[7] N.G. Makagonov, A.I. Posyagin. "Development of a prototype neural network analog-to-digital converter", Science of the present and future. 2017. Vol. 1. pp. 217-218 (in Russian).

[8] E. V. Katsko, A. I. Posyagin, A. A. Yuzhakov. "An autocontrol system based on neurons in a self-routing analogue converter", Russian Electrical Engineering. 2014. Vol. 85. № 11. pp. 703-707.

[9] A.A. Naborshikov, A. A. Yuzhakov "On increasing the reliability of a self-routing analog-to-digital converter based on a neural network", Mathematical Modeling : Abstracts of the II International Conference. 2021. pp. 67-68 (in Russian).

Adaptive LUT - Decoder LUT for FPGAs

Nikita E. Oputin
Department of Automatic and Telemechanic
Perm National Research Polytechnic University
Perm, Russia
oputin1906@mail.ru

Sergey F. Tyurin
Department of Automatic and Telemechanic
Perm National Research Polytechnic University
Perm, Russia
tyurinsergfeo@yandex.ru

Abstract- **Known adaptive logic modules (ALM) based on LUT elements, which, by setting up and connecting several LUTs, can implement the logic functions of five, six, seven, and even partially eight variables. However, to configure some of the important devices, for example n variables decoder, too many ALMs are needed.**

Decoder LUT was previously developed, but it cannot be used in normal LUT mode. Therefore, it is proposed to implement two modes of operation on one device using one LUT. The first is the implementation of one logical function, the second is the implementation of the input variables decoding. This has the effect of reducing the number of transistors compared to decoder implementation by multiple LUTs.

Keywords — Adaptive LUT, Decoder LUT, FPGA, logic function, logic element.

I. INTRODUCTION

Logic Cell or logic elements LE are the basis of programmable logic integrated circuits such as FPGA (field-programmable gate array) [1] – and are read-only memory devices LUT (Look Up Table).

Linear representation of the 1-LUT's logic function [2], [3], [4] is the next:

$$z_{out}(x) = d_1 \bar{x} \cup d_0 x \qquad (1)$$

where d_1, d_0 are configuration data of the one argument (n=1) $z_{out}(x)$ function. Combining d_1, d_0 we can get 2^2 functions (Fig1).

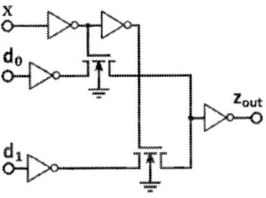

Fig. 1. 1-LUT tree.

Figure 1. shows 1-LUT according to tree representation of (1), with two MOS-p pass transistors two configuration inputs d_1 d_0 one input variable (x) and single output function (z). NOT gates (invertors) are an amplifier, the signal's restoration elements, one x-invertor realizes NOT(x) signal.

The disadvantages of the device taken as a prototype [5], [6] are: 1) large hardware costs for the implementation of one logical function; 2) large hardware costs for the implementation of systems of logical functions from the same variables implementing decoding of the input signal.

Disadvantages exist for the following reasons. The technical means of the prototype are not adaptable

and are focused on the implementation, depending on the configuration of only systems of logical functions from the same n variables in perfect disjunctive normal form (SDNF) using blocks for calculating functions that are redundant when implementing decoding the input signal.

In this regard, hardware costs increase significantly in the number of transistors for large n – 6, 7, 8. Therefore, it is advisable to implement dual-mode operation:

1) implementation of only one logical function;

2) implementation of decryption of input variables.

In mode 2, other, external programmable logic devices can play the role of function calculation blocks

II. MODELING ADAPTIVE LUT

The developed logical element for two variables (2-LUT) implements any logical function up to two variables (X1, X2) by appropriately setting the SRAM configuration memory cells (d0, d1, d2, d3). The mathematical description of the developed logic element for two variables 2-LUT is presented below:

$$z_{OUT}(x_2 x_1) = d_0 \overline{x_2}\, \overline{x_1} \cup d_1 \overline{x_2}\, x_1 \cup d_2 x_2 \overline{x_1} \cup d_3 x_2 x_1 \qquad (2)$$

Modeling Adaptive LUT for two variables was performed in the NI Multisim circuit modeling system of the National Instruments Electronics Workbench Group. The simulation result is shown in Figure 2.

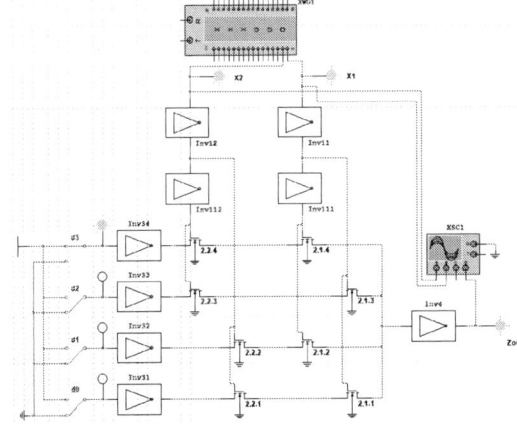

Fig. 2. Simulation of a LUT for two information (address) inputs configured to implement the function $x_1 \cap x_2$

Adaptive LUT is implemented on transistors 2.1.1-4 and 2.2.1-4. Settings Adaptive LUT are made with the keys d0, d1, d2, d3, the status of the keys is displayed on the corresponding indicators. These keys simulate the state of the SRAM configuration memory cells into which the required truth table is loaded. A word generator (XW G1) models the values of variables X1, X2. Blocks Inv11, Inv12, Inv111,

979-8-3503-8358-4/23 $31.00 © 2023 IEEE

Inv112, Inv31, Inv32, Inv33, Inv34, Inv4 implement inverter blocks, which are necessary elements of the LUT. The circuit of the inverter blocks is shown in Figure 3. The result of the calculation is formed at the Z_{out} output.

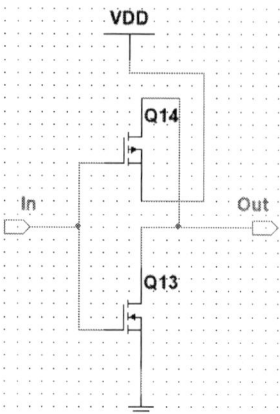

Fig. 3. Inverter block

An oscilloscope (XSC1) is used to analyze the operation of the circuit. Using an oscilloscope (XSC1), the output indicators of the circuit are taken. Figure 4 shows the dependence of the output of the general indicator configured for the implementation of the function $x_1 \cap x_2$ on the input variables.

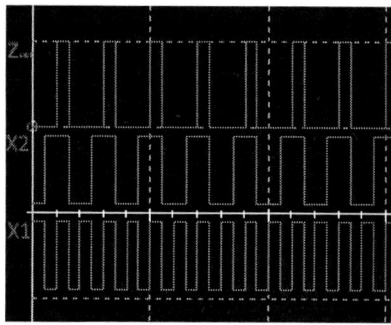

Fig. 4. Diagram of the dependence of the output of the general indicator configured for the implementation of the function $x_1 \cap x_2$ on the input variables

For X1=1 and X2=1, Z_{out} takes the value 1. For other variable values, Z_{out} takes the value 0.

III. MODELING ADAPTIVE LUT – DECODER LUT

To decrypt the input signal, it is necessary to upgrade the existing circuit. To do this, inverters Inv51, Inv52, Inv53, Inv54 are installed at the outputs, it is necessary to amplify the signal that has passed through the transmitting transistors. Additionally, diodes are connected so that a short circuit does not occur in the decryption mode.

In order to prove the accuracy of the address decryption at the inputs X1, X2, we will set the set 11. In Figure 5 it can be seen that when dealing this set, the third output, Y3, remains not activated.

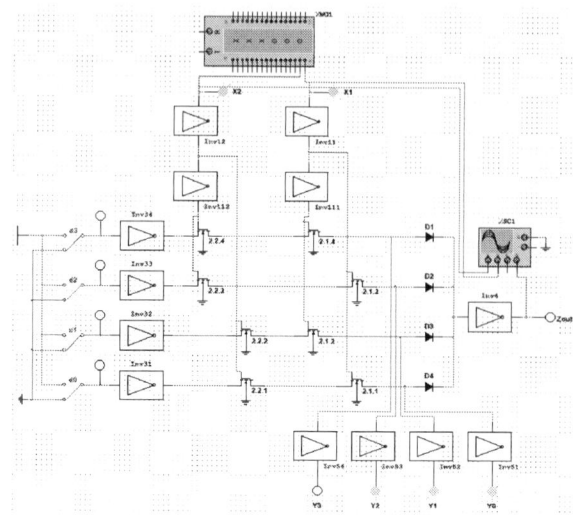

Fig. 5. LUT modeling for two information (address) inputs, decryption of the input set 11

In this case, the circuit can also work in the mode of implementing a logical function (Fig. 6).

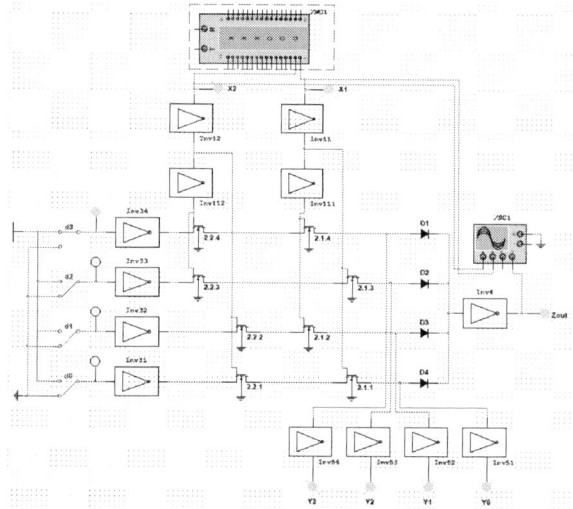

Fig. 6. Operation in logical function mode

The disadvantage of the simulated circuit is the lack of a choice of operating modes, as well as circuit breakage in the decryption mode.

IV. IMPROVED ADAPTIVE LUT – DECODER LUT

Let's model a logical element into three 3-LUT variables (Fig. 7). We will additionally introduce the possibility of selecting operating modes using the Mode key and two rows of transistors 16.1-16.16.

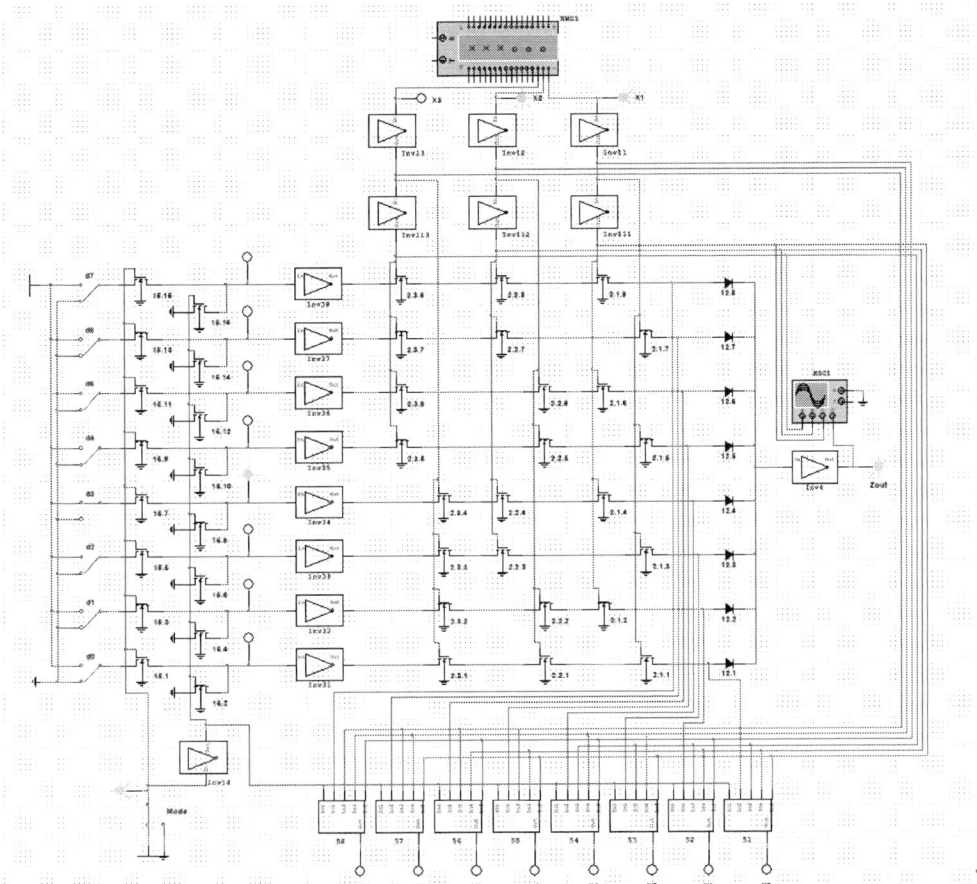

Fig. 7. Fig. 7. Simulation of a LUT for three information (address) inputs configured to implement the function $\overline{x_3} \cap x_2 \cap x_1$

Blocks 51-58 implement decryption blocks, their scheme is shown in Figure 8. Transistors 5.i.1.1, 5.i.1.2, 5.i.1.3 perform the function of constituents of zero, necessary to prevent circuit breakage.

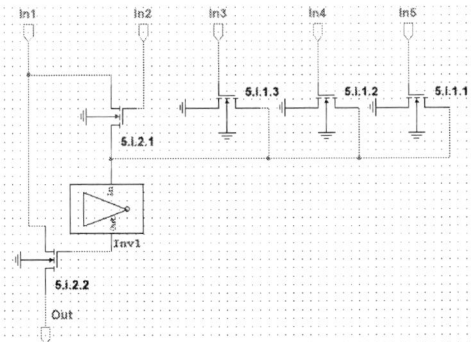

Fig. 8. Decryption Block

An oscilloscope (XSC1) is used to analyze the operation of the circuit. Using an oscilloscope (XSC1), the output indicators of the circuit are taken.

Figure 9 shows the dependence of the output of the general indicator configured for the implementation of the variable function $\overline{x_3} \cap x_2 \cap x_1$ on the input variables when the circuit is operating in the implementation mode of the logical function.

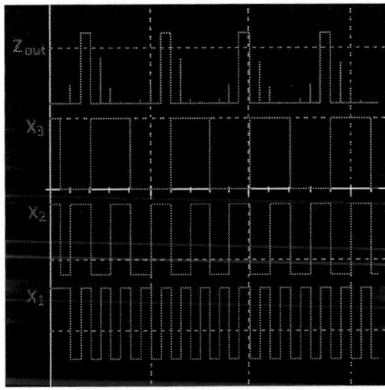

Fig. 9. Diagram of the dependence of the output of the general indicator configured for the implementation of the function $r_3 \cap x_2 \cap x_1$ on the input variables

In order to prove the accuracy of the address decryption mode, we will set a set of 100 at the inputs X3, X2, X1. Figure 10 shows that with this set, the fifth output remains inactive – Y5.

979-8-3503-8358-4/23 $31.00 © 2023 IEEE

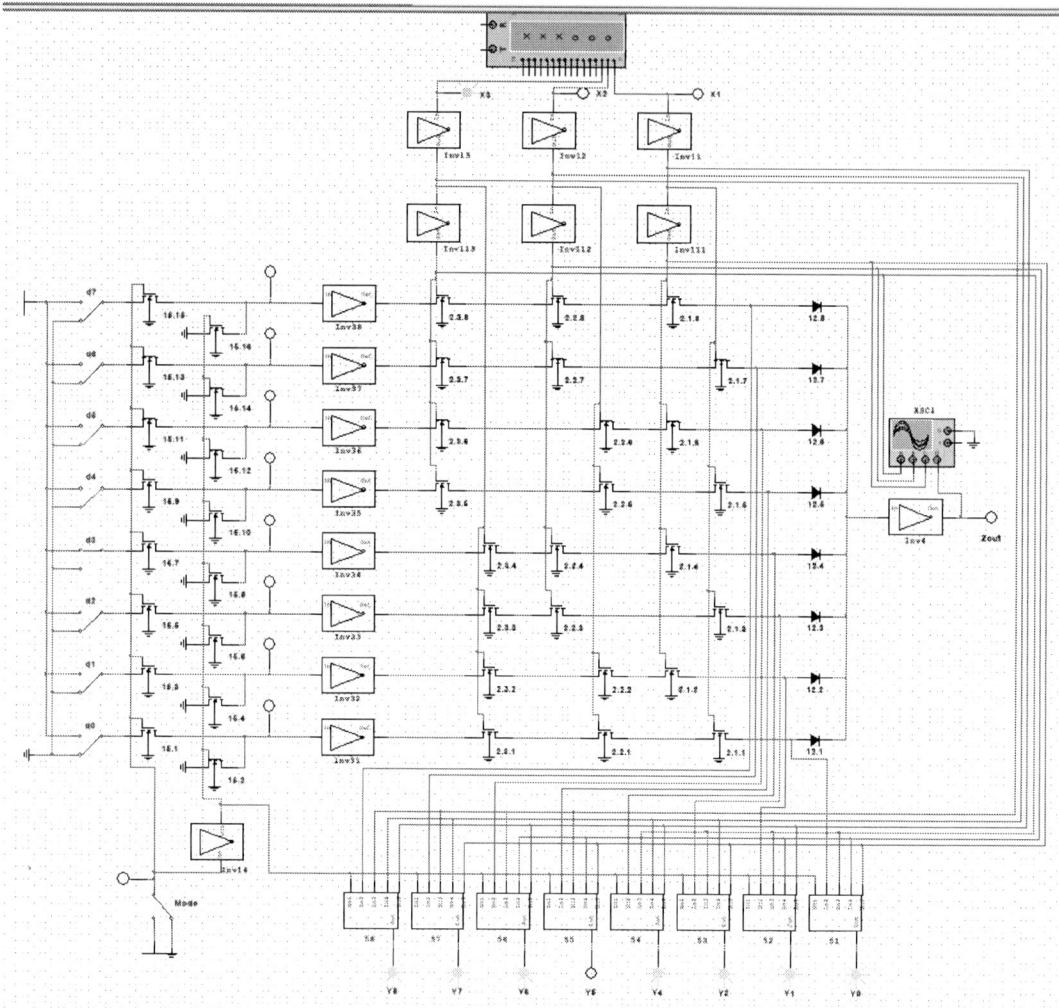

Fig. 10. LUT modeling for three information (address) inputs, decryption of the input set 100

V. CONCLUSION

The proposed device allows to implement two modes of operation: 1) the implementation of only one logical function; 2) the implementation of the decryption of input variables. This significantly reduces hardware costs due to the reduction in the number of transistors. The simulation confirms the operability of the proposed advanced LUT-based device. In addition, this device can be used as universal logic elements.

ACKNOWLEDGMENT

Great thanks to Senior lecturer Elena Leonidovna Kavardakova (Perm National Research Polytechnic University) and Candidate of Sciences Skornyakova Alexandra Yurievna

REFERENCES

[1] Brown S. Architecture of FPGAs and CPLDs: A Tutorial. Available at:https://www.ece.iastate.edu/~zambreno/classes/cpre583/documents/BroRos96A.pdf (accessed: 7 November 2022).

[2] Field Programmable Gate Arrays. Available at: http://www.eng.auburn.edu/~nelson/courses/elec4200/FPGA/FPGAoverview.pdf (accessed 5 November 2022)

[3] FPGA Architecture White Paper – Altera. Available at: https://www.altera.com/en_US/pdfs/literature/wp/wp-01003.pdf (accessed 7 November 2022)

[4] 7 Series FPGAs Data Sheet: Overview. Available at: https://www.xilinx.com/support/documentation/data_sheets/ds180_7Series_Overview.pdf (accessed 2 November 2022)

[5] Stroganov A., Tsybin S. Programmable switching in FPGAs: an inside look. *Komponentyitekhnologii* [Components and Technologies], 2010, no. 11, pp. 56-62. *(in Russian)*

[6] Tyurin S.F. Vikhorev R.V. *Programmiruemoelogicheskoeustrojstvo* [Programmable Logic Device] Patent RF,no. 2573732, 2016.

Charge Relaxation in Polylactide/Montmorillonite Composite Materials

Andrey A. Pavlov
*Higher School of High Voltage
Engineering*
*Peter the Great St.Petersburg
Polytechnic University*
Saint-Petersburg, Russia
pavlov.aa.hv@mail.ru

Almaz M. Kamalov
*Higher School of High Voltage
Engineering*
*Peter the Great St.Petersburg
Polytechnic University*
Saint-Petersburg, Russia
spb.kamalov@gmail.com

Margarita E. Borisova
*Higher School of High Voltage
Engineering*
*Peter the Great St.Petersburg
Polytechnic University*
Saint-Petersburg, Russia
vladimirl.borisov@gmail.com

Mikhail A. Kovalenko
*V.A. Belyi Metal-Polymer
Research Institute
of National Academy of Sciences
of Belarus*
Gomel, Belarus
mikhailkovalenko1991@gmail.com

Victor A. Goldade
*V.A. Belyi Metal-Polymer
Research Institute
of National Academy of Sciences
of Belarus*
Gomel, Belarus
victor.goldade@gmail.com

Sergey V. Zotov
*V.A. Belyi Metal-Polymer
Research Institute
of National Academy of Sciences
of Belarus*
Gomel, Belarus
zotov-1969@mail.ru

Abstract—This study focuses on charge relaxation process in a new composite material based on polylactide (PLA) and montmorillonite (MMT), which was used as a filler. The samples with different mass percentages of MMT were charged in a corona discharge and studied using the compensation method with a vibrating electrode. Time dependences of the electret potential differences were achieved. Thermally stimulated depolarization currents (TSDC) were used to identify the relaxation processes. The TSDC spectra were analyzed in terms of the superposition method.

Keywords—charge, electret, TSD, polylactide, montmorillonite

I. INTRODUCTION

The problem of recycling polymer waste, which since the 1960s has become a constant source of pollution, is caused by a significant increase in the level of consumption of polymer films used as packaging materials. Until recently, about 40% of the total volume of polymer materials produced in Europe was used for packaging. More than 40% of the polymer film waste is made from polyolefins. They decompose in natural conditions for decades. [1]. Development of biodegradable polymer materials (BPM) with a set "lifetime" created an environmentally optimal approach to solving the problem of polymer waste disposal.

One of the popular biopolymers that is currently used in the packaging industry is polylactide (PLA). PLA is the most widely studied and used renewable and biodegradeable polyester, which is also thermoplastic. It is used either in industrial packaging or as a biocompatibleble medical material. Lactic acid, which is a component of PLA, is a chiral molecule and exists in the form of two enantiomers (L- or D-) depending on the position of the methyl group (Fig. 1).

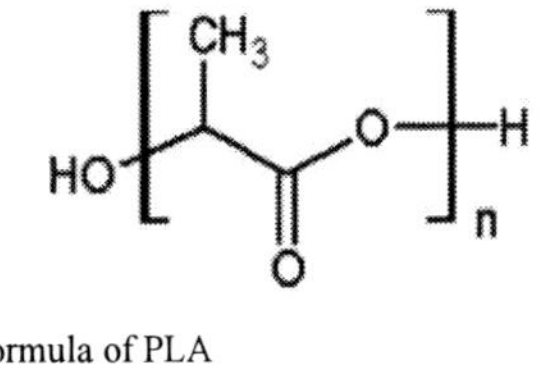

Fig. 1. Structural formula of PLA

Polylactide has stereoisomers such as L-PLA, D-PLA and LD-PLA. Processing temperature, annealing time and molecular weight are the factors that define the properties of PLA. Stereochemistry and thermal history have a direct impact on the crystallinity of PLA and, consequently, on its properties. The crystallinity degree is an indicator of the amount of the crystalline phase in the polymer in relation to the amorphous one. Crystallinity affects many properties of the polymer, including modulus of elasticity, tensile strength, etc. Therefore, when choosing a polymer, its crystallinity plays a primary role. There are three modifications of PLA crystals: α, β and γ. They have different conformations and cell parameters that are formed under various thermal and mechanical influences.[2].

The U.S. Food and Drug Administration (FDA) classified PLA as safe: it is recommended for food packaging. The approval for medical use gave a huge boost to PLA researches. At the same time, the use of industrial polymers based on lactic acid has increased dramatically.

In order to improve the consumer qualities of packaging, polymer materials might gain active properties. Currently, there are three main approaches to creating biologically active packaging:

1) natural antioxidants modification,

2) development of nanocomposite films with functional fillers,

3) development of polymer materials with a stable electret state.

In the first approach, the active substances in packaging films absorb oxidative radicals from food or migrate into food, increasing the quality and shelf life of the product [3]. Antioxidants obtained from water and alcohol extracts of plants, concentrates obtained from various bio-sources are used. These packaging materials can release active substances into food at a stable preset rate, thereby reducing lipid oxidation and maintaining the nutritional value and quality of food.

The second approach implies no direct effect on the food product, but a change in the properties of the film. In this case, nanofillers (metals, silicates, natural fibers, etc), which affect the mechanical, thermal and barrier properties of

979-8-3503-8358-4/23 $31.00 © 2023 IEEE

biopolymer matrix. In [4], food films based on PLA with the addition of zinc oxide nanoparticles were created. These composite films were characterized by reduced water vapor permeability, and zinc oxide particles added antibacterial properties.

The third approach involves the formation of a stable electret state in a polymer package [5]. Under the influence of an electric field, the energy parameters of microbial cells decrease, and, consequently, metabolic processes slow down. A number of studies have been conducted that confirmed that the polymer film, which has gained a stable electret state, significantly increases the shelf life of various food products [6].

Polylactide is a poor electret material. Therefore, one of the promising methods for improving the electret properties of PLA is the creation of composite materials [7]. With the introduction of a filler, the electret properties in polylactide films increased regardless of the sign of the corona in which the samples were polarized. In study [8], polymer composites were made from PLA with barium titanate, titanium dioxide, nanoscale aerosil and diatomite. The highest values of the electret potential difference were observed in composites with 2-4% content of BaTiO3 and TiO2 (2.5 – 3.5 times higher than that of the original PLA). This is explained by the fact that new defects appear in the polymer structure, acting as charge traps. Absorption of polymer macromolecules on the filler surface reduces their mobility and slows down relaxation processes.

The analysis of the mentioned research results suggests that composite materials with stable electret properties can be obtained by adding nanofillers into a PLA matrix. The study of relaxation mechanisms in the initial PLA films and PLA composites is of particular interest, since the nature of electrical processes in such films has not been fully studied. The aim of the work is to develop composite films based on polylactide and montmorillonite and to identify the optimal filler content that will provide a stable electret state.

II. MATERIALS AND METHODS

The object of the study is a composite material based on polylactide. Matrix: polylactide brand "Biopolymer 4043D". Montmorillonite (MMT) was used as a filler. The mass percentage of filler was from 0 to 3%.

Montmorillonite is a widespread clay mineral from the smectite group of the layered silicate subclass. The relevance of using this material as a filler for polymer electrets is due to its properties: fine dispersion (it contains nanodisperse particles up to 100 nm in size), high dielectric properties and low cost.

Montmorillonite is a mineral that has been used for a long time due to its adsorbing properties. It has an unstable chemical composition, which largely depends on the percentage of water. Approximate composition: aluminum oxide (11-22%), magnesium oxide (4-9%), iron oxide (5%), water (12-24%); the mineral can also contain calcium oxide (up to 3.5%). The structure of montmorillonite is characterized by a large distance between the bundles of layers. A wet piece of MMT swells strongly due to water permeating the gaps between the layers of the structure.

The samples were activated in a corona discharge (Fig. 2). The needle potential was +7 kV. The value of the set

electret potential difference was regulated using a grid with a +500 V potential. The samples were charged at room temperature for 2 minutes.

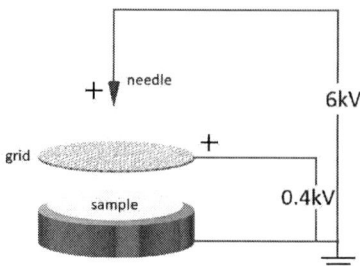

Fig. 2. Charging scheme

The electret potential difference was measured using the electrostatic induction method [9].

The analysis of the spectra of thermally stimulated depolarization currents (TSDC) makes it possible to determine the mechanisms of charge relaxation in the composite. TSD currents were measured at a constant heating rate of 1.5 °K/min. An insulating layer (polytetrafluoroethylene (PTFE) film with a thickness of 30 microns and with a resistivity of 10^{16} ohm·m) was placed between the sample surface and the electrode. The resistance of the insulating layer is several orders of magnitude higher than the resistance of the PLA+MMT film. In this case, the direction of the TSD current corresponds to the movement of charge carriers through the thickness of the sample and has the same direction as the charging current. Using the compensation method, it was found that a homocharge accumulated in the sample after charging in a corona discharge.

The TSDC measurement scheme is shown in Fig. 3.

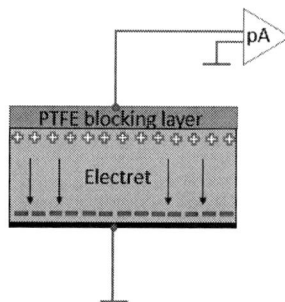

Fig. 3. TSDC measurement scheme

The spectra of TSDC are analyzed according to the superposition method [9]. The value of the current density J_{TSD} for each maximum can be determined using equation (1).

$$J_{TSD} = J_m \exp\left[\frac{W}{k}\left(\frac{1}{T_m}-\frac{1}{T}\right)\right]\exp\left\langle-\frac{W}{kT_m^2}\int_{T_m}^{T}\exp\left[\frac{W}{k}\left(\frac{1}{T_m}-\frac{1}{T'}\right)\right]dT'\right\rangle \quad (1)$$

The current density at the maximum J_m, according to the model of a two-layer dielectric, is expressed in equation (2).

$$J_m = \frac{\varepsilon\varepsilon_0\varepsilon_{lay}U_{e0}}{\left(\varepsilon h_{lay}+\varepsilon_{lay}h\right)}\exp\left\{-\frac{W}{kT_m^2}\int_{T_0}^{T_m}\exp\left[\frac{W}{k}\left(\frac{1}{T_m}-\frac{1}{T'}\right)\right]dT'\right\},$$
(2)

where h_{lay} and ε_{lay} are material thickness and permittivity of the insulating layer, respectively; T_0 is the initial temperature; h and ε are the thickness of the sample and dielectric constant; U_{e0} is the initial value of the compensating voltage; ε_0 is the dielectric constant.

The activation energies of the experimentally measured maxima were determined by the "fitting" method. The calculated and experimentally obtained curves match by varying the activation energy W.

The conductivity of the samples was measured as in [10].

III. RESULTS AND DISCUSSION

Table 1 shows the conductivity of the samples.

TABLE I.   SAMPLE CONDUCTIVITY

Mass percentage of MMT	Conductivity, 10^{-14} S/m
0	1.27
1	1.42
1.5	0.21
3	0.61

The stability of the electret state was studied by measuring the electret potential difference from time at room temperature. The time dependences of the relative values of U_e/U_0 are shown in Fig. 4.

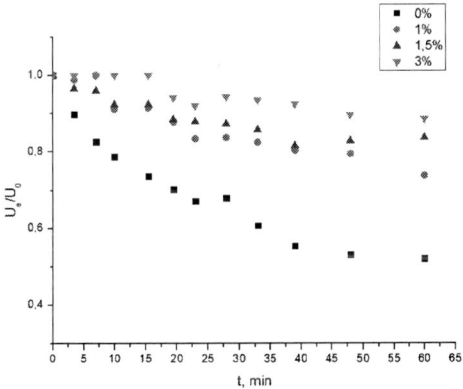

Fig. 4.   Time dependences of the electretic potential differences

Fig. 4 shows that with an increase in the percentage of filler, the stability of the electret state increases. After 50 minutes, a stabilization of the charge is observed in samples with 1.5 and 3% MMT.

The TSCD spectra are shown in Fig. 5.

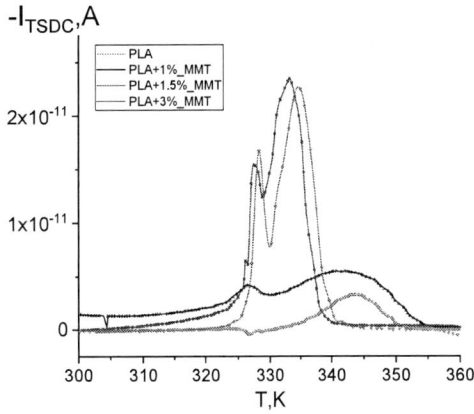

Fig. 5.   TSDC spectra of the PLA+MMT samples

Fig. 5 suggests that with an increase in the percentage of filler, the curves do not change their character: the introduction of montmorillonite leads to an additional maximum in the glass-transition temperature region.

The high-temperature maximum becomes more acute and shifts to the low temperature region with an increase in the percentage of filler (>1%).

Table 2 shows the charge calculated from the TSD currents.

TABLE II.   SAMPLE CHARGE

Mass percentage of MMT	Charge, C/cm²
0	0.2×10^{-2}
1	0.8×10^{-2}
1.5	1.1×10^{-2}
3	1.1×10^{-2}

The "fitting" was used to identify the activation energies (Fig. 6).

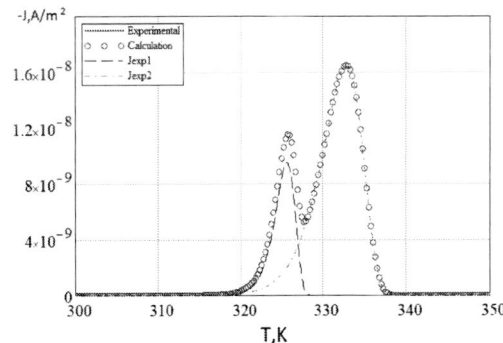

Fig. 6.   "Fitted" TSDC curves

TABLE III. THE ACHIEVED ACTIVATION ENERGIES. TSDC MAXIMA PROPERTIES

Maxima	$J_m \times 10^{-8}$, A/m²	T_m, °K	W, eV
Initial PLA			
J_1	2.43	342.0	2.3
PLA+1%MMT			
J_1	1.2	326.7	5.0
J_2	2.7	342.0	1.3
PLA+1.5%MMT			
J_1	6.5	328.5	13.0
J_2	13.0	334.7	3.2
PLA+3%MMT			
J_1	3.0	327.5	13.0
J_2	11.6	333.0	3.2

Such high values of the activation energy (5-13 eV) are not characteristic of the intrinsic conductivity of a composite dielectric. At the same time, with an increase in the filler concentration, the conductivity of the sample decreases, and the amplitude of the TSD currents and the sample charge increase. Therefore, it can be assumed that the sharp low-temperature maxima may appear due to the release of charge carriers from traps that are localized at the interface between PLA and MMT or in MMT, which was caused by activation of the segmental motion in PLA.

The introduction of filler leads to a shift of high-temperature maxima to the low-temperature region. The maxima become narrower and the activation energy increases. This effect may be associated with a change in the structure of the material and the formation of traps at the interface between the PLAN and MMT or in MMT.

IV. CONCLUSION

Composite films based on polylactide and montmorillonite were developed and produced. The composite film with 3% MMT at a temperature of 25 °C has the best electret properties.

REFERENCES

[1] Gerald S. Abiotic Control of Polymer Biodegradation. Trends in polymer science, 1997, no. 11, pp. 361-368.

[2] di Lorenzo M. L. Crystallization behavior of poly (L-lactic acid). European Polymer Journal, 2005, vol. 41, № 3, pp. 569–575.

[3] Barbosa-Pereira L. et al. Development of antioxidant active films containing tocopherols to extend the shelf life of fish. Food Control, 2013. vol. 31, № 1, pp. 236–243.

[4] Pantani R. et al. PLA-ZnO nanocomposite films: Water vapor barrier properties and specific end-use characteristics. European Polymer Journal, 2013, vol. 49, № 11, pp. 3471–3482.

[5] Goldade V., Kovalenko M., Zotov S. Electret charge in nanocomposites based on polyethilene. ISJ Theoretical & Applied Science, 2021, vol. 11 (103), p. 759–765.

[6] Gorohovatskij Ju. A. Electret effect and its application. *Sorosovskij obrazovatel'nyj zhurnal* [Soros' educational journal], 1997, № 8, pp. 92 – 98. (in Russian)

[7] Galikhanov M. F., Zhigaeva I. A. Changes of polyethylene electrets properties when filled with barium titanate. Proceedings of 14 th International Symposium on Electrets, Montpellier, France, 2011, pp. 161 – 162.

[8] Guzhova A. A. *Jelektretnye kompozicionnye materialy na osnove polilaktida.* PhD Diss. [Electret composite materials based on polylactide]. Kazan, 2016, 132 p.

[9] Borisova M., Galukov O. *Fizika dielektricheskih materialov. Elektroperenos i nakoplenie zaryada v dielektrikah* [Physics of dielectric materials. Charge transfer and preservation in dielectrics]. Saint-Petersburg, SPbPU Publ., 2004. 106 p.

[10] A. Pavlov, M. Borisova, A. Kamalov, K. Malafeev, V. Yudin, "The Effect of Hydroxyapatite on the Electrical Properties of a Polylactide-Based Composite," 2020 IEEE Conference of Russian Young Researchers in Electrical and Electronic Engineering (EIConRus), St. Petersburg and Moscow, Russia, 2020, pp. 1042-1044, doi: 10.1109/EIConRus49466.2020.9039381.

Experimental Study of Self-Healing Characteristics and Insulation Resistance of Metallized Polypropylene Film under Various Inter-Layer Pressure

Alexey V. Pechnikov
Higher School of High Voltage Engineering
Peter the Great St. Petersburg Polytechnic University
Saint Petersburg, Russia
pechnikov.av@edu.spbstu.ru

Ahmet A. Hojamov
Higher School of High Voltage Engineering
Peter the Great St. Petersburg Polytechnic University
Saint Petersburg, Russia
hodzhamov_aa@spbstu.ru

Abstract—The self-healing of metallized film capacitor's performance after dielectric breakdown is a unique ability making them high reliable, attractive for application under high voltage stresses, including pulse voltages typical for power electronics applications. The actual metallized film capacitors are tightly wounded rolls with inter-layer pressure declining from peak values counting tens of bars around the mandrel to units of bars in top layers. This work presents experimental findings of self-healing process characteristics such as energy, duration and micro-arc discharge time-dependent resistance under actual inter-layer pressures typical for metal-film capacitors. The insulation resistance of stacked metallized polypropylene films after self-healing under high pressures were found to be decreasing, resulting in current leakages to occur. Current leakages origination was associated with feasible carbon deposition near the breakdown channel due to polymer decomposition in high temperature arc discharge. Obtained results show the proper choice of film winding tension importance for capacitor's reliability.

Keywords—self-healing, metallized film capacitor, insulation resistance, breakdown strength, inter-layer pressure

I. INTRODUCTION

Metallized polypropylene film capacitors (MPPFCs) are one of the crucial components in modern power electronics due to their low loss, superior reliability and specific volumetric characteristics [1]. At the moment they have been almost substituted electrolytic and ceramic capacitors in such applications as filtering, DC-linking and snubbering for electric vehicle, renewable energy sources and power supplies [2].

Typical design of MPPFCs includes two polypropylene (PP) films coated with vacuum sprayed thin metal layer [3]. The PP films could be wrapped on the mandrel, forming cylindrical construction called capacitor section, or could be stacked one to each other making up multilayer sandwich-like construction. The thickness of metal electrodes in both cases are units and tens of nano-meters, the material is zinc, aluminium and their compositions [4]. Thin electrodes enable MPPFCs to be more material-saving than film/foil capacitors and, in addition, provide self-healing (SH) ability. In general, SH ability lies in the fact that metallized film capacitor stays operable after single local dielectric breakdown. Thanks to electrodes' small thickness it is rapidly destructed around the breakdown point by electric explosion and micro-arc discharge [5].

This work was supported by the Russian Science Foundation under Project 19-79-10075.

SH process mainly consists of capacitors' self-discharge quickly extinction caused by micro-arc elongation in radial direction and cooling by adjacent film layers [6]. The voltage across the capacitor decreases during SH, the current through the micro-arc discharge firstly rises, has peak, and then drops [7]. The current and voltage product determines the power dissipated near the breakdown channel. Depending on SH energy level different scenarios of capacitor's failure are possible. Enhancing of the dissipated power can lead to relatively moderate undesirable sequences like accelerated capacitance decrease caused by more intensive SH acts, or to severe like the catastrophic failure inducted by adjacent polymer layers heating, their breakdown, and further, resulting in avalanche-like SH [8]. Thereby the key characteristics of SH process are SH duration and energy, which is the time integral of the product of SH current and voltage across the capacitor. Many intensive studies have been conducted on clarifying the effect of various factors on SH energy, such as applied voltage, capacitors' capacitance, interlayer pressure, metallization resistivity [9 – 11].

As it was established earlier by a number of investigators SH energy and breakdown strength tend to decrease with inter-layer pressure growth, with slightly changing after inter-layer pressure reaching a value of 1 MPa (i.e. 10 atm) [8,9]. First of these is presumably associated with air gap between adjacent films reducing leading to more intensive interaction between micro arc plasma and PP film, decreasing overall micro arc temperature and thus plasma conductivity, resulting in faster arc extinguishing. The temperature of micro arc plasma can reach up to 7000 K [12], that provides a significant local heating of polymer film which involves the possibility of its thermal decomposition and carbon deposition. Currently, the deposited near the breakdown area carbon is considered as the most likely reason of capacitors' insulation resistance decline during its exploitation [13,14]. Such viewpoint is not new and originates from early studies in the late 70s [15]. Nonetheless, despite the decades of vigorous studies this topic is still challenging and a matter of interest. The one of issues is to associate the SH energy with insulation resistance decrease considering influence of inter-layer pressure on SH characteristics. Previous studies used equivalent parallel resistance as indicative parameter of capacitors' ageing caused by SH acts [13,16]. However, it is frequency dependent and much sensitive to measuring frequency choice. In authors' opinion, the insulation resistance would be more appropriate to use.

The present paper reports results of experimental study of SH process characteristics on samples of PP film with overall zinc metallization, and the electrical characteristics of capacitor samples before and after SH. The objective of study is to clarify the effect of inter-layer pressure on SH characteristics and capacitors' insulation resistance.

II. EXPERIMENTAL

A. Experimental samples

The experimental samples were PP films from Birkelbach Kondensatortechnik GmbH with all-over zinc one side metallization. Thickness of the metallized films was $d_f = 6.8$ μm, sheet resistance was $R_s = 21.5$ Ω/sq, thickness of metallization was around $d_m = 10$ nm. The film thickness was additionally controlled by means of opto-mechanical thickness gauge with accuracy of 0.5 μm.

B. Experimental setup

Experimental setup for the study consists of laboratory hydraulic press, two cylindrical bases for uniform distribution of inter-layer pressure on the surface of experimental sample, a flat platform for installing a pair of experimental samples, two rubber gaskets to ensure a tight fit of the film to the hard part of the installation, a polypropylene or polyimide spacer to form a margin, an upper flat cover for additional pressing of samples, as well as two non-metallized PP films of 8 μm thick to prevent sticking of experimental samples to rubber gaskets. To supply high voltage and provide a reliable contact with thin metallization layer a spring-loaded foil brass polished electrodes with thickness of 8 μm were used. Rectangular samples of metallized film of fixed size were cut out to assemble a capacitor element, and a new spacer was cut out as the aging progress. The geometry of the capacitor element system can be seen in Fig.1.

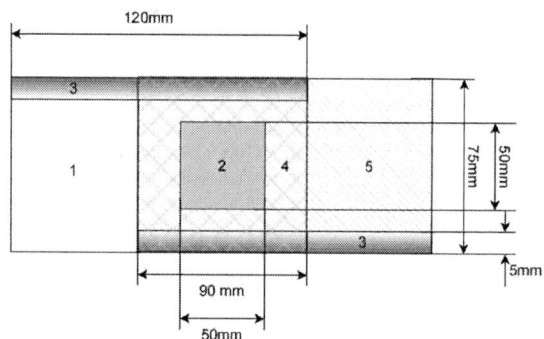

Fig. 1. Capacitor element

1 - the film under potential, 2 - active zone, 3 - margin, 4 - the spacer (located between two films), 5 - the grounded film

As it is seen in the Fig. 1 the area of active zone is $S_{active} = 25$ cm², maintaining the constancy of this parameter it is possible to achieve the same primary capacitance of the system. It depends on the inter-layer pressure and was changing from 7.73 nF to 14.33 nF, dependence will be presented further. The general scheme of the entire installation is shown in the Fig. 2.

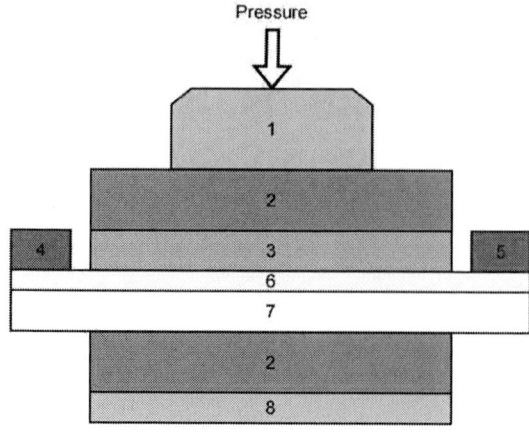

Fig. 2. Full scheme of experimental setup

1 – upper piston of the hydraulic press, 2 – cylindrical bases, 3 – upper flat cover, 4 – electrode under potential, 5 – grounded electrode, 6 – capacitor element, 7 – flat platform for installing samples and electrodes, 8 – lower piston of the hydraulic press

C. Primary characteristics measurement method

Measurement part of the experimental setup contains a combination of measurement electric scheme and equipment for getting primary parameters of SH and capacitor element like: current and voltage waveform, insulation resistance, capacitance and loss tangent. In this study were used the GwINSTEK GDS-72204 digital storage oscilloscope, ProfKiP E6-14M terraohmmeter and HIOKI IM3533-01 LCR-meter. The electrical scheme for measuring of current and voltage waveform can seen in the Fig. 3.

The voltage was supplied by high-voltage DC source with output voltage up to 20 kV. The voltage was raised manually, with rate about 200 V·s⁻¹. Capacitor element is depicted as a capacitance C_1. Resistors R_1=1 GΩ, $R_2 = 50$ MΩ, $R_3 = 1$ MΩ are a resistive divider for monitoring the voltage supplied from a high-voltage DC source. The choice of exactly 3 resistors is due to the insufficient insulation gap between terminals of resistors R_1 and R_2, in order to avoid measuring resistor flashover. After supplying a high voltage parallel capacitance $C_{charge} = 0.1$ μF started charging through resistor $R_{charge} = 10$ MΩ.

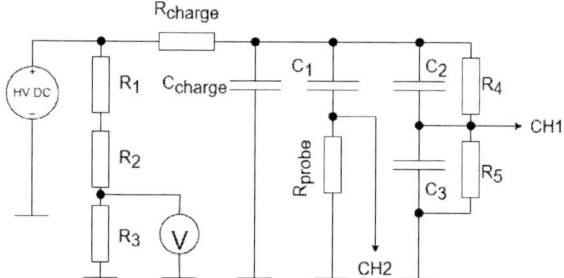

Fig. 3. Electrical scheme for measurement of SH voltage and current

When a breakdown of experimental sample occurs C_{charge} discharges on breakdown channel. The value of C_{charge} was chosen of one order higher than C_1 to provide sufficient amount of energy for SH micro arc development. SH current was measured using the current shunt $R_{probe} = 0.1$ Ω by oscilloscope via the CH2 channel (the oscilloscope is triggered via the CH2 channel by current rise). The voltage across C_1, hereinafter referred as SH voltage was measured by the capacitance-resistive divider with division ratio $k = 1000$, via the channel CH1. The capacitances and resistances values were chosen so that the product $R_4 C_2 = R_5 C_3$, also regarding oscilloscope input impedance.

D. Inter-layer air gap measurement

Inter-layer air gap size, along with polymer film thermophysical characteristics and chemical composition determines the conditions in which SH micro arc develops. In order to obtain the air gap size a preliminary experiment was carried out. The capacitance of capacitor element shown in the Fig. 2 was measured at 1 kHz frequency at different inter-layer pressures up to 25 atm. Then according the series equivalent scheme of air gap and polymer film connection the air gap size was calculated in accordance with Eq.1.

$$ d_{air} = \frac{\varepsilon_0 \cdot \varepsilon_f \cdot S_{act} - C_{meas} \cdot d_f}{C_{meas} \cdot \varepsilon_f}, \qquad (1) $$

where d_{air} is a size of air gap, ε_0 is electric constant, ε_f − relative permittivity of polymer film, S_{act} − active area, C_{meas} − measured capacity.

In the calculations, the minimum size of the air gap, based on the typical polymer film roughness [17] was assumed as $d_{min} = 0.1\mu m$. The dependencies of measured capacitance and calculated air gap on inter-layer pressure are shown in the Fig. 4. Additionally, it was established that after applying of $10 - 15$ atm of inter-layer pressure the air gap size took the constant value. Hence, in following studies the upper limit of applied inter-layer pressure was set as 10 atm.

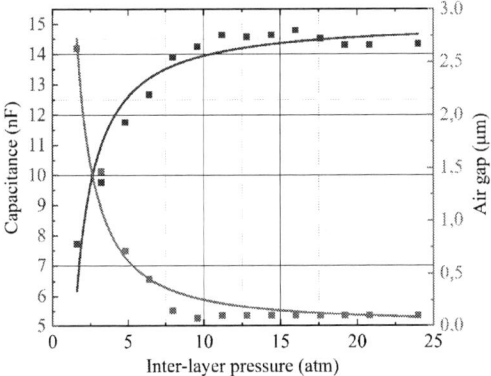

Fig. 4. . Dependencies of capacitance and calculated air gap size on inter-layer pressure of studied experimental samples

III. RESULTS

A. Electrode current density and resistance computation

The SH current amplitude is determined by time-dependent resistance of micro arc discharge and resistance of metallized electrodes. The last one cannot be easily measured. Therefore, to calculate it we developed computer model of the capacitor element. In Electric current module of software package COMSOL Multiphysics® we simulated the surface distribution of current density in the capacitor element electrodes and calculated surface resistance depending on the position of breakdown channel which resistance was neglected. The equation system (2) was solved

$$ \nabla J = 0, $$
$$ J = \sigma E, \qquad (2) $$
$$ E = -\nabla V, $$

where J is electric current density, σ is electrode conductivity, E is electric field, V is electric potential.

Results of current density modeling are shown in the Fig. 5. Potential electrode is located on top and current density lines converge to a breakdown point located in the center of active zone. As can be seen in the picture, the highest current density is in the vicinity of the breakdown channel. It is believed that due to high current density metallization is initially exploded and then evaporated by micro arc emerged in metal vapor.

The electrode resistance R_{el} was calculated for different locations of breakdown channel in the active area, then obtained values were divided on sheet resistance giving relative electrodes' resistance $R'_{el} = R_{el}/R_s$. The results are presented in the Fig. 6. The average electrodes' resistance, as one can see is $\approx 4 R_s$. This value is used by evaluation of correct micro arc time dependent resistance.

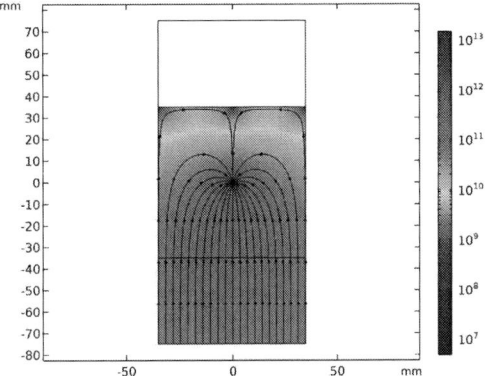

Fig. 5. Current density (A/m^2) surface distribution in the capacitor element potential electrode at typical breakdown voltage of 3700 V

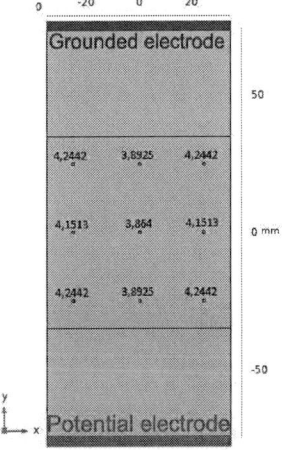

Fig. 6. Relative electrodes' resistance depending on the breakdown channel position

B. Self-healing characteristics as a function of inter-layer pressure

In total, dependencies of three main SH characteristics on different inter-layer pressure were obtain. The main characteristics of the SH process are breakdown strength, SH energy and duration. Breakdown strength was calculated based on breakdown voltage and film thickness. The study was conducted for 6 different inter-layer pressure ratings and 10 breakdowns were obtained for each inter-layer pressure. Typical waveforms of SH voltage (U_{SH}), current (I_{SH}) and calculated breakdown channel resistance (R_{SH}) are shown in the Fig. 7.

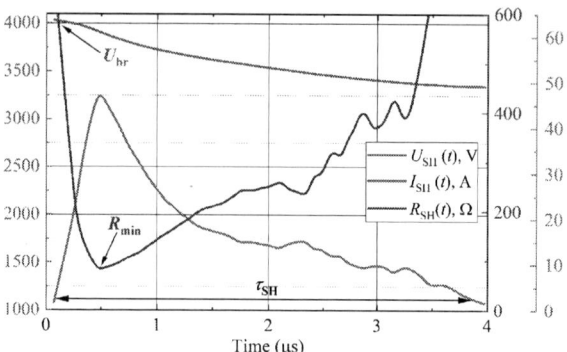

Fig. 7. Typical waveforms of SH voltage, current and resistance of micro arc discharge

The breakdown channel resistance was calculated as follows

$$R_{SH} = U_{SH} / I_{SH} - R_{el}. \quad (3)$$

The SH duration τ_{SH} is the time during which SH current drops to zero value (depicted in the Fig.7). The breakdown voltage U_{br} was determined from SH voltage waveforms as initial peak value of voltage on capacitor element. After breakdown SH voltage drops as capacitances C_1 and C_{charge} discharge. The breakdown strength was calculated from simply dividing $E_{br} = U_{br}/d_f$. As one can see in the Fig. 8 the breakdown strength of PP film has almost constant value at different inter-layer pressures, slightly deviating from average value about 550 kV/mm. Some larger deviations are likely caused by strong film tension, further thinning and local fracture on spacer edge by application of pressure. Generally, the obtained E_{br} values are in agreement with known data on breakdown strength of similar PP films [18].

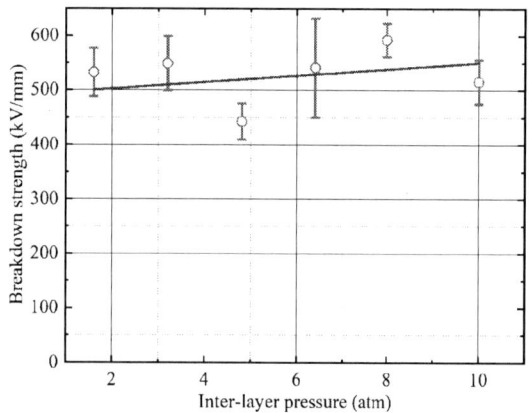

Fig. 8. Dielectric strength as a function of inter-layer pressure

SH energy was obtained by integrating of the product of SH voltage and current to an upper limit of τ_{SH} subtracting energy loss in electrodes as follows

$$W_{SH} = \int_0^{\tau_{SH}} U_{SH} I_{SH} dt - R_{el} \times \int_0^{\tau_{SH}} I_{SH}^2 dt. \quad (4)$$

The dependence of the self-healing energy on the inter-layer pressure is presented in the Fig. 9.

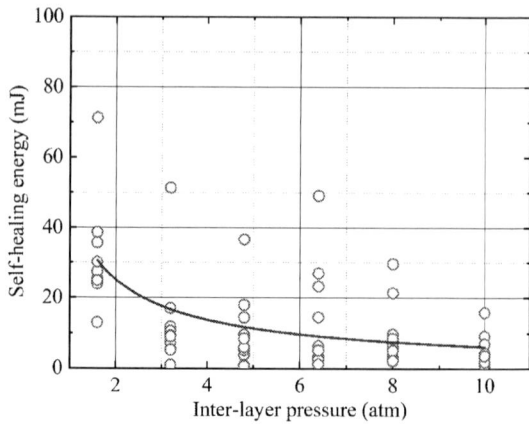

Fig. 9. Self-healing energy as a function of inter-layer pressure

The SH energy approximately follows a power law decreasing as inter-layer pressure grows. These experimental observations are in good agreement with previous data [10] and could be simply explained by SH duration decrease due to more intensive heat transfer from micro arc discharge to PP film. Nevertheless, the SH duration (see Fig. 10) surprisingly has no evident tendency to decrease. It has a significant dispersion, however an average value remains roughly constant, approximately $\tau_{SH} \approx 1.5$ μs. Presumably, the explanation is somehow associated with the fact that there is also a huge electrostatic pressure affecting film layer before and after breakdown which could be estimated as

$$P_{el-stat} \approx \varepsilon_0 \varepsilon_f E_{br}^2 / 2, \quad (5)$$

where $\varepsilon_f = 2.3$ for our case. Substituting experimental breakdown strength value gives $P_{el-stat} \sim 30$ atm, which rating considerably exceeds inter-layer pressure.

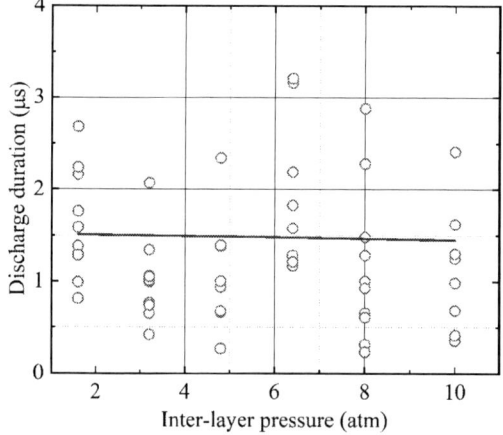

Fig. 10. Discharge duration as a function of inter-layer pressure

Electrostatic pressure is obviously absent in the demetallized zone because of vanished electrode, as the inter-layer pressure is applied to all zones of metallized film independent of electrode presence. The micro arc plasma after breakdown tends to expand, creating its own pressure about 100 – 300 atm [12]. Thus, the inter-layer pressure limits micro arc plasma expansion in the direction transverse the film layers, yet apparently doesn't affect the expansion alongside the layers. Probably, it depends on electrostatic pressure only, which is nearly stable by all interlayer pressures (see Fig. 8). Investigation of this issue is quite important for SH phenomena theory, however is out of a scope of present paper.

The minimal resistance R_{min} of micro arc discharge (see Fig.7) determined from experimental waveforms of SH current and voltage is shown in the Fig. 11. Minimal resistance corresponds to the current peak value and increases with inter-layer pressure. This dependence also could be explained taking into account theoretical explanations made above about micro arc plasma conductivity decrease with more intensive cooling caused by free volume decrease. Micro arc resistance dependencies on pressure are useful for simulation of temperature distribution of MPPFCs in various areas of wound capacitor section.

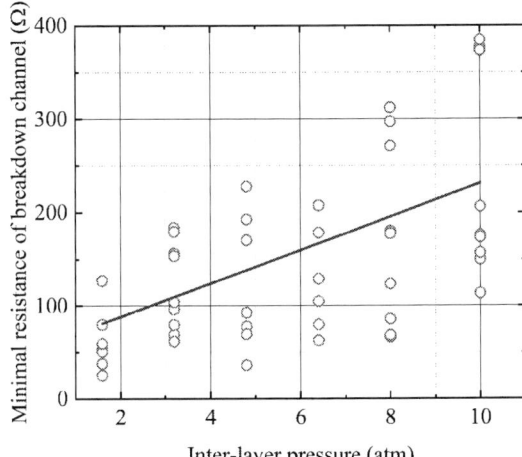

Fig. 11. Minimal resistance of micro arc as a function of inter-layer pressure

C. Capacitor element insulation resistance measurement

During the research the insulation resistance, capacitance and loss tangent measurements were carried out before and after 10 SH acts at the same inter-layer pressure. The most notable are results of insulation resistance measuring presented in the Table 1. The measuring voltage of teraohmmeter was 10 V for all samples. As seen from the table the insulation resistance dramatically decreases at $3-4$ orders of magnitude.

TABLE I.   INSULATION RESISTANCE MEASUREMENT RESULTS

Insulation resistance measurement results		
P, atm	R_{ins} before 10 SH acts, $G\Omega$	R_{ins} after 10 SH acts, $G\Omega$
3,2	100	0,058
4,8	17	0,0032
6,4	1000	0,034
8	100	0,013

We attribute this tendency to the carbon deposition in the demetallized zone on both film sides. In order to check this assumption a microphotograph of SH areas was taken (see Fig. 12).

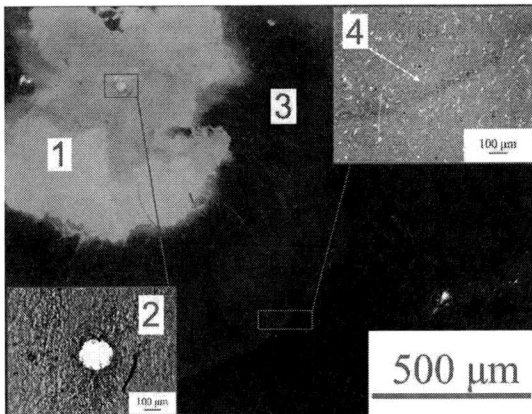

Fig. 12. Microphotograph of SH area. 1 – demetallized zone, 2 – breakdown pinhole, 3 – zinc metallization, 4 – deposited carbon trace

On the microphotograph are visible demetallized zones, breakdown pinhole and intact metallization. It is also seen that PP film is very rough there, having hollows likely caused by polymer melting by micro arcs. The darkening of demetallized zone is assumed to be a graphite layer which provides low resistance of capacitor element after SH acts. It should be noted that much less amount of SH energy was needed to form a conductive layer, than it was observed earlier – tenths of joules in our case contrary units of joules in [14]. Unfortunately, due to small sample amount authors did not managed to find a relation between SH energy and insulation resistance of capacitor element. The relation between inter-layer pressure and insulation resistance, or at least probability of conductive layer occurrence was not also found out. Remarkable are circular traces, which being inspected closer are clusters of black particles of units and tens micron size. They are supposedly graphite particles, thrown out from breakdown channel by expanding micro arc plasma. To credible establish the conductive carbon presence in the SH areas by different interlayer pressures the detailed chemical analysis of demetallized zone surface is demanded.

IV. CONCLUSION

The characteristics of SH depending on the inter-layer pressure were studied at zinc metallized PP films. A significant decrease in insulation resistance of capacitor elements after tens of SH acts independent on applied inter-layer pressure were found out. It was established that relatively small amounts of SH energy ~ 0.1 J are sufficient to considerable reduce the insulation resistance. The obtained results should be taken into account by analysis of capacitors' performance after SH and by choosing winding tension.

REFERENCES

[1] R. Ramos, "Film Capacitors in Power Applications: Choices and Particular Characteristics Needed," *IEEE Power Electron. Mag.*, vol. 5, no. 1, pp. 45–50, 2018, doi: 10.1109/MPEL.2017.2782401.

[2] H. Wang and F. Blaabjerg, "Reliability of capacitors for DC-link applications in power electronic converters - An overview," *IEEE Trans. Ind. Appl.*, vol. 50, no. 5, pp. 3569–3578, Sep. 2014, doi: 10.1109/TIA.2014.2308357

[3] V. O. Belko, O. A. Emelyanov, I. O. Ivanov, A. P. Plotnikov, and E. G. Feklistov, "Application of Numerical Simulation for Metallized Film Capacitors Electrodes Design," IEEE Access, vol. 9, pp. 80945–80952, 2021, doi: 10.1109/ACCESS.2021.3085695.

[4] V. O. Belko and O. A. Emelyanov, "Self-healing in segmented metallized film capacitors: Experimental and theoretical investigations for engineering design," J. Appl. Phys., vol. 119, no. 2, Jan. 2016, doi: 10.1063/1.4939954.

[5] O. A. Emelyanov and V. O. Belko, "Pattern formation in electrical exploding of thin metal films," IEEE Trans. Plasma Sci., vol. 41, no. 4, pp. 961–966, 2013, doi: 10.1109/TPS.2013.2251008.

[6] S. Qin, S. Ma and S. A. Boggs, "The mechanism of clearing in metalized film capacitors," 2012 IEEE International Symposium on Electrical Insulation, 2012, pp. 592-595, doi: 10.1109/ELINSL.2012.6251539.

[7] V. O. Bel'ko, P. N. Bondarenko, and O. A. Emel'yanov, "The dynamic characteristics of self-healing processes in metal film capacitors," Russ. Electr. Eng., vol. 78, no. 3, pp. 138–142, 2007, doi: 10.3103/S106837120703008X.

[8] V. Belko, O. Emelyanov, and I. Ivanov, "Critical Parameters of Metallized Film Capacitor's Failure," in 2018 IEEE 2nd International Conference on Dielectrics (ICD), Jul. 2018, vol. 2018-Janua, pp. 1–3. doi: 10.1109/ICD.2018.8514586.

[9] I. Rytöluoto and K. Lahti, "Effect of inter-layer pressure on dielectric breakdown characteristics of metallized polymer films for capacitor applications," Proc. IEEE Int. Conf. Solid Dielectr. ICSD, pp. 682–687, 2013, doi: 10.1109/ICSD.2013.6619787.

[10] Y. Chen et al., "Study on self-healing and lifetime characteristics of metallized-film capacitor under high electric field," IEEE Trans. Plasma Sci., vol. 40, no. 8, pp. 2014–2019, 2012, doi: 10.1109/TPS.2012.2200699.

[11] J. H. Tortai, A. Denat, and N. Bonifaci, "Self-healing of capacitors with metallized film technology: experimental observations and theoretical model," J. Electrostat., vol. 53, no. 2, pp. 159–169, Aug. 2001, doi: 10.1016/S0304-3886(01)00138-3.

[12] J. H. Tortai, N. Bonifaci, A. Denat, and C. Trassy, "Diagnostic of the self-healing of metallized polypropylene film by modeling of the broadening emission lines of aluminum emitted by plasma discharge,"

J. Appl. Phys., vol. 97, no. 5, p. 053304, Feb. 2005, doi: 10.1063/1.1858872.

[13] V. O. Belko, O. A. Emelyanov, I. O. Ivanov, and A. P. Plotnikov, "Self-Healing Processes of Metallized Film Capacitors in Overload Modes—Part 1: Experimental Observations," IEEE Trans. Plasma Sci., vol. 49, no. 5, pp. 1580–1587, May 2021, doi: 10.1109/TPS.2021.3071187.

[14] V. O. Belko and O. A. Emelyanov, "Self-Healing Processes of Metallized Film Capacitors in Overload Modes—Part II: Theoretical and Computer Modeling," IEEE Trans. Plasma Sci., vol. 49, no. 6, pp. 1898–1905, Jun. 2021, doi: 10.1109/TPS.2021.3076007.

[15] H. Heywang, "Physikalische und chemische Vorgänge in selbstheilenden Kunststoff-Kondensatoren," Colloid Polym. Sci. 1976 2542, vol. 254, no. 2, pp. 139–147, Feb. 1976, doi: 10.1007/BF01517025.

[16] V. Belko, O. Emelyanov, I. Ivanov, and V. Starovoytenkov, "The Diagnostics of Metallized Film Capacitors under Soft Training Test," in 2020 International Conference on Diagnostics in Electrical Engineering (Diagnostika), Sep. 2020, pp. 1–4. doi: 10.1109/Diagnostika49114.2020.9214585.

[17] I. Rytöluoto, A. Gitsas, S. Pasanen, and K. Lahti, "Effect of film structure and morphology on the dielectric breakdown characteristics of cast and biaxially oriented polypropylene films," Eur. Polym. J., vol. 95, pp. 606–624, Oct. 2017, doi: 10.1016/J.EURPOLYMJ.2017.08.051.

[18] S. J. Laihonen, U. Gäfvert, T. Schütte, and U. W. Gedde, "DC breakdown strength of polypropylene films: Area dependence and statistical behavior," IEEE Trans. Dielectr. Electr. Insul., vol. 14, no. 2, pp. 275–286, Apr. 2007, doi: 10.1109/TDEI.2007.344604.

979-8-3503-8358-4/23 $31.00 © 2023 IEEE

Study of the Sensitive Element of a Resonant Pressure Sensor with Membranes of Various Shapes

Phyo Win Tun
Department of Nano and Micro System Technology
National Research University of Electronic Technology, "MIET"
Moscow, Russia
kophyowinhtun0@gmail.com

Sergey P. Timoshenkov
Department of Nano and Micro System Technology
National Research University of Electronic Technology, "MIET"
Moscow, Russia
spt@mail.ru

Boris M. Simonov
Department of Nano and Micro System Technology
National Research University of Electronic Technology, "MIET"
Moscow, Russia
serbosel@mail.ru

Abstract— **In this paper, the various shapes of membrane of resonant pressure sensor with a pressure range of 100-1000 kPa, using the growing requirements of industrial gas pressure calibration equipment. The presented resonance pressure sensor design consists of an insulated silicon layer (SOI) and a silicon to glass (SOG) layer. The design of the sensing element (SE) was created in the SolidWorks and modeling was carries out in the ANSYS program. The deformations of SE diaphragms with various shapes under applied pressure range of 100-1000 kPa, mechanical stresses arising in SE structures with diaphragms of various shapes under applied pressure, and natural frequencies were calculated. It has been established that the natural oscillation frequency shifts depending on the applied pressure. The analysis showed that the studied SE designs have a high sensitivity and an optimized shape of the membranes and the parameters of the pressure sensors.**

Keywords— *resonators, pressure sensor, diaphragm of the sensitive element.*

I. Introduction

Resonant pressure sensors are widely used in weather, spacecraft, aircraft, etc. because of their high precision, stable long-term parameters and high sensitivity compared to other pressure sensors. They are also distinguished by high linearity and resolution characteristics [1-3]. In comparison to other types of pressure sensors, such as capacitive pressure sensors, piezo-electric pressure sensors and piezoresistive pressure sensors, resonance pressure sensors are characterized by high precision, high resolution, quasi-digital outputs and long-term stability. Resonant pressure sensors operate with intrinsic frequency shift from resonators, due to changes in axial stresses. The high long-term stability of such sensors is due to the fact that their resonant oscillation frequency practically does not depend on unstable or drifting electric signals, but which are rather determined by the mechanical characteristics of the structure [4-5].

In this work, the design of the resonant pressure sensor and the designs of membranes of various shapes of this sensor with a pressure measurement range of 100-1000 kPa are studied. And then maximum deformations and maximum stresses of different shape of membranes were calculated under applied pressure and obtained frequency responses of the resonant pressure transducer using the ANSYS program.

II. The Design of the Resonant Pressure Sensor and the Principle of Its Operation

The resonance pressure sensor design consists of an insulated silicon layer (SOI) as sensitive elements and a layer of silicon on glass (SOG) with a cavity for vacuum packing (Fig. 1a). In the layer of SOI, two H-form resonators were mounted on a pressure-sensitive diaphragm. The resonator I is located closer to the middle area of the diaphragm, and the resonator II is closer to the border of the diaphragm, respectively. In addition, the electrical connections are formed by a plating layer using eight through silicon vias. Figure 1 shows its 3D image, create in SolidWorks. When the pressure is applied on a pressure sensitive diaphragm, it deforms and modulates the axial stress of the H-form resonators. The measured pressure is related to an increase in the resonant frequency of the central resonator and a decrease in that of the lateral resonator, which leads to the appearance of a differential frequency output signal.

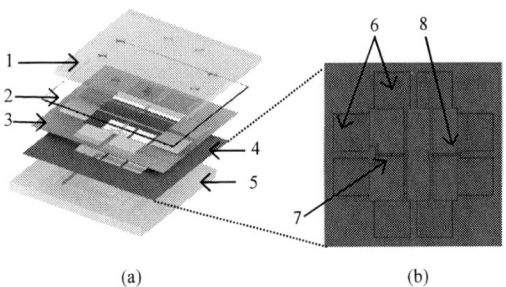

(a)   (b)

Fig. 1. Design of a resonant pressure sensor (a), the device layer topology (b): 1 - silicon layer (Si), 2 - glass layer, 3 - device layer (Si), 4 - oxide layer (SiO₂) , 5 - diaphragm (Si), 6 - metallization (contacts), 7 - central resonator (resonator I), 8 - side (edge) resonator (resonator II)

The parameters of the resonant pressure sensor: dimensions $10 \times 10 \times 1$ mm³, is composed by a glass cap with eight vias (diameter 600 microns) with a cavity (5 millimeters × 5 millimeters × 120 microns) and a SOI plate with sensitive elements. The sensitive element consists of a pressure sensitive diaphragm with dimensions of 5 millimeters × 5 millimeters × 300 microns of the SOI structure and the two H-form resonators (two single conductors (beams) 1400 microns × 18 microns × 40 microns with a (60 microns × 18 microns connection × 40 microns) in the Si layer of the device in the SOI wafer. Double-clamped H-form resonators deployed in the central regions (resonator I) and the border (resonator II) of the diaphragm, respectively, are connected to the diaphragm through anchor structures within the oxide layer of the SOI plate.

III. Study of Pressure Sensitive Membrane Design and Finite Element Modeling (FEM)

Finite element modeling based on the ANSYS program was used to design and optimize the balanced pressure in the linear elasticity range of a resonant micro pressure sensor. Structural and modal static analysis is used to compute natural frequency variations in response to pressure applied to the transducer. In the simulation, the mesh size of geometric structures of 100 μm was used and the meshing method of tetrahedral elements was used to construct a grid of resonators of the resonant pressure sensor. After mesh was created, the total number of elements of the modeled geometric structures was approximately 300,000 elements (Fig 2 (b)). Materials used for FEM simulations are presented in Table 1.

TABLE I. MATERIAL PROPERTIES OF PROPOSED DESIGN IN FEM SIMULATION.

Parameters	Materials		
	Silicon (Si)	Glass	Silicon dioxide (SiO₂)
Material's density (kg/m³)	2330	2500	2200
Young's Modulus (GPa)	169	75	73
Possion's Ratio	0.28	0.23	0.17

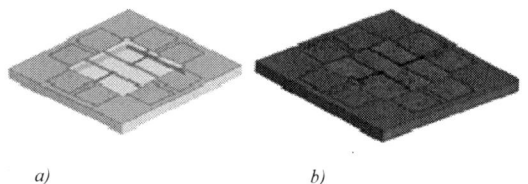

a)   b)

Fig. 2. Resonant pressure sensor structure in SolidWorks, (a) 3D model, (b) tetrahedral method of the proposed model

The boundary conditions at the simulation of the pressure sensor were defined in such a way that the edges of the pressure sensitive diaphragm were fixed to avoid unintended movements. In the static analysis of the structure, the pressure of 100 kPa to 1000 kPa has been used as the load. Deformations and mechanical stresses in pressure-sensitive diaphragms of square and round shape were calculated (Table 2, Fig. 3).

TABLE II. RESULTS OF SIMULATION OF MECHANICAL STRESS IN PRESSURE SENSOR DIAPHRAGMS

Pressure (kPa)	Diaphragm design			
	Square diaphragm		Circular diaphragm	
	Deformation (μm)	Mechanical Stress (MPa)	Deformation (μm)	Mechanical Stress (MPa)
100	0.21855	22.356	0.18951	12.178
200	0.43597	44.734	0.37901	24.354
300	0.65412	67.108	0.56852	36.53
400	0.87195	89.486	0.75802	48.826
500	1.09112	111.78	0.94753	60.91
600	1.3081	134.16	1.137	73.15
700	1.52614	156.52	1.3265	85.232
800	1.74389	178.92	1.516	97.423
900	1.96215	201.35	1.7055	109.62
1000	2.1765	223.72	1.8958	121.18

From Table 1 it can be seen that at an operating pressure in the range from 100 to 1000 kPa on the pressure sensing

diaphragm, the mechanical stress in the circular diaphragm was lower than in the square diaphragm.

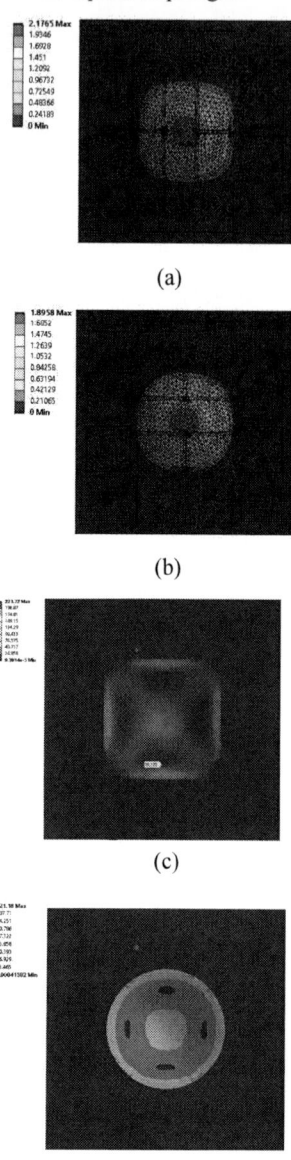

(a)

(b)

(c)

(d)

Fig. 3. Modeling results: deformations in square (a) and circular (b) diaphragms, mechanical stresses in square (c) and circular (d) diaphragms under applied pressure of 1000 kPa.

From the obtained results, we can conclude that in a circular diaphragm, tensile mechanical stress occurs in the central resonator (resonator I), and compressive stress occurs in the side resonator (resonator II) when the diaphragm is under the applied pressure. Then we used these stress distributions in the structures as loads for the modal part in the ANSYS program, and the simulation results were the natural frequency shifts of the resonators responding to the applied pressure (Fig. 4, 5). Figure 4 (a) shows the tensile stress and compressive stress in the resonators and 4 (b) the results of modeling the natural frequencies of the resonators in the first mode (operating mode).

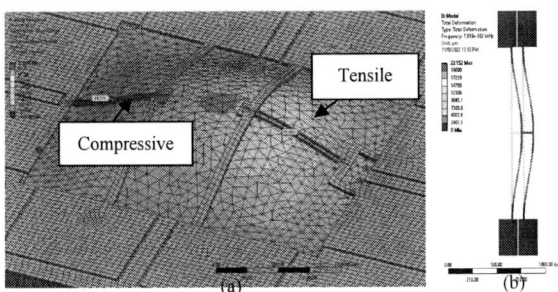

Fig. 4. Simulation results of compressive and tensile stresses (a) appearing under the action of pressure and natural frequencies of oscillations of resonators (b) in the first (working) mode

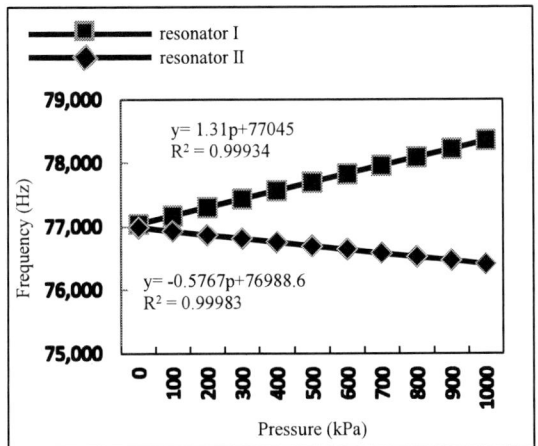

Fig. 5. The relationship between resonant frequencies of the resonator beams and applied pressure in a range of 100 to 1000 kPa.

In figure 5 shows the relationship between the resonant frequency and the applied pressure. The pressure sensitivity of the central resonator is 1.31 Hz/kPa and the linear coefficients are 0.99934, and the pressure sensitivity of the side resonator is 0.5767 Hz/kPa and the linear coefficients are 0.99983 in the pressure range from 100 kPa to 1000 kPa.

IV. CONCLUSIONS

A design has been developed for a resonant pressure sensor with various forms of a pressure-sensitive diaphragm for the pressure measurement range from 100 to 1000 kPa. using ANSYS simulations, deformations and mechanical stresses of circular and rectangular diaphragms were calculated. The results of calculations of deformations, mechanical stresses, pressure sensitivity and linear coefficient of two resonators located on the diaphragm of a resonant pressure sensor, with circular and square shape diaphragms in the pressure measurement range of 100 - 1000 kPa are presented.

REFERENCES

[1] Eaton, W.P., Smith, J.H.// Micromachined pressure sensors: Review and recent developments// 1997, P- 530 - 539.

[2] Beeby S.P., Ensel G., Kraft M.//MEMS Mechanical Sensors// London, UK, 2008, P- 97–112.

[3] L. Zhu, Y.H. Xing, B. Xie, C. Xiang, Y.L. Lu, D.Y. Chen, J.B. Wang, and J. Chen// A High- Quality Resonant Pressure Micro Sensor with Through-silicon-via Electrical Interconnections// Transducers 2017. Kaohsiung, Taiwan- P - 82-85.

[4] Chao Xiang, Yulan Lu, Pengcheng Yan, Jian Chen, Junbo Wang, Deyong Chen// A Resonant Pressure Microsensor with Temperature Compensation Method Based on Differential Outputs and a Temperature Sensor// Micromachines 2020. P- 1-14.

[5] Yulan Lu, Bo Xie, Qiuxu Wei, Yadong Li, Xiaoqing Shi, Chao Xiang, Deyong Chen// A Resonant Pressure Microsensor with a Stress Isolation Layer// IEEE SENSORS JOURNAL, VOL. 19, NO. 18, 2019, P – 7875 -7883.

Investigation of the Aging of Electrical Insulating Paper in Polydimethylsiloxane Liquid

Alexandr S. Reznik
Higher School of High Voltage Engineering
Peter the Great St. Petersburg Polytechnic University
Saint Petersburg, Russia
alexxxandr2803@mail.ru

Natalia M. Zhuravleva
Higher School of High Voltage Engineering
Peter the Great St. Petersburg Polytechnic University
Saint Petersburg, Russia
natalia_zhurav@mail.ru

Dmitry V. Kiesewetter
Higher School of High Voltage Engineering
Peter the Great St. Petersburg Polytechnic University
Saint Petersburg, Russia
dmitrykiesewetter@gmail.com

Pavel A. Poyarkov
Higher School of High Voltage Engineering
Peter the Great St. Petersburg Polytechnic University
Saint Petersburg, Russia
Pavel.poyarkov.99@gmail.com

Ekaterina G. Smirnova
Department of Paper and Cardboard Technology
St. Petersburg University of Technology and Design
Saint Petersburg, Russia
smirnovalta@gmail.com

Abstract—Polydimethylsiloxane liquids used in electrical engineering are a promising dielectric liquid material, however, the aging of electrical insulating paper, including modified with biopolymers, in such dielectrics has not been studied in sufficient detail. The work presents the results of the study of the mechanical and optical properties of various types of papers during the process of artificial thermal aging. It is shown that the deterioration of mechanical parameters of the paper in polydimethylsiloxane liquids approximately corresponds to the similar indicator of paper in petroleum oils. It was revealed that the composite cellulose with the biopolymer material is characterized by increased resistance to thermal aging compared with widely used standard paper and a mutual influence of paper with the organosilicon impregnating medium does not lead to an additional decrease in the light transmission of the liquid caused by the destruction of the cellulose material compared to petroleum oil.

Keywords—electrical insulating paper, artificial aging, polydimethylsiloxane, bacterial cellulose, mechanical strength

I. INTRODUCTION

Paper-impregnated insulation (PII) is widely used in medium and high power transformers (PT). The reliability of transformers and, consequently, of the entire power system as a whole depends on the parameters of electrical insulation (EI). Therefore, improving an insulation properties of PT – electrical insulating paper (EIP) and liquid dielectric, is still an urgent technical task. In particular, one of the promising areas of development is the use of paper with the cellulose macromolecules of which linked with molecular of a biopolymer [1-4], as well as the use of synthetic esters and transformer silicone liquids (TSL) in power transformers, the most common of which is polydimethylsiloxane (PDMS) liquid dielectric. Despite the higher cost of such liquids, compared with petroleum oils, their use may be appropriate, as it reduces the maintenance costs of power transformers. It is known that petroleum oils and silicone liquids are compatible substances [5] ("questionable compatibility"). This makes it possible to replace the transformer's petroleum oil in the operated transformer with a synthetic liquid, in particular, with TSL in the form of a polydimethylsiloxane liquid dielectric. However, the presence of a certain amount of petroleum oil in TSL, as well as a certain amount of PDMS in paper-impregnated insulation, affects the electrical properties of insulation [6], which must be taken into account when servicing transformers.

An interest in the use of power electric cables with PII is returning with the improvement of materials and technologies. Taking into account the accumulated experience of successful operation of cables with PII of the first generation (in particular, cables with PII at the nominal voltage of 6 kV have been operated in St. Petersburg for more than hundred years [7]). It can be assumed that modern cables with PII will have better performance characteristics.

Many scientific articles are devoted to investigation of an aging of paper-impregnated insulation. For example, in [8-11] the artificial aging of standard insulating paper in petroleum oil is considered, in [12-15] – in synthetic and natural ether.

The aging of EI paper in polydimethylsiloxane has not been examined in sufficient detail, but such data are necessary for the design and successful operation of power transformers. This research is devoted to the investigation of the aging of the components of PII in polydimethylsiloxane liquids.

II. MATERIALS AND METHODS

A. Materials

The investigation was fulfilled for the samples of EI paper of industrial production from pine pulp (reference sample No. 1, three-layer, calandered), laboratory castings of electrical insulating cellulose (EIC) pine (sample No. 2), as well as from a composite of plant (PC) and bacterial cellulose (BC) (samples No. 3 and 4). Modification of EIC from pine pulp of traditional sulfate boiling was carried out using bacterial cellulose (BC CKR) synthesized by the culture of *Komagataeibacter rhaeticus* bacteria. Grinding of the nano-gel film of bacterial cellulose was fulfilled in two ways: using of the roll of Valley (such BC was introduced into the composite sample No. 3) and using the industrial disintegrator (such BC was introduced into the composite sample No. 4). As a result of this modification of the EIC, a composite with the following composition was obtained from plant cellulose: 95% pine EIC + 5% BC CKR and presumably an isotropic structure.

979-8-3503-8358-4/23 $31.00 © 2023 IEEE

Petroleum oil "GK" brand and transformer silicone liquid "Sofexil" (analog PDMS-50) were used as a dielectric liquid, in the medium of which thermal aging of EIP samples was fulfilled and the light transmission parameters of which were measured in the aging process.

B. Measurement technique

Artificial aging of EIP samples was performed using a technology close to the standard. The materials under study, with the exception of sample No 2, were cut into strips measuring 15 x 60 mm and placed in containers with two types of liquid dielectric: widely used petroleum oil "GK" and organosilicon liquid "Sofexil". After that, the containers with samples and copper wire (the catalyst for the destruction process) were placed in a thermostat and subjected to thermal aging at the temperature of 140 °C for 350 hours. The sampling of the material (5 fragments per point) was carried out periodically for measurements. The impregnating liquid was removed from the surface of the strips with a filter material, after which the mechanical tensile strength was determined.

The measurement of the mechanical tensile strength of the EIP, as well as the deformation curves of materials, was carried out using the ES M301/ESM301L "Mark-10" test bench.

Photographs demonstrating the morphological features of the samples were reached using electronic scanning microscope "Supra 55VP-25-78".

The relative light transmission coefficient (K_{oc458} (%)) to assess the degree of destruction of the liquid, the aging of which was fulfilled jointly with the studied EIP variants, was measured at wavelength of 458 nm using the MKMF-1 micro colorimeter.

III. DATA OBTAINED

The main parameters of the EI paper samples measured in the original state (before thermal aging, after preparation under standard conditions) are shown in the Tab. 1. In the Tab. 1: h is the thickness of the sample, ρ is the specific density, σ_0 is the mechanical tensile strength, E_b is the breakdown strength of the electric field.

An understanding of the morphological features of the composite material made of PC and BC gives the image shown in Fig. 1. In particular, Fig. 1 shows plant cellulose fibers linked by cellulose formations of bacterial cellulose and also the difference in the thickness of the fibers is clearly visible.

The obtained dependences of the mechanical tensile strength on the duration of thermal aging are shown in Fig. 2. Laboratory castings (No 2 – No 4) have density almost two times less than the sample industrially manufactured (No 1), which is due to their single-layer nature and lack of sealing (calandering).

TABLE I.   BASIC PARAMETERS OF EIP

Type of EIP	100% EIC			
	h (mm)	ρ (g/cm³)	σ_0 (MPa)	E_b (kV/mm)
100% EIC, industrially manufactured EIP, three-layer, calandered (reference sample, No. 1)	0.177	0.780	76.8	8.1
100% EIC, laboratory casting EIP (sample, No. 2)	0.215	0.423	15.7	6.1

Type of EIP	100% EIC			
	h (mm)	ρ (g/cm³)	σ_0 (MPa)	E_b (kV/mm)
	single-layer composite			
EIP laboratory castings, BC grinding using roll of Valley, 95% EIC + 5% BC (sample No 3)	0.201	0.453	33.5	8.7
EIP laboratory castings, BC grinding with the disintegrator, 95% EIC + 5% BC (sample No 4)	0.180	0.542	39.3	8.9

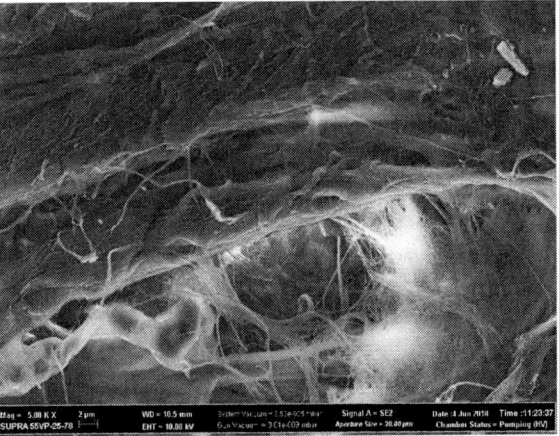

Fig. 1.   SEM-image of the composite material surface.

This reason also explains the fact that the mechanical tensile strength of samples No 2 - No 4 is significantly lower than that of industrially manufactured paper.

The relative light transmission coefficient of the dielectric liquid after artificial aging together with EIP and

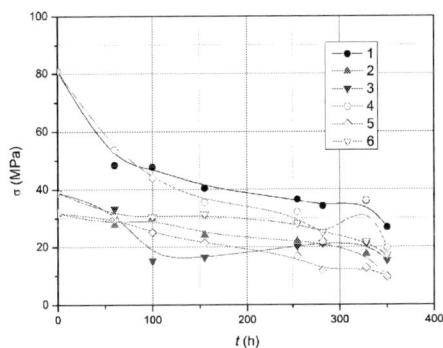

Fig. 2.   Dependence of the mechanical tensile strength on the duration of thermal aging: 1 – reference sample, 2-4 – laboratory castings (Table 1), 1-3 – aging in the petroleum oil, 4-6 – in "Sofexil" (PDMS-50).

TABLE II.   MAXIMUM ELONGATION OF THE SAMPLE (MM)

Type of EIP	Type of liquid dielectric	
	Petroleum oil "GK"	TSL "Sofexil"
Standard EIP (Reference sample, No 1)	0.70	0.50
Composite material, EIC was milled in the roll of Valley (No 3)	0.45	0.39

Type of EIP	Type of liquid dielectric	
	Petroleum oil "GK"	TSL "Sofexil"
Composite material, EIC was milled using the disintegrator (No 4)	0.55	0.45

the catalyst for 350 hours for "GK" oil decreased from 100% to 7%-17% and for TSL it practically did not change (100%).

The deformation curves of the materials were measured both in the initial state and in the aging process. In particular, the maximum elongation of the samples is determined. Table 2 shows the average elongation values of the samples after 350 hours of thermal aging. It was found that the electrical insulating paper impregnated with polydimethylsiloxane dielectric liquid has a maximum elongation of the samples in the case of application of a tensile load somewhat less than when impregnated with petroleum oil.

IV. CONCLUSION

It was found that during the tests in the oil medium, the mechanical strength of the industrial EIP (sample No 1) decreased by 67% from the initial value, and for composites No 3 and No 4 by 70% and 60%, respectively. The decrease in the tensile strength of these samples for the same duration of aging in the TSL "Sofexil" was: 76%; 70% and 56%, respectively.

It is obvious that the influence of the considered impregnating media on the mechanical strength of the experimental EIP in the process of thermal aging is similar. However, it was revealed that the development of destructive processes in composite No 4 is slowed down in comparison with composite No 3 and even with paper of traditional manufacture (reference sample No 1): the decrease in mechanical strength of the specified type of modified EIP was 60% in the petroleum oil medium (in sample No 1 – 67%), and in TSL "Sofexil" medium – 76% and 56%, respectively. The resulting effect is due to the structuring of the EIP web with bacterial cellulose. There is a crosslinking of plant fibers with the biopolymer, which is confirmed by the parameters of the deformation curves.

REFERENCES

[1] Zhuravleva N.M., Reznik A.S., Kiesewetter D.V., Stolpner A.M., Smirnova E.G., Khripunov A.K. Improving the efficiency of power transformers insulation by modifying the dielectric paper with bacterial cellulose. IOP Journal of Physics: Conf. Series, 2019, V. 1236 (1), 012002, pp. 1-4.

[2] Zhuravleva N.M., Reznik A.S., Kiesewetter D.V., Stolpner A.M., Smirnova E.G. Study of the Electrophysical Properties of the Composite of Plant and Bacterial Cellulose. Proceedings of the 2019 IEEE conference of Russian young researches in electrical and electronics engineering, St. Petersburg, 2019, p. 838-842.

[3] Zhuravleva N., Reznik A., Kiesewetter D., Stolpner A., Smirnova E., Budaeva V. Improvement of properties of cellulose dielectrics by their structure modification with nanocellulose produced of wastes of agricultural crops. IOP Journal of Physics: Conf. Series, 2019, V. 1410 (1), 012068, pp. 1-4.

[4] Zhuravleva N., Reznik A., Kiesewetter D., Stolpner A., Khripunov A. Possible applications of bacterial cellulose in the manufacture of electrical insulating paper. IOP Journal of Physics: Conf. Series, 2018, V. 1124, 031008, pp. 1-4.

[5] Fernandez I., Ortiz A., Delgado F., Renedo C., Perez S. Comparative evaluation of alternative fluids for power transformers. Review. Electric Power Systems Research, 2013, V. 98, pp. 58–69.

[6] Kiesewetter D.V., Reznik A.S., Zhuravleva N.M., Litvinov D.V. Research of dielectric properties of the mixtures of petroleum transformer oils and silicone liquids. Proceedings of the 2021 IEEE Conference of Russian Young Researchers in Electrical and Electronic Engineering, St. Petersburg, 2021, pp. 1224-1227.

[7] Barinov V.M., Kiesewetter D.V. Experience in Enhancing the Reliability of Operation of Power Cable Lines in St. Petersburg. Proceedings of the 10th Electric Power Quality and Supply Reliability Conference, Tallinn, 2016, pp. 55-58.

[8] Zhuravleva N., Reznik A., Kiesewetter D., Tukacheva A., Smirnova E. On the increasing of the sorption capacity and temperature resistance of cellulosic insulation dielectrics. Proceedings of the 11th International Conference "ELEKTRO 2016", High Tatras, 2016, pp. 649–653.

[9] Zhuravleva N., Reznik A., Tukacheva A., Kiesewetter D., Smirnova E. The study of thermal aging components paper-impregnated insulation of power transformers. Proceedings of the 2016 IEEE North West Russia Section Young Researchers in Electrical and Electronic Engineering Conference, St. Petersburg, 2016, pp. 747–751.

[10] Sari M.I, Suwarno I., Kinkeldey T., Werle P. Effect of Thermal Aging on the mechanical Characteristic of Insulating Paper Impregnated with Different Insulating Oils. Proceedings of the conference Condition Monitoring and Diagnosis, 2018, pp. 1-4.

[11] Fu Q., Wang M., Chen T., Li X., Li J., Zhang J. The relationship between carbon oxides in oil and thermal aging degree of oil-paper insulation. Proceedings of the 2016 IEEE International Conference on High Voltage Engineering and Application (ICHVE), 2016, pp. 1-4.

[12] Madkour W. A., Mansour D. -E. A., Abosheiasha H. F. Thermal Aging Influence on Relaxation Time of Transformer Insulation Paper Impregnated in Natural and Synthetic Ester Oils. Proceedings of the 22nd International Middle East Power Systems Conference (MEPCON), 2021, pp. 407-411.

[13] Fernandez O. H. A., Fofana I., Jalbert J., Gagnon S., Rodriguez-Celis E., Duchesne S., Ryadi M. Aging characterization of electrical insulation papers impregnated with synthetic ester and mineral oil: Correlations between mechanical properties, depolymerization and some chemical markers. IEEE Transactions on Dielectrics and Electrical Insulation, 2018, V. 25 (1), pp. 217-227.

[14] Meira M., Alvarez R. E., Verucchi C. J., Catalano L. J., Ruschetti C. R. Thermal aging analysis of mineral oil and natural ester immersed windings. Proceedings of the 2020 IEEE Electrical Insulation Conference (EIC), 2020, pp. 132-135.

[15] Cai S., Chen C., Li H., Shao M., Chen J., Yin J., Cai J. Research on electrical properties of natural ester-paper insulation after accelerated thermal aging. Proceedings of the 1st International Conference on Electrical Materials and Power Equipment (ICEMPE), 2017, pp. 397-401.

The Research of GaN HEMT Transistor Input Gate Capacitance Dependence as a Function of Operating Mode in GHz Band

Denis V. Rodionov
National Research University of Electronic Technology
Moscow, Zelenograd, Russia
denis.rodionov@gmail.com

Alexander I. Khlybov
National Research University of Electronic Technology
Moscow, Zelenograd, Russia
alex1818@yandex.ru

Evgeny Yu. Kotlyarov
SC MERI;
National Research University of Electronic Technology
Moscow, Zelenograd, Russia
ORCID: 0000-0002-6314-3178

Pavel V. Timoshenkov
National Research University of Electronic Technology
Moscow, Zelenograd, Russia
cator@yandex.ru

Nikolay V. Guminov
National Research University of Electronic Technology
Moscow, Zelenograd, Russia
gummi.qdn@gmail.com

Abstract— **Authors propose new methodic for GaN field transistor gate capacitance extraction based on high-frequency S_{21}-parameter measurement and transmission line theory. The experimental research of the GaN field transistor (W=2x80 um, L_g=0.25 um) fabricated on silicon substrate was done according to described methodic. Authors have got the input gate capacitance as function of drain voltage (1.0 – 20.0 V) at the input signal frequency 15 GHz. Measured capacitance demonstrates 460-340 fF values respectively for pointed voltage band. Obtained results were compared with time-domain reflectometry data. The comparition shows good matching.**

Keywords— *GaN; HEMT; S-parameters, time-domain reflectometry, maximum gain frequency, GHz, gate-drain capacitance, gate-source capacitance.*

I. INTRODUCTION

The determination of the GaN field transistor input capacitance as function of operating mode in GHz band is important problem for understanding the device's physical condition and small-signal and large-signal models development. Ordinary methods for transistor input capacitance determination are quasistatic and high-frequency C-V approaches. But "high"-frequency methods perform measurements at the maximum frequency 5 MHz. Capacitance value measured by these approaches can be different with high-frequency (> 1.0 GHz) capacitance value. It causes incorrect transistor behavior description. There are approaches for transistor parameters determination in GHz band: - time-domain reflectometry [1] measurement based on analyze of the interaction between picoseconds pulse and device under test; - the extraction of the transistor parameters from measured s-matrix [2, 3]. Authors propose methodic for the GaN field transistor input gate capacitance (gate-source and gate-drain) determination based on transmission line theory and s-parameters high-frequency measurement. Obtained results were compared with TDR measurement results.

II. THEORETICAL RESEARCH

Test circuit for transistor research is shown on figure 1.

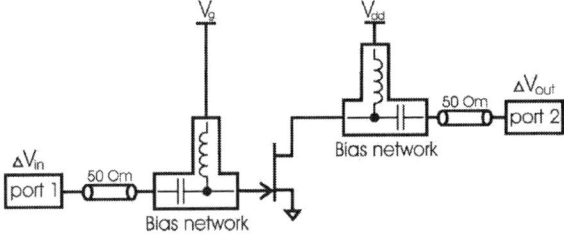

Fig. 1. Test circuit for transistor research using vector network analyzer. Here Vg – gate bias voltage, Vdd – power supply voltage.

Device under test is used with common source circuit. Harmonic signal is going from one of the VNA port through matched 50 Ohm waveguide to transistor's gate. Other VNA port is connected to transistor's drain.

Equivalent circuit for input and output signals analysis according to equivalent wave rule is shown on Fig. 2.

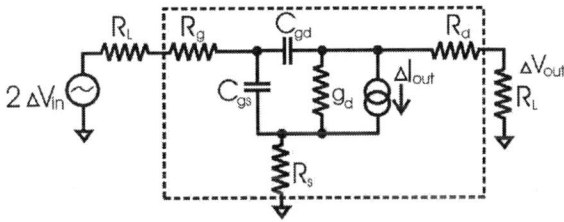

Fig. 2. Equivalent circuit for input and output signals analysis. RL = 50 Ohm – port input impedance and transmission line impedance.

According to circuit on figure 2, input current change ΔI_{in} is determined:

$$\Delta I_{in} = \frac{2\square \Delta V_{in}}{R_L + Z_{in}} \quad (1)$$

where ΔV_{in} – signal source voltage change, R_L = 50 Ohm RF path impedance, Z_{in} – transistor input impedance determined:

$$Z_{in} = R_g + \frac{Z_{gs}\square Z_{gd}}{Z_{gs} + Z_{gd}} \quad (2)$$

$$Z_{gs} = R_s + \frac{1}{j\omega C_{gs}}, \quad Z_{gd} = R_d + R_L + \frac{1}{j\omega C'_{gd}},$$

979-8-3503-8358-4/23 $31.00 © 2023 IEEE

where R_s, R_d, R_g – source, drain and gate resistance respectively; g_d – output conductance, C_{gs}, - gate-source capacitance, C'_{gd} - gate-drain effective capacitance.

Effective (due to Miller effect) gate-drain capacitance C'_{gd} for circuit shown on figure 1 determined as:

$$C'_{gd} = C_{gd}(1 + K_V) \qquad (3)$$

where K_V – voltage gain, C_{gd} – gate-drain capacitance.

Authors take into account that $\dfrac{1}{\omega C_{gs}}$ is much more than R_s and R_g, $\dfrac{1}{\omega C'_{gd}}$ is much more than R_d, R_g and R_L, and output resistance $R_{out}=1/g_d \gg R_d$, and R_L, thus input impedance is determined as:

$$Z_{in} = \frac{1}{j\omega(C_{gs} + C'_{gd})} \qquad (4)$$

Thus, current change in gate-source network determined as:

$$\Delta I_{gs} = \frac{\Delta I_{in} C_{gs}}{(C_{gs} + C'_{gd})} \qquad (5)$$

and gate-source voltage change ΔV_{gs}:

$$\Delta V_{gs} = \frac{\Delta I_{gs}}{j\omega C_{gs}} = \Delta I_{in} \frac{(\frac{C_{gs}}{C_{gs} + C'_{gd}})}{j\omega C_{gs}} = \frac{\Delta I_{in}}{j\omega(C_{gs} + C'_{gd})} \qquad (6)$$

Output current change value using equations (1) and (6) is:

$$\Delta I_{out} = g\square\Delta V_{gs} = \frac{g\square\Delta I_{in}}{j\omega(C_{gs} + C'_{gd})} =$$
$$= \frac{g}{j\omega(C_{gs} + C'_{gd})}\square\frac{2\square\Delta V_{in}}{R_L + Z_{in}} \qquad (7)$$

here g – transistor transconductance. Value for Z_{in} (equation 4) has been substituted in (7), than output current change is determined as:

$$\Delta I_{out} = \frac{2g\square\Delta V_{in}}{j\omega\square R_L\square(C_{gs} + C'_{gd}) + 1} \qquad (8)$$

Then voltage gain can be written by equation:

$$K_V = \frac{\Delta V_{out}}{\Delta V_{in}} = \frac{\Delta I_{out}\square R_L}{\Delta V_{in}} = \frac{2\square g\square R_L}{j\omega\square R_L\square(C_{gs} + C'_{gd}) + 1} \qquad (9)$$

Now authors determine voltage gain module $|K_V|$:

Numerator and denominator in equation (9) should be multiplied on complex conjugate factor:

$$K_V = \frac{2gR_L}{(1 + j\omega R_L(C_{gs} + C'_{gd}))} \frac{(1 - j\omega R_L(C_{gs} + C'_{gd}))}{(1 - j\omega R_L(C_{gs} + C'_{gd}))} =$$
$$= \frac{2gR_L(1 - j\omega R_L(C_{gs} + C'_{gd}))}{1 + [\omega R_L(C_{gs} + C'_{gd})]^2} \qquad (10)$$

Now voltage gain equation should be written in complex form with separate real and imagine parts:

$$K_V = \frac{2gR_L}{1 + \omega^2 R_L^2(C_{gs} + C'_{gd})^2} - j\frac{2gR_L^2\omega(C_{gs} + C'_{gd})}{1 + \omega^2 R_L^2(C_{gs} + C'_{gd})^2} \qquad (11)$$

Voltage gain module is formulated as:

$$|K_V| = \sqrt{[\mathrm{Re}(K_V)]^2 + [\mathrm{Im}(K_V)]^2} =$$
$$= \frac{1}{1 + \omega^2 R_L^2(C_{gs} + C'_{gd})^2} \times$$
$$\times \sqrt{(2gR_L)^2 + [2g\omega R_L^2(C_{gs} + C'_{gd})]^2} =$$
$$= \frac{2gR_L}{\sqrt{1 + \omega^2 R_L^2(C_{gs} + C'_{gd})^2}} \qquad (12)$$

It's known that input power is $P_{in} = \dfrac{(\Delta V_{in})^2}{R_L}$, output power is $P_{out} = \dfrac{(\Delta V_{out})^2}{R_L} = (\Delta I_{out})^2 R_L$, then power gain ($K_P$) equation can be written as:

$$K_P = \frac{P_{out}}{P_{in}} = \frac{(V_{out})^2}{(V_{in})^2} = |K_V|^2 =$$
$$= (\frac{2gR_L}{\sqrt{1 + \omega^2 R_L^2(C_{gs} + C'_{gd})^2}})^2 \qquad (13)$$

Then S_{21} parameter (power gain in dB) has been formulated as:

$$S_{21} = 10\square\lg(K_P) = 20\square\lg(|K_V|) =$$
$$= 20\square\lg(\frac{2gR_L}{\sqrt{1 + \omega^2 R_L^2(C_{gs} + C'_{gd})^2}}) \qquad (14)$$

It's known that

$$C_{in} = C_{gs} + C'_{gd} = C_{gs} + C_{gd}(1 + K_V) \qquad (15)$$

Input capacitance as function of S_{21} parameter has been formulated as:

$$C_{in} = C_{gs} + C_{gd}(1 + 10^{\frac{S_{21}}{20}}) = \frac{1}{\omega R_L}\sqrt{\frac{(2gR_L)^2}{10^{\frac{S_{21}}{20}}} - 1} =$$
$$= \frac{1}{\omega R_L}\sqrt{(\frac{2gR_L}{K_V})^2 - 1} = \frac{1}{2\pi f R_L}\sqrt{(\frac{2gR_L}{K_V})^2 - 1} \qquad (16)$$

Each capacitances in equation (16) – C_{gs} and C_{gd} can be determined by performing S_{21} measurement at the tow neighbor frequencies:

$$\begin{cases} C_{in}(f_1) = C_{gs} + C_{gd}(1 + K_V(f_1)) \\ C_{in}(f_2) = C_{gs} + C_{gd}(1 + K_V(f_2)) \end{cases} \qquad (17)$$

There is capacitance C_{gs} from the first equation in system (17):

$$C_{gs} = C_{in}(f_1) - C_{gd}(1 + K_V(f_1)) \qquad (18)$$

This equation has been substituted in to the second equation in system (17). After that equation for the C_{gd} capacitance is:

$$C_{gd} = \frac{C_{in}(f_2) - C_{in}(f_1)}{K_V(f_2) - K_V(f_1)} \quad (19)$$

Then C_{gd} equation should be substituted to equation (18) to determine C_{gs} capacitance. There are two equations for capacitances C_{gs} and C_{gd}:

$$\begin{cases} C_{gd} = \dfrac{C_{in}(f_2) - C_{in}(f_1)}{K_V(f_2) - K_V(f_1)} \\[3mm] C_{gs} = C_{in}(f_1) - \dfrac{C_{in}(f_2) - C_{in}(f_1)}{K_V(f_2) - K_V(f_1)}(1 + K_V(f_1)) \end{cases} \quad (20)$$

Input capacitance determination by time-domain reflectometry is described in [1-4].

III. EXPERIMENTAL RESEARCH

Gallium-nitride field transistor used in this work was fabricated on the silicon substrate. Transistor has two parallel gates with W=160 um overall width and L_g=0.25 um (figure 3).

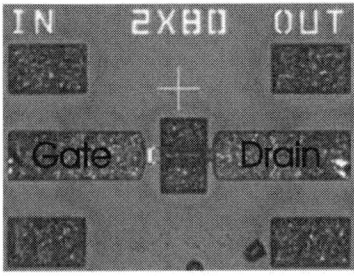

Fig. 3. Gallium-nitride field transistor layout (W = 2x80 um, L_g = 0.25 uм, L_{gs} = 2 uм, L_{gd} = 4 uм).

Scientific equipment used in this work allows perform the measurement of the transistor static parameters in pulse mode. Output IV-curves for GaN transistor under test (W = 2x80 um, L_g = 0.25 um) is shown on figure 4.

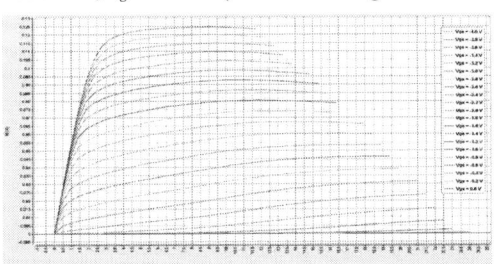

Fig. 4. Output IV-curves for Gallium-nitride field transistor (W = 2x80 um, L_g = 0.25 um).

Test transistor was mounted on the PCB between two coplanar lines (figure 5).

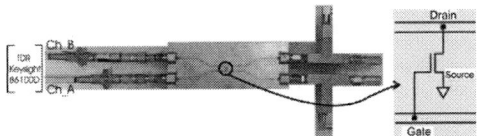

Fig. 5. The printed circuit board and transistor measurement circuit for C_{gs} and C_{gd} TDR measurement.

The measurement of S-parameters was done the first (figure 6).

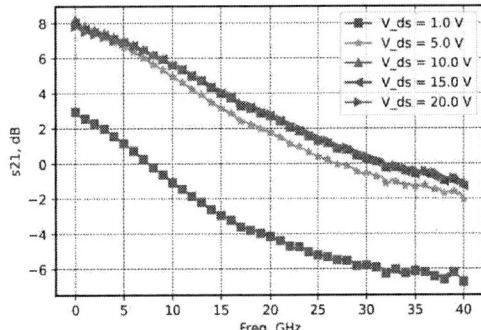

Fig. 6. Measured S-parameters for gallium-nitride field transistor (W = 2x80 um, L_g = 0.25 uм, L_{gs} = 2 uм, L_{gd} = 4 uм).

Pads geometry allows perform the measurements using mw-probes. After that drain and source of the transistor under test were connected by wires to coplanar lined to perform time-domain reflectometry (TDR) measurements. Gate was connected to low line, drain – to the upper line (figure 5). All measurements was done for continues mode. Measurement methodic using time-domain reflectometry approach was detailed in [1, 4]. Transistor parameters were measured in saturation region: V_{ds}=1.0 – 20.0 V, V_{gs}=-3.0 V. Measurement methodic contains input capacitance (C_{in}) measurements. Rise time at the transistor gate was 23 ps.

Transistor measurement circuit, connection to metrological equipment, distribution of incident, reflected and transition signals for input capacitance measurement are shown in figure 7.

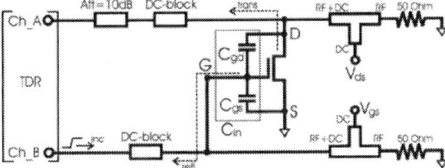

Fig. 7. Transistor measurement circuit for C_{in} TDR measurement.

Input capacitance was determined by reflected signal analysis and can be formulated as:

$$C_{in} = C_{gs} + (1 + K_V)C_{gd} + C_P \quad (22)$$

where C_{gs} – gate-source capacitance, C_{gd} – gate-drain capacitance, K_V – voltage gain on the 25 Ohm ($R_L/2$) load.

Reflected and transition signals during input capacitance (C_{in}) and transconductance (g) measurements are shown in figure 8.

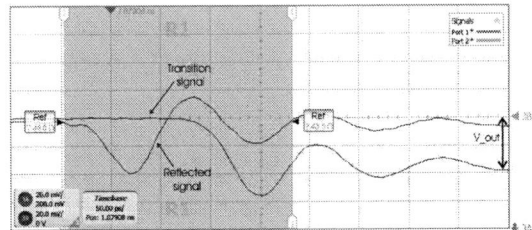

Fig. 8. Reflected and transition signals for input capacitance (C_{in}) and transconductance (g) measurement.

979-8-3503-8358-4/23 $31.00 © 2023 IEEE

Input capacitance was calculated using the following equation:

$$C_{in} = -\frac{2}{R_L V_{in}} \int_{T_1}^{T_2} V_{\text{Refl}} dt \qquad (23)$$

where T1-T2 – time interval for integration (gray color on figure 6), V_{in} – input signal voltage amplitude V_{in} = 200 mV, V_{refl} – reflected signal and load impedance R_L = 50 Ohm.

Transistor transconductance in the selected operating point can be determined using output signal amplitude. It's V_{out} = 126.5 mV and V_{in} = 200 mV for transistor under test. Transistor transconductance g is determined using the following equation:

$$g = \frac{V_{out}}{(\frac{R_L}{2}) V_{in}} \qquad (24)$$

Measured value for transconductance was used in equation (16) for the determination of the input capacitane based on S-parameters.

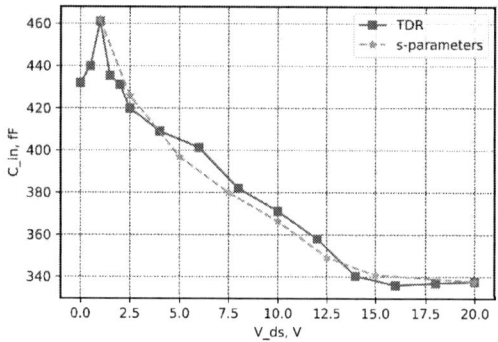

Fig. 9. TDR and S-parameters based results.

Input capacitance as function of drain-source voltage calculated using measured S_{21}-parameter (at the 15 GHz) and using TDR approach (incident wave rise time 23 ps) are shown on figure 7. Obtained functions demonstrate good matching.

IV. Conclusion

The authors declare that the following provisions and results are new in this work:

- Authors propose new methodic for field transistor gate capacitance extraction based on S21 parameter measurement. This paper contains analytical equations for input capacitance, gate-source and gate-drain capacitances as functions of S21 parameter;

- Experimental research was done for GaN field transistor according to proposal methodic. The dependence of the input capacitance as function of drain voltage (1.0 – 20.0 V) at the frequency 15 GHz was presented.

- Gate-source capacitance (Cgs) typical value is 15 times more than gate-drain capacitance (Cgd) value in described voltage band (1.0 – 20.0 V).

- Input capacitance demonstrates nonlinearity with increasing of the drain voltage. Measurements were done in 1.0 – 20.0 V band. Capacitance demonstrates values 460 – 340 fF respectively.

- The results from S21 measurements was compared with time-domain reflectometry results. The comparition demonstrates good matching.

Acknowledgment

The study was supported by a grant from the Russian Science Foundation (project No. 20-19-00521)

References

[1] Timoshenkov V.P., Khlybov A.I., D.V. Rodionov, N.A.Shelepin, A.V.Seletsky Microwave reflectometry of n-MOSFET SoI transistors // Republic of Crimea, Alushta, September 30 – October 05, 2019, Collection of abstracts of articles 359-353.

[2] Sokolov A.A., Babak L.I. Method of constructing a low-signal model of a microwave transistor with high electron mobility Sokolov A.A., Babak L.I. //Reports of TUSUR, No. 2 (22), Part 1, pp. 153-156 December 2010.

[3] Jiong-Guang Su, Shyh-Chih Wong, Chun-Yen Chang, Tiao-Yuan Huang //The extraction of MOSFET gate capacitance from S-parameter measurements//Solid-State Electronics 46 (2002) 1163–1167.

[4] Timoshenkov V., Rodionov D., Khlybov A. TDR method for determination of IC's parameters // Proceedings of the International Conference on Micro- and Nano-Electronics 2016 conference. SPIE Digital Library. Proc. of SPIE Vol. 10224 1022427-1.

Concentric Topology for Sensitive Elements of Surface Acoustic Wave Devices

Maria A. Sorvina
PhD student of Laser and Navigation System Department
Saint Petersburg Electrotechnical University "LETI"
Saint Petersburg, Russia
masorvina@etu.ru

Abstract — **This article presents modern requirements for navigation sensors. Benefits and disadvantages of micromechanical devices are shown. A new concentric topology of sensitive elements of surface acoustic waves sensors is described. Proposed concept is analyzed by full finite element method.**

Keywords — *surface acoustic waves, anisotropic material, topology, sensitive element, interdigitated transducer, finite element method*

I. INTRODUCTION

Micromechanical accelerometers (MMA) are inertial sensors that are used to measure the acceleration of an object. Compared with traditional sensors, they have many advantages such as light weight, small size, low cost, relatively high accuracy, and easy integration. These advantages have led to wide application in many areas, including various dynamic objects, consumer and industrial electronic devices, robotics, automotive, military, medicine, aerospace, etc. [1].

Sensitive element design and manufacturing technology are key factors influencing a sensor's measurement capability, performance, and applications. At present, the leader among sensors of a tactical accuracy class is an accelerometer manufactured using the technology of microelectromechanical systems, which is mass-produced [2]. However, in many applications, such as military technology, the navigation system must not only be compact, but also resistant to shock and vibration. Conventional microelectromechanical accelerometers cannot meet the modern requirements put forward for the development of inertial navigation systems [3], which include impact resistance up to 50000 g and an average vibration resistance value up to 50 g, due to the presence of moving parts in their design - elastic suspensions. It is possible to ensure resistance to such impacts at high sensitivity of the device using accelerometers on surface acoustic waves (SAW). The main elements of such devices are delay lines and resonators whose properties depend on the measured motion parameters. The exceptional simplicity of the kinematic scheme and a high level of constructive integration create the prerequisites for improving the accuracy characteristics, reducing the overall dimensions and reducing the cost of production [4].

Devices based on surface acoustic waves have already proved their significant superiority over their counterparts in various electronic equipment due not only to high electrical parameters, small dimensions, reliability, but also the possibility of their mass production based on microelectronic technology. Currently, in radio equipment, especially in devices for receiving and processing signals, SAW delay lines, SAW filters, SAW convolvers, SAW decoders and phase shifters are widely used [5-6].

The purpose of this work is to consider a new concentric design of SAW-based MMA sensing elements and to propose a method for modeling such devices.

II. A NEW DESIGN CONCEPT FOR SAW-BASED MMA SENSING ELEMENTS

SAW-based solid-state electronic devices are usually a substrate of a piezoelectric material, on one or more flat surfaces of which a periodic interdigitated structure of conducting electrodes is formed. The electrodes are designed both for electrical excitation of SAW and for their reception (reverse conversion into an electrical signal). Under bending deformations, the phase velocity of the SAW and the pitch of the periodic structure change [7] (Fig. 1).

The process of converting acceleration into a change in the frequency of the microaccelerometer output signal consists of three key steps. First, the measured acceleration deforms the piezoelectric substrate. Secondly, the resulting internal stresses change the oscillation frequency of the self-oscillator in the feedback circuit of the delay line. Thirdly, the shift of the phase-frequency characteristic of the sensor reflects the change in the speed of the object.

Recently, much attention has been paid to non-standard ways of organizing the sensing element. For example, in 2019 the ring topology of interdigital transducers (IDT) was proposed [8]. The general view of the sensitive element of the membrane is a console attached to the body with silicone glue. The resonator consists of two ring IDT and a piezoelectric crystal located between the transducers [9].

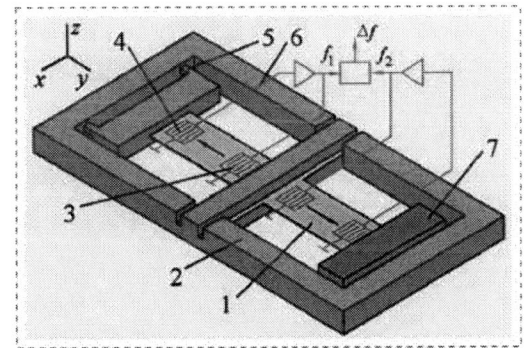

Fig. 1. The generalized scheme of a microaccelerometer with a frequency output: 1 - cantilever beams, 2 - frame housing, 3 - input IDT, 4 - output IDT, 5 - jumpers, 6 - conductors, 7 - inertial weights.

The proposed model of the sensitive element of a micromechanical accelerometer with a concentric shape of the IDT is shown in fig. 2. Here, two groups of electrodes are deposited on the piezoelectric substrate of 128° YX LiNbO$_3$: excitation and receiving, made in the form of ellipses. The shape of the electrodes is supposed to be elliptical, since the

979-8-3503-8358-4/23 $31.00 © 2023 IEEE

speed of wave propagation in an anisotropic material depends on the direction. The process of converting a physical quantity into the frequency of the output signal of a sensor on surface acoustic waves can be represented as a sequence of transformations: the measured mechanical quantity is converted into a deformation of the piezoelectric substrate, which, accordingly, causes a change in the phase-frequency characteristic of the SAW accelerometer and leads to a change in the oscillation frequency of the self-oscillator containing in the feedback circuit delay line communication.

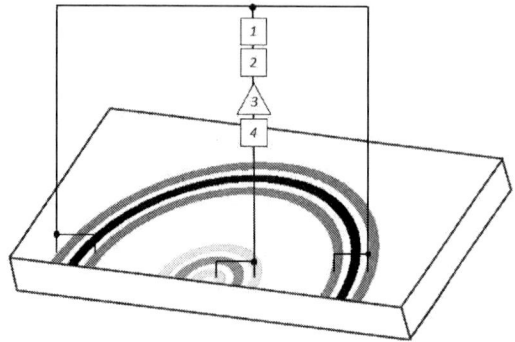

Fig. 2. SAW sensor model with concentric electrode topology.

It is stated in [11] that with a decrease in the SAW natural frequency, the scale factor increases. Currently, the frequency range for electronic components of SAW devices is from 5 MHz to 6 GHz. Based on this, and also taking into account the limitation on the calculated power, the value of the calculated frequency was chosen to be 5 MHz.

Compared with the existing classical IDT placement concepts, it is hypothesized that the concentric shape of the sensing element topology will have higher quality factors, and hence better energy efficiency and, possibly, sensitivity.

The main expected advantage of a concentric design over linear topologies is higher energy efficiency due to the concentration of energy in the center. The concentric topology of the electrodes will avoid the scattering and reflection of the acoustic wave, which is typical for linear structures. The acoustic wave propagates not only in the forward direction from the exciting IDT to the receiving one, but also in other directions, as shown in Fig. 3. The propagating wave is reflected from the boundaries of the device, which leads to the appearance of excess noise on the amplitude-frequency characteristic.

Another advantage of the concentric shape of the electrodes is that it takes into account the anisotropy of the piezoelectric material. In paper [12], R. Woods criticized linear topologies.

Fig. 3. Scattering of an acoustic wave in a linear construction.

Monograph [13] presents a graph of the dependence of the SAW velocity on the free surface for lithium niobate YZ-cut (Fig. 4). Along the crystallographic axis X, the velocity is 4080 m/s. Along the Z axis - 3488 m/s. The speed depends on the angle, that is why it is important to model different cases of wave propagation.

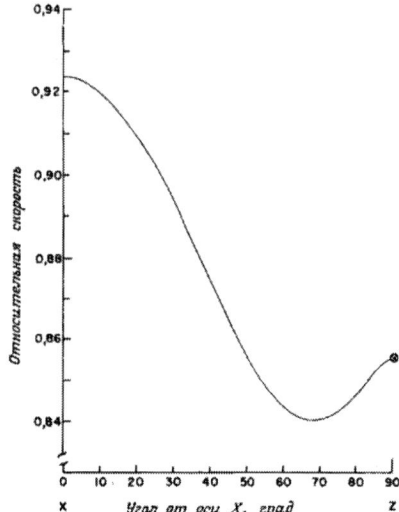

Fig. 4. Surface wave velocity for propagation on the surface of the Y-cut LiNbO₃

III. SENSOR SIMULATION

To test the proposed hypothesis, it is planned to conduct simulations in the *OOFELIE::Multiphysics* environment. To do this, the following boundary conditions are specified in the model: an alternating voltage with an amplitude of 1 V is applied to the even electrodes of the IDT 2, and the odd ones are grounded. Also, a perfectly matched layer 5 (PML) has been added to the model. The PML layer defines a quasi-infinite space that allows the wave to propagate freely beyond the formal boundaries of the model without experiencing reflections. This representation is due to the fact that in a real device, the transducers are located at a significant (with respect to the wavelength) distance from the boundaries of the sound duct and, therefore, it attenuates significantly before reaching them.

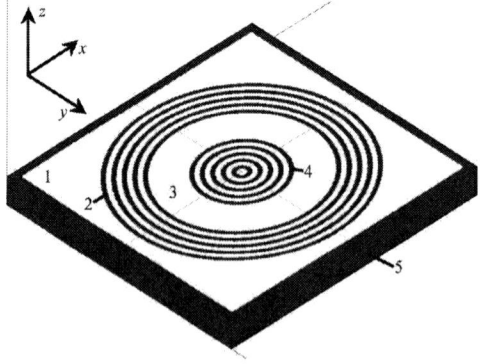

Fig. 5. SAW sensor model with concentric electrode topology.

To study the anisotropy of the material, a delay line model was built, shown in Figure 6. The model was rotated around the Z axis with a step of 10 degrees to obtain information about the wave propagation velocity. The data obtained are summarized in Table 1 and on the basis of it the graph in Figure 7 is built.

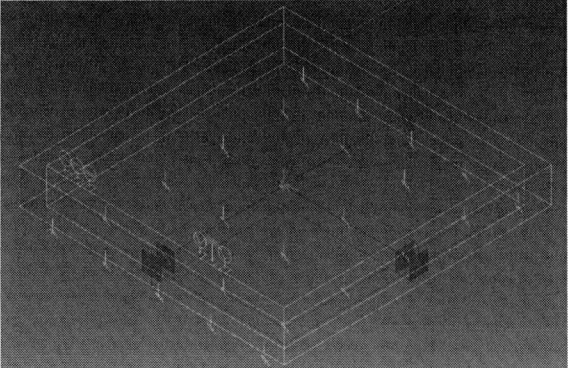

Fig. 6. SAW sensor model with concentric electrode topology.

TABLE 1. PROPOGATION SPEED

Angle	Speed, m/s	Angle	Speed, m/s	Angle	Speed, m/s
0°	3749,74	120°	5222,85	240°	5042,75
10°	3749,74	130°	5222,85	250°	5222,85
20°	3400,93	140°	4178,28	260°	5222,85
30°	2867,45	150°	2658,9	270°	5222,85
40°	2924,8	160°	3249,77	280°	5042,75
50°	4062,22	170°	3400,93	290°	5042,75
60°	5624,61	180°	4178,28	300°	4874,66
70°	5416,29	190°	4178,28	310°	3400,93
80°	5416,29	200°	3400,93	320°	3952,43
90°	5416,29	210°	2867,45	330°	3249,77
100°	5416,29	220°	3952,43	340°	2867,45
110°	5416,29	230°	5042,75	350°	2812,3

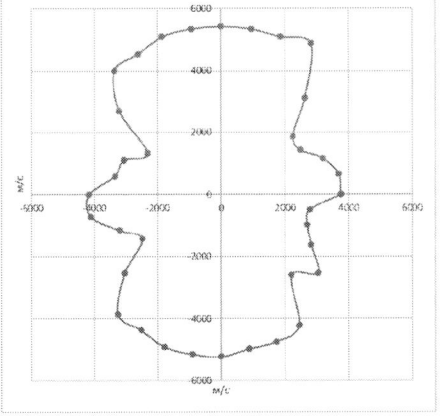

Fig. 7. Graph of the propagation speed of SAW in the polar coordinate system.

IV. CONCLUSION

The advantages and disadvantages of modern MMA were considered. The concentric topology of a SAW microaccelerometer which is able to withstand high shock loads was presented. The implementation of the sensing element model in the OOFELIE::Multiphysics software environment was described. Direction for further research: it is planned to optimize the design of a concentric SAW sensor. For this, the influence of the distance between the exciting and receiving IDTs, as well as the shape of the sensitive element substrate on the amplitude-frequency characteristic of the sensor, will be studied.

REFERENCES

[1] Mandal D. and Banerjee S. Surface Acoustic Wave (SAW) Sensors: Physics, Materials, and Applications, USA, 2021

[2] G. Girardin. Sensors for Cellphones and Tablets 2016 // YOLE Developpement Market & Technology Report, 2016. P. 1 - 4.

[3] Precise Robust Inertial Guidance for Munitions (PRIGM): Advanced Inertial Micro Sensors (AIMS), Microsystems Technology Office, DARPA-BAA-15-38, May 29, 2015

[4] Shevchenko S.Yu. Development of a microaccelerometer based on surface acoustic waves. Dissertation for the degree of candidate of technical sciences. 2007. 132 p.

[5] Orlov, V. S. Filters on surface acoustic waves / V. S. Orlov, V. S. Bondarenko. - M. : Radio and communication, 1984. - 272 p.

[6] Rechitsky, V. I. Acoustoelectronic radio components / V. I. Rechitsky. - M. : Radio and communication, 1987. - 192 p.

[7] Lukyanov D.P., Luchinin V.V., Skvortsov V.Yu. Microaccelerometer on surface acoustic waves. M.: Microsystem technique, 2001. P. 2.

[8] Shevchenko S.Yu., Mikhailenko D.A., Lukyanov D.P. The choice of material for sensitive elements of accelerometers based on SAW. Izvestiya vuzov Rossii. Radioelectronics. 2020. V. 23, No. 6. S. 70-83.

[9] Shevchenko S.Yu., Mikhailenko D.A., Nyamveru B. Optimization of the design of an interdigital converter of a ring resonator on surface acoustic waves. Izvestiya vuzov Rossii. Radioelectronics. 2021. V. 24, No. 6. pp. 51-62.

[10] Kukaev A.S. Multiphysics modeling of sensitive elements of microgyroscopes on surface acoustic waves. Dissertation for the degree of candidate of technical sciences. 2016. 130 p.

[11] Lao B.Y. Gyroscopic Effect in Surface Acoustic Waves. Proceeding of IEEE ultrasonic symposium, In, 1980. P 687.

[12] Woods R.C., Kalami H., and Johnson B. Evaluation of a Novel Surface Acoustic Wave Gyroscope, USA, 2002

[13] Oliner A. Surface acoustic waves. 1981

Multi-Function LUT for FPGAs

Stanislav I. Sovetov
Department of Automatic and Telemechanic
Perm National Research Polytechnic University
Perm, Russia
fizikoz@gmail.com

Sergey F. Tyurin
Department of Automatic and Telemechanic
Perm National Research Polytechnic University
Department of Software Computing Systems
Perm State University
Perm, Russia
tyurinsergfeo@yandex.ru

Abstract— **FPGA's basic element is the Look Up Table (LUT). Known LUTs implement only one logic function for a given configuration in canonical normal form (CDNF). In this case, there is always a part of inactive pass transistors (PT). Previously developed LUT implements two functions simultaneously. A further modification of this method is proposed with the implementation of more than two functions of the same variables for one configuration. This provides the effect of reducing the number of transistors in comparison with the implementation in several LUTs.**

Keywords—FPGA, LUT, pass transistors

I. Introduction

A. Motivation

Recently, in widely used FPGAs, the basic element is a configurable logic block. The main element of this block is the LUT, which implements a certain logical function. Modern LUTs are configurable multiplexers for realizing logic functions with 2^n inputs and one output for n variables [1,2]. Using the existing LUTs for n variables requires 2^n memory cells (SRAM) and $2^{n+1}-2$ pass transistors (PT) when only one logic function is implemented. For implementing simultaneously n logical functions, it is necessary to use the same number of LUTs. However, when using single function in the LUT, the second half of the tree of pass transistors in the amount of 2^n-1 remain inactive [3-5]. In previous work, it was proposed to implement simultaneously 2 logical functions on one tree of pass transistors [10,12].

B. State of the Art

The linear representation of the 3-LUT logic function is given below:

$$z(x_3 x_2 x_1 d) = d_0 \cdot \overline{x}_3 \overline{x}_2 \overline{x}_1 \vee d_1 \cdot \overline{x}_3 \overline{x}_2 x_1 \vee d_2 \cdot \overline{x}_3 x_2 \overline{x}_1 \vee$$
$$d_3 \cdot \overline{x}_3 x_2 x_1 \vee d_4 \cdot x_3 \overline{x}_2 \overline{x}_1 \vee d_5 \cdot x_3 \overline{x}_2 x_1 \vee \quad (1)$$
$$\vee d_6 \cdot x_3 x_2 \overline{x}_1 \vee d_7 \cdot x_3 x_2 x_1$$

where $d_0, d_1, d_2, d_3, d_4, d_5, d_6, d_7$ - configurations data of the three arguments function $z(x_3 x_2 x_1)$.

The 3-LUT (Fig. 1) consists of three stages of NMOS pass transistors. For any one of the 8 values of the inputs, exactly one path from a single SRAM cell to the output is enabled, and all 7 other paths have at least one NMOS pass transistor is turned off.

Combining $d_0, d_1, d_2, d_3, d_4, d_5, d_6, d_7$ we can get 2^8 functions.

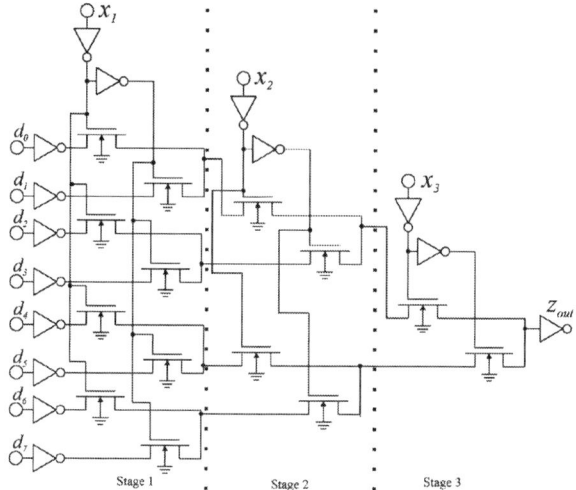

Fig. 1. 3-LUT tree.

Each branch of the tree (Fig. 1) $x^{\sigma_3}_3 x^{\sigma_2}_2 x^{\sigma_1}_1$, where $\sigma_i \in \{0,1\}$ is indicator of the negation presence (=1) or negation absence (=0) is orthogonal to another branches. So only one branch activates.

In this way, all n-LUTs produce only single logic function of n arguments in the minterm canonical form (MCF) or canonical disjunctive normal form (CDNF).

At the same time, each minterm can activate other logic functions of the same arguments (for example sum and carry functions). Combining this minterms by OR we can get multi-outputs logic element.

C. Objectives and structure

This article proposes an advanced circuit that uses the inactive branches of the pass transistor tree. By adding an additional 1-LUT to stage 3, it is possible to use the second function on the inactive half of the tree.

The new solution makes it possible to increase the number of simultaneously implemented functions of the same variables, which is important when implementing, for example, code transformations. To solve this problem, the authors perform

- synthesis and analysis of the proposed multi-functional n-LUT design,

- simulation of the proposed multi-functional LUT,

- comparing the complexity in the number of transistors of the obtained solution with the known

979-8-3503-8358-4/23 $31.00 © 2023 IEEE

II. Multi-functional n-LUT design method

To implement an additional logic function, the LUT design has been supplemented with the output NOT gate of the second function, the pass transistors, the outputs of which are combined and connected to the NOT gate, the group of input pass transistors of setting SRAM, the outputs of which are combined and connected to the input NOT gate. Additional elements are marked in Fig. 2.

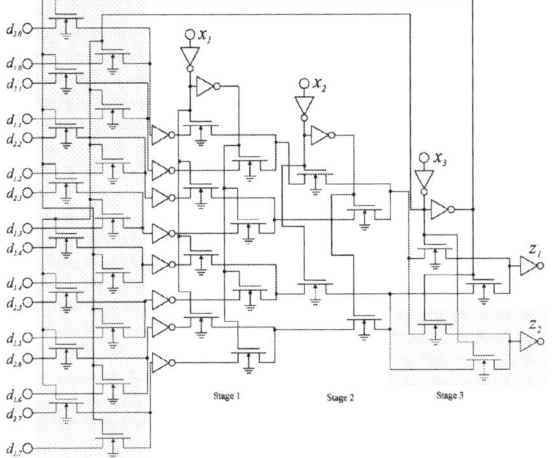

Fig. 2. Two-functions 3-LUT tree.

Computing two logic functions at the same time requires setting logic levels corresponding to logic functions of n variables.

The implementation of the two functions assumes the following decomposition of the logical device setting in terms of the highest variable on stage 3 (x_3).

For the first function:

$$z_1(x_3 x_2 x_1 d) = \bar{x}_3[d_{1.0} \cdot \bar{x}_2 \bar{x}_1 \vee d_{1.1} \cdot \bar{x}_2 x_1 \vee d_{1.2} \cdot x_2 \bar{x}_1$$

$$\vee\, d_{1.3} \cdot x_2 x_1] \vee x_3[d_{1.4} \cdot \bar{x}_2 \bar{x}_1 \vee d_{1.5} \cdot \bar{x}_2 x_1 \qquad (2)$$

$$\vee\, d_{1.6} \cdot x_2 \bar{x}_1 \vee d_{1.7} \cdot x_2 x_1]$$

For the second function:

$$z_2(x_3 x_2 x_1 d) = x_3[d_{2.4(0)} \cdot \bar{x}_2 \bar{x}_1 \vee d_{2.5(1)} \cdot \bar{x}_2 x_1 \vee$$

$$d_{2.6(2)} \cdot x_2 \bar{x}_1 \vee d_{2.7(3)} \cdot x_2 x_1] \vee \bar{x}_3[d_{2.0(4)} \cdot \bar{x}_2 \bar{x}_1 \vee \qquad (3)$$

$$d_{2.1(5)} \cdot \bar{x}_2 x_1 \vee d_{2.2(6)} \cdot x_2 \bar{x}_1 \vee d_{2.3(7)} \cdot x_2 x_1]$$

where $d_{i.j}$; $i = 1, 2(^v)$; $j = 1, 2, 3, 4 \ldots 2^3(2^n)$. the tuning constant is specified in the format $d_{i.j(k)}$, where k means the real function set number, and j – the number of the input that is used to connect it. In this case, k from j is obtained by inverting the most significant bit on stage 3. So, the top half of the SRAM is swapped with the bottom half to use the inactive part of the pass transistors and now two paths from a double SRAM cell of two functions to the output is enabled.

The following logical levels are set at SRAM for the implementation of two logical function: XOR (z_1) and majority function (z_2) shown in Table 1:

TABLE. I. Initial SRAM data

Input port	Configurable SRAM cell for z_1	Configurable SRAM cell for z_2
1	0	0
2	1	1
3	1	1
4	0	1
5	1	0
6	0	0
7	0	0
8	1	1

To implement four logical functions, we added 2-LUT tree for each additional function and tuning transistors controlled by second variable x_2 (Fig. 3).

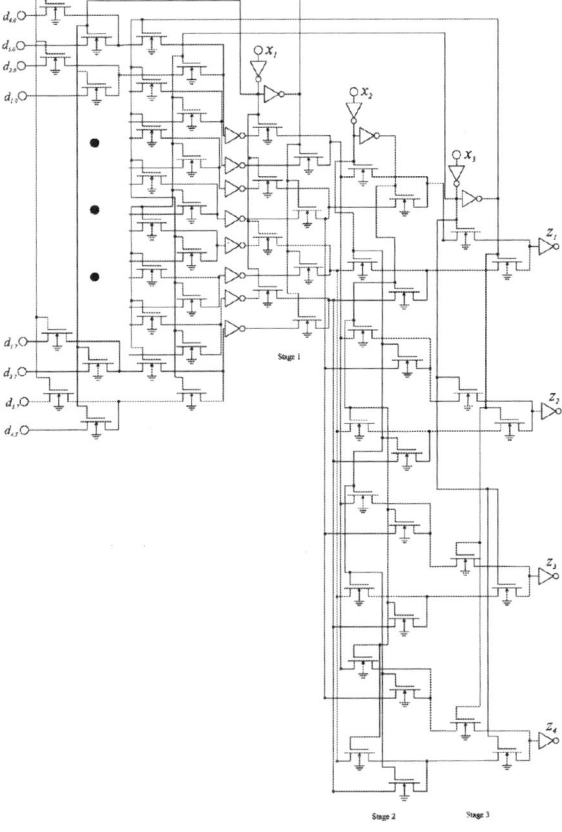

Fig. 3. Four-functions 3-LUT tree.

The implementation of the four functions assumes the following decomposition of the logical device setting in terms of x_3 and x_2 variables.

For the first function:

$$z_1(x_3 x_2 x_1 d) = \bar{x}_3 \bar{x}_2[d_{1.0} \cdot \bar{x}_1 \vee d_{1.1} \cdot x_1] \vee$$

$$\bar{x}_3 x_2[d_{1.2} \cdot \bar{x}_1 \vee d_{1.3} \cdot x_1] \vee x_3 \bar{x}_2[d_{1.4} \cdot \bar{x}_1 \vee d_{1.5} \cdot x_1] \vee$$

$$x_3 x_2[d_{1.6} \cdot \bar{x}_1 \vee d_{1.7} \cdot x_1]$$

For the second function:

$$z_2(x_3x_2x_1d) = \bar{x}_3x_2[d_{2.2(0)}\cdot\bar{x}_1 \vee d_{2.3(1)}\cdot x_1] \vee$$

$$x_3\bar{x}_2[d_{2.4(2)}\cdot\bar{x}_1 \vee d_{2.5(3)}\cdot x_1] \vee x_3x_2[d_{2.6(4)}\cdot\bar{x}_1 \vee d_{2.7(5)}\cdot x_1] \vee$$

$$\bar{x}_3\bar{x}_2[d_{2.0(6)}\cdot\bar{x}_1 \vee d_{2.1(7)}\cdot x_1]$$

For the fourth function:

$$z_3(x_3x_2x_1d) = x_3\bar{x}_2[d_{3.4(0)}\cdot\bar{x}_1 \vee d_{3.5(1)}\cdot x_1] \vee$$

$$x_3x_2[d_{3.6(2)}\cdot\bar{x}_1 \vee d_{3.7(3)}\cdot x_1] \vee \bar{x}_3\bar{x}_2[d_{3.0(4)}\cdot\bar{x}_1 \vee d_{3.1(5)}\cdot x_1] \vee$$

$$\bar{x}_3x_2[d_{3.2(6)}\cdot\bar{x}_1 \vee d_{3.3(7)}\cdot x_1]$$

For the third function:

$$z_4(x_3x_2x_1d) = x_3x_2[d_{4.6(0)}\cdot\bar{x}_1 \vee d_{4.7(1)}\cdot x_1] \vee$$

$$\bar{x}_3\bar{x}_2[d_{4.0(2)}\cdot\bar{x}_1 \vee d_{4.1(3)}\cdot x_1] \vee \bar{x}_3x_2[d_{4.2(4)}\cdot\bar{x}_1 \vee d_{4.3(5)}\cdot x_1] \vee$$

$$x_3\bar{x}_2[d_{4.4(6)}\cdot\bar{x}_1 \vee d_{4.5(7)}\cdot x_1]$$

The following logical levels are set at SRAM for the implementation of two logical function: XOR (z_1), majority function (z_2), disjunction (z_3) and conjunction (z_4) shown in Table 2:

TABLE. II. INITIAL SRAM DATA

Input port	Configurable SRAM cell for z_1	Configurable SRAM cell for z_2	Configurable SRAM cell for z_3	Configurable SRAM cell for z_4
1	0	0	1	0
2	1	1	1	1
3	1	0	1	0
4	0	0	1	0
5	1	1	0	0
6	0	1	1	0
7	0	0	1	0
8	1	1	1	0

III. SIMULATION MULTI-FUNCTIONAL LUT

The simulation of the multifunctional LUT was carried out in the Multisim dynamic simulation system from National Instruments [6]. The NMOS transistor model was chosen as the pass transistors BSIM 4.8.0 [6]. The NOT gate are implemented on two MOSFET transistors according to the circuit shown in Fig. 4.

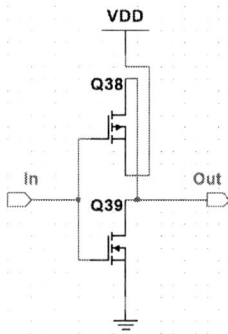

Fig. 4. NOT Gate Design.

Consider the implementation of a multifunctional LUT for a different number of variables.

A. 1-LUT

The diagram 1-LUT implementation of two functions simultaneously (NOT x and x) is shown in Fig. 5.

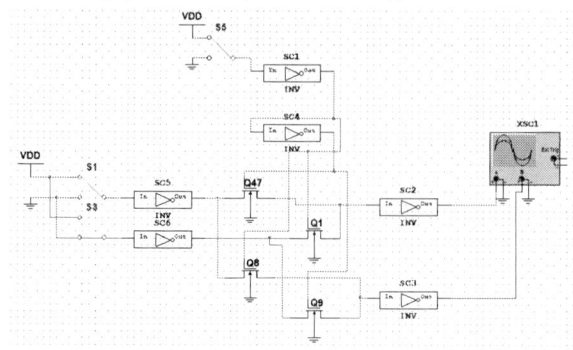

Fig. 5. 1-LUT design implementation of two functions simultaneously.

B. 2-LUT

The diagram 2-LUT implementation of two functions simultaneously (XOR, majority) is shown in Fig. 6.

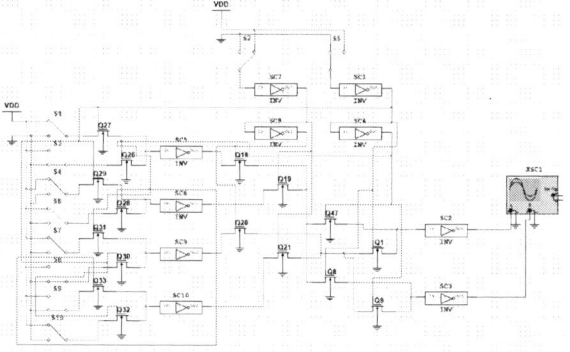

Fig. 6. 2-LUT design implementation of two functions simultaneously.

C. 3-LUT

The diagram 3-LUT implementation of two functions simultaneously (XOR, majority) is shown in Fig. 7.

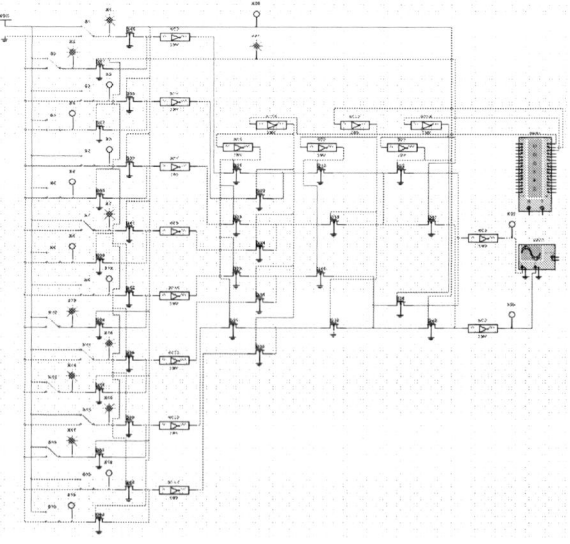

Fig. 7. 3-LUT design implementation of two functions simultaneously.

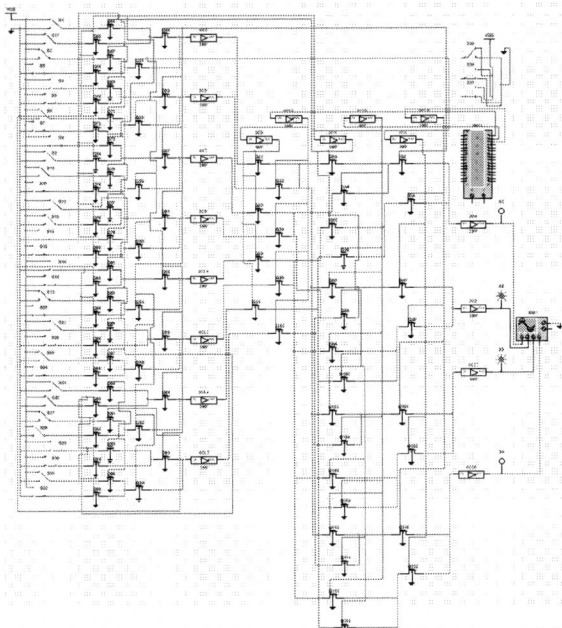

Fig. 8. 3-LUT design implementation of four functions simultaneously.

Setup input signals are set using dynamic voltage switches according to Table 1. A high voltage level (VDD 5V) corresponds to a logic 1, and a low voltage level (GND 0V) corresponds to a logic 0.

The variable signals are set using the WordGenerator, which has a 3-bit Gray code Fig. 9. The word generator operates at a frequency of 1 kHz, so it takes 8 ms to iterate through all the Gray code bits.

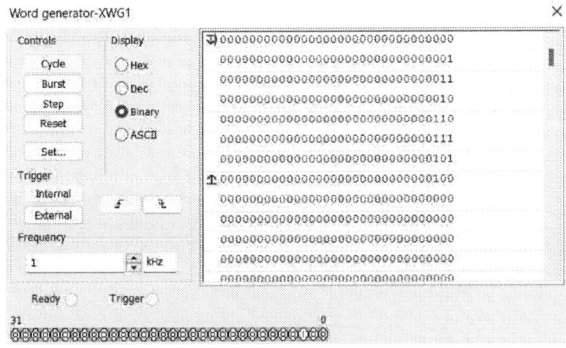

Fig. 9. Gray code in WordGenerator XWG1.

Function outputs connected to a two-channel oscilloscope XSC1.

IV. SIMULATION RESULTS

Simulation results are presented on Fig. 10.

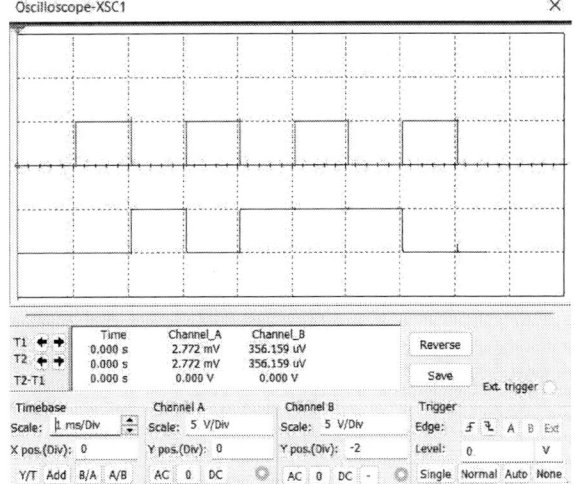

Fig. 10. Waveform of two-function 3-LUT.

The upper waveform corresponds to the XOR function (z_1), the lower waveform corresponds to the majority function (z_2). Each time division corresponds to one state of the Gray code in WordGenerator.

Simulation of four functions proved corresponding to truth tables of the declared functions.

V. CONCLUSION

Unlike the basic 3-LUT design which uses one inactive half of the transmitting NMOS, the proposed new design uses the inactive parts of both halves of the pass transistors over the value of several 2^n-1 high-order variables.

Thus, it is possible to simultaneously implement 2^{n-1} logic functions on the inactive parts of the pass transistor groups, which increases the functionality of the device. It takes 2^{n-2} LUTs to implement the same number of logic functions. Despite the additional cost, there is a gain in complexity relative to the number of required transistors.

Thus, the complexity in the number of LUT transistors depending on the number of variables n ($n=1,2,3,4...$) is estimated as:

$$L_1(n) = (2+2) \cdot 2^n + (2^{n+1} - 2) + 2n + 2 + 2 =$$
$$2^{n+2} + 2^{n+1} + 2n + 2 = 3 \cdot 2^{n+1} + 2n + 2$$

To calculate 2^v functions in the proposed design, we get the complexity:

$$L_v(v,n) = (2^{v+1} - 2 + 2) \cdot 2^n + 2^{n+1} + 2n +$$
$$(2^{v+1} - 2)(2^v - 1) + 2(2^v - 1) =$$
$$2^{v+1} \cdot 2^n + 2^{n+1} + 2n + 2^{v+1} \cdot (2^v - 1)$$

So, for example, to implement four logical functions ($v=2$) from the same variables (for $n=4$), two LUTs are required with a total use of 212 transistors, and in the proposed new LUT design, which implements all four functions, the number of necessary transistors equals 192.

The complexity diagrams which represent the usual LUT $L_{v1}(v)$ and the multi-functional LUT $L_v(v)$ in the implementation of logical functions depending on v with the number of variables $n=8$ are shown in Fig. 11.

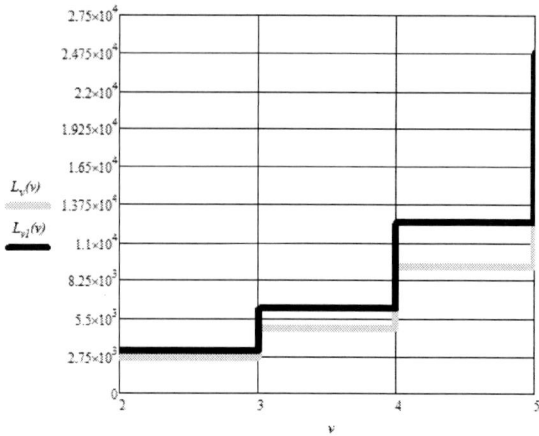

Fig. 11. Comparing the complexity of usual LUT and multifunctional LUT.

As can be seen from the diagram, there is a gain in complexity in terms of the number of transistors for a multifunctional LUT in relation to the corresponding number of usual LUTs.

REFERENCES

[1] Drozd O., Perebeinos I., Martynyuk O., Zashcholkin K., Ivanova O., Drozd M. Hidden fault analysis of FPGA projects for critical applications. *15th Conference on Advanced Trends in Radioelectronics, Telecommunications and Computer Engineering,* IEEE, 2020, pp. 128–132. DOI: 10.1109/TCSET49122.2020.235591

[2] Drozd O., Nikul V., Antoniuk V., Drozd M. Hidden faults in FPGA-built digital components of safety-related systems. *14th Conference on Advanced Trends in Radioelecrtronics, Telecommunications and Computer Engineering,* IEEE, 2018. pp. 805–809. DOI: 10.1109/TCSET.2018.8336320

[3] Vikhorev R. Universal logic cells to implement systems functions. *Conference of Russian Young Researchers in Electrical and Electronic Engineering.* IEEE, 2016. pp. 404–406. DOI: 10.1109/EIConRusNW.2016.7448197

[4] Vikhorev R. Improved FPGA logic elements and their simulation. *Conference of Russian Young Researchers in Electrical and Electronic Engineering,* IEEE, 2018. pp. 275–280. DOI: 10.1109/EIConRus.2018.8317080

[5] Skornyakova A.Yu., Vikhorev R.V. Self-Timed LUT Layout Simulation. *Conference of Russian Young Researchers in Electrical and Electronic Engineering,* IEEE, 2020. pp. 176–179. DOI: 10.1109/EIConRus49466.2020.9039374

[6] *National Instruments.* Available at: http://www.ni.com/multisim/ (accessed 5 November 2022).

[7] *Microwind & Dsch Version 3.5.* Available at: https://www.yumpu.com/en/document/view/40386405/microwind-manual-lite-v35pdf-moodle (accessed 5 November 2022).

[8] Mead C. A., Conway L. Introduction to VLSI Systems. Available at: https://www.researchgate.net/publication/234388249_Introduction_to _VLSI_systems (accessed 5 November 2022).

[9] Tyurin S. F. LUT's Sliding Backup. *IEEE transactions on device and materials reliability,* 2019, vol. 19, pp. 221-225. DOI: 10.1109/TDMR.2019.2898724

[10] Tyurin S. F., Prokhorov A.S. Programmiruemoe logicheskoe ustroystvo [Programmable Logic Device]. Patent RF, no. 2637462, 2017 (In Russian).

[11] Tyurin S.F. Green Logic: Green LUT FPGA Concepts, Models and Evaluations. *Green IT Engineering: Components, Networks and Systems Implementation,* 2017, vol. 105, pp. 241-261. DOI: 10.1007/978-3-319-55595-9

Arc Spot Movement and Its Effect on Flow in an AC Plasma Torch

Alexander Surov
National Center for Cognitive Research
ITMO University
St.Petersburg, Russian Federation
avsurov@ieeras.ru

Nikita Obraztsov
Institute of Energy
Peter the Great St Petersburg Polytechnic University
St.Petersburg, Russian Federation
nikita.obrazcov@mail.ru

Nikolay Bykov
National Center for Cognitive Research
ITMO University
St.Petersburg, Russian Federation
nbykov2006@yandex.gh

Ghennady Nakonechny
Laboratory for Plasma Technologies
IEE RAS
St.Petersburg, Russian Federation
ghennady@mail.ru

Abstract— AC electric arc plasma torches are promising thermal plasma generators for a wide range of applications. New models of plasma torches are used both for heating the plasma-forming gas, and as independent models of plasma-chemical reactors with the required composition of the gas mixture at the output. In the electric discharge chambers of such devices, processes of various physical nature occur simultaneously. Computer simulation of the combination of these processes provides an opportunity to optimize existing and develop new plasma technologies. One of the important processes is the movement of the arc spot in the channel of the plasma torch. Moving the arc spot over the electrode surface is used as one of the main techniques to ensure good performance and increase non-stop operation of devices. To date, there is no complete mathematical model of the AC plasma torch considering the dynamics of the arc binding. The main goal of the work is to analyze the possibility of using experimental data on the dynamics of the movement of arc spot in the channels of an AC plasma torch to build an appropriate data-driven model. Data on the arc spot movement is obtained using high-speed video recording. The reconstruction of the equations describing t h e dynamics of the motion of the electric arc spot on the electrode surface involves the subsequent use of generative design methods.

Keywords— AC plasma torch, electric arc spot, high-temperature flow dynamics, plasma flow simulation

I. INTRODUCTION

Thermal plasma generators are in demand for the development of new plasma chemical technologies [1–5]. The most studied and widely used electric arc plasma torches are DC plasma torches [6], while AC plasma torches [7–9] have a number of advantages for industrial applications. Computer simulations ca n accelerate the development of new devices and allow for a better understanding of the processes occurring in the modules of existing plasma torches that are difficult to diagnose. At present there are models that describe interaction of an electric arc with a plasma-forming gas, comprehensive studies of the electric arc interaction with electrode surface are conducted by the researchers from different scientific teams [11–15], however, there are no models that can fully describe the processes occurring in the discharge chambers of AC plasma torches. At the same time, the development of such models can allow us to move to the creation of digital counterparts for powerful

This research is financially supported by The Russian Science Foundation, Agreement No. 21-11-00296, https://rscf.ru/en/project/21-11-00296/.

devices and significantly reduce the costs of development due to the need for long-term energy - consuming resource tests. The processes occurring in the discharge chambers and jets of AC plasma torches are non-stationary, and the computational models being developed also require significant time to perform calculations, even on powerful supercomputers. The methods of generative design of analytical models can solve the difficult part of the problem - to construct differential equations that characterize the motion of the electric arc spot on the electrode surface, using experimental data. To be successful in creation and realization of these methods is necessary to achieve the following goals. The preparation and analysis of such experimental data is one of the goals and the aim of this work. The second goal is to evaluate the influence of the arc spot dynamics on the gas-dynamic and electrical parameters of the plasma torch. The latter task involves performing a series of calculations of the plasma flow in an alternating electric field for various parameters describing the motion of the electric arc spot on the electrode surface.

II. METHODS AND APPROACHES

A. Features of AC plasma torches

A wide range of AC plasma torches was developed at the IEE RAS [8, 16]. The most promising are high-voltage devices that have the longest continuous operation time (up to 2000 hours with a power of up to 500 kW). A characteristic design feature of such plasma torches is the separate arrangement of the electrodes in cylindrical channels. This allows the burning of long arcs (from several cm to 2.5 m) with a high voltage drop (up to 3 kV) at relatively low current values (from units to 100 A). Studying plasma torches of this type, characteristic zones with different geometric and dynamic parameters of the arc column were noted [17]. So, it is possible to distinguish one zone of the arc column, stabilized by the gas flow in the channel, and another, intensively moving in the transverse gas flow at a distance from the exit from the channels. When considering the energy characteristics of high-voltage AC plasma torches, as a rule, little attention is paid to near-electrode processes. This is due to the fact that the role of each electrode (cathode or anode) changes cyclically with a frequency determined by the period of current change, and the near-electrode voltage drops are hundreds of times lower than the voltage drop over the entire long arc. Nevertheless, new models of plasma torches with multi-gas plasma-forming media can be used as independent plasma-chemical reactors. Information about all the non-stationary processes

979-8-3503-8358-4/23 $31.00 © 2023 IEEE

occurring in them can be crucial for the development of a new technology. Thus, the process of moving the arc spot over the surface of the electrode, which ensures the distribution of the thermal load to increase the electrode resource, should also be considered as a source of the parameter gradient.

B. Operating modes of the experimental plasma torch

The high-speed video material obtained for various operating modes of the electrode system of the experimental AC plasma torch was used for the research. To control the movement of the arc spot on the surface of the electrode, the swirling gas flow is used. Solenoids with different characteristics can be also used to enhance the arc spot motion. The parameters corresponding to these modes are shown in Table 1. Fig. 1 shows a schematic of a cylindrical electrode with characteristic points for which the magnetic field parameters are determined. The table shows the values of the field density in the vicinity of point 1 (Fig. 2) at the moments of the current maximum. Zero values are specified for the mode without applying an external magnetic field. For each of the modes, the current and voltage drop on the arcs with a frequency of 10 kHz were recorded. Data on the motion of the electric arc spot was recorded using a Citius Imaging C100 CENTURIO high-speed digital video camera with a frame rate of 4000 fps.

When processing the data using specially developed software, the image is filtered for each video recording. The contours described by the arc spot, the coordinates of the arc spot position at each frame, and the corresponding tangential and linear speeds are determined.

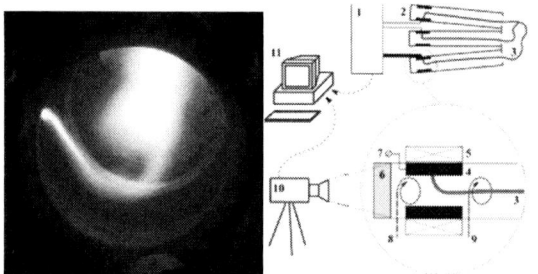

Fig. 1. Photo of an AC electric arc spot on the electrode surface (exposure 1/4000 s) and schematic of the experimental installation [18]. 1 - power supply, 2 - plasma torch, 3 - arc column, 4 - hollow electrode, 5 - solenoid, 6 - window for photo and video shooting, 7 - terminal for connection of the power supply, 8 and 9 - plasma forming gas load, 10 - high-speed video camera, 11 – data acquisition system.

TABLE. I. THE ARC CURRENT OF THE PLASMA TORCH AND THE PARAMETERS OF THE MAGNETIC FIELD IN THE VICINITY OF THE CHARACTERISTIC POINT OF THE ELECTRODE SURFACE

Operating mode, №	Arc current (rms), A	Field density in the vicinity of the point 1 (Fig. 2), $\times 10^{-3}$ T
1	46.2	0
2	47.3	10.7
3	84.3	0
4	85.0	7.2
5	83.4	14.1
6	85.0	15.2
7	83.6	19.8

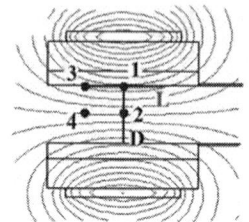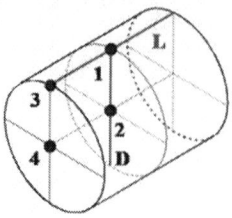

Fig. 2. Schematic diagram of the hollow cylindrical electrode with solenoid. 1 – 4 - characteristic points in the electrode chamber; L and D – lines showing the area of the working surface in the longitudinal direction and the diameter, respectively.

C. Modeling of processes in the channel of a high-voltage plasma torch

High-speed video data will allow us to determine the time change in the position of the radius vector of the reference point of the centre of the arc electrode spot r, as well as to calculate the corresponding time derivatives $\dot{r}$ and $\ddot{r}$. Using the generative design method, the equation of motion of the arc spot will be restored,

$$(\ddot{r}, \dot{r}, r, t, u) = 0,$$

where t is time, and u is a function that considers the influence of external factors.

The equation is restored under the assumption of the continuity of the process of moving the spot along the electrode. In the first approximation, it is assumed that the dimensions of the arc spot are known and do not change during movement

The equation of motion will be integrated into the general algorithm for calculating the processes in an ac plasma torch with air used as a plasma-forming gas [19, 20]. This algorithm involves the joint calculation of the equation of motion of partially ionized air and the Maxwell equations of electrodynamics. Air is modeled as an isobaric non-isothermal single-phase medium, where the molar mass, electrical conductivity and thermophysical parameters depend on temperature.

During the simulation, it is assumed to establish the influence of the movement of the arc spot point both on the parameters of the medium in the vicinity of the electrode, and on the integral parameters, such as the VAC of the ac plasma torch.

III. RESULTS AND DISCUSSION

Video clips taken for different operating modes of the plasma torch (see Table 1) were processed. Examples of captured frames are shown in Fig. 3. Eight frames recorded sequentially with an interval of 1.25 ms during one half-cycle of current change (10 ms) are presented at the top row. The images show the movement of the near-electrode section of the arc column. One can notice an increase and subsequent decrease in the brightness of the arc column, corresponding to an increase and decrease in the arc current. The figure below shows the same frames (4 in a row) with markers of the arc spot positions applied to the image.

As a result, for each considered mode, data on the position of the arc binding spot at each fixed time (angular coordinate) were obtained, and the tangential and linear speeds of the arc spot movement along the electrode surface were calculated.

979-8-3503-8358-4/23 $31.00 © 2023 IEEE

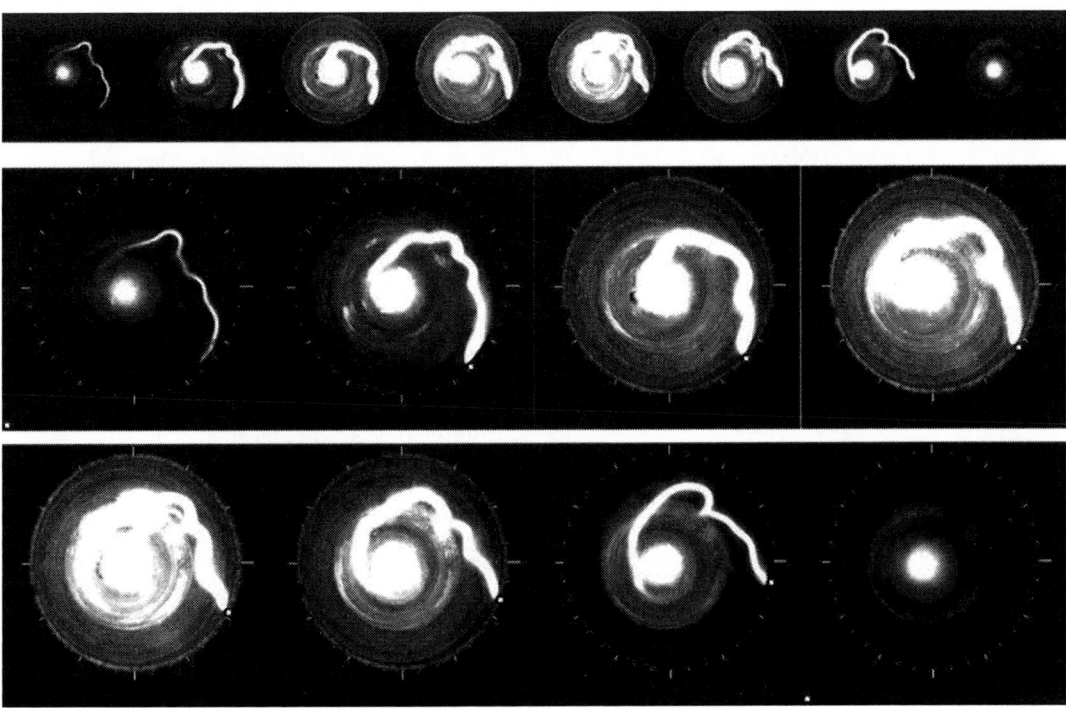

Fig. 3. Frames of a video clip of an electric arc with movement of the arc spot along the electrode surface.

The movement of the arc during each half-period is unique, it is noticed that electric arcs move, leaving a continuous trace on the surface, although, so-called jumps are also observed, repeating with different frequency in different operating modes. However, for each mode, patterns can be determined by considering data for a long period of time (a large number of periods of current change). Fig. 4 shows the graphs of the dependence of the tangential component of the arc spot speed vs time, obtained for the operation of a plasma torch with a current of 83.6 A (Tab.1, operating mode 7). The field is switched on in antiphase with the current, i.e. it "slows down" the arc from the moment of its initiation to the current drop. The graphs show that the spot moves more intensively only after 5 ms from the moment of arc initiation, when the modulus of the value of the current changing a long the sinusoid would already have a falling branch. The curve of the arc spot speed for each moment of time is averaged over 174 half-periods (recorded time is 1.74 s). The curves of arc speed for even and odd half-periods, averaged over 87 half-periods, respectively, are also presented. These curves have different shapes, which should be due to the difference in the interaction of the arc with the electrode surface in the cathode and anode phases.

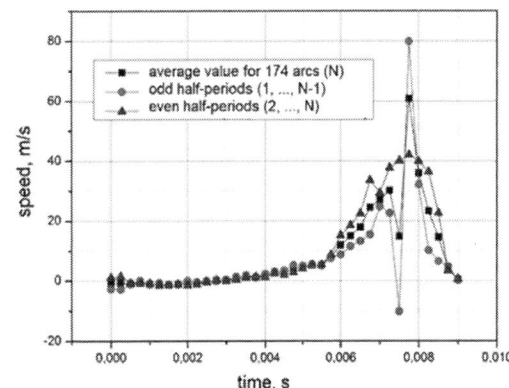

Fig. 4. The change of the speed of the arc spot moving along the surface of the hollow electrode during the lifetime of the arc (one half-cycle of the current).

TABLE. II. THE AVERAGE SPEED OF THE ELECTRIC ARC SPOT MOVEMENT ALONG THE SURFACE OF HOLLOW CYLINDRICAL ELECTRODE FOR DIFFERENT OPERATION MODES OF THE PLASMA TORCH

№	Arc current (rms), A	Flow rate, g/s	Flow rate proportion, upstream : downstream	Tangential gas velocity in the arc spot area, m/s	Field density in the arc spot area, ×10⁻³ T	Speed of the arc spot, m/s
1		5		14.33		4.50
2	47.3	9.4	4 : 5	26.82	10.7	8.80
3		12.2		34.63		10.72
4			3 : 5	12.08		10.20
5	83.5	4.5	1 : 1	12.35	14.1	11.10
6			5 : 3	11.43		9.80

Table 2 shows the data on the average speed of the electric arc spot for different opera ting modes of the plasma torch. The flow proportions are given for the two gas inputs in the near-electrode region, shown in the diagram Fig.1 (8 - upstream and 9 - downstream). The values of the tangential velocity of the gas flow in the near-electrode region are calculated for the cold flow without considering the electric arc. Data on the flow rate and magnetic field density are given for a certain region in the plasma torch channel near the electrode surface (point 1 in Fig. 2). The average speed of the electric arc was determined for each of the modes for the time during which the arc spot passed 10 complete revolutions along the cylindrical surface of the hollow electrode.The movement of the arc spot in the considered electric arc systems can be controlled by changing the plasma forming gas flow rates, as well as by changing the parameters of the magnetic field. For an AC arc having a sinusoidal current waveform, the application of an alternating magnetic field with a synchronous change in the current in the solenoid allows either to accelerate the movement of the arc spot with an increase in the arc current, or to slow it down when the opposite polarity is turned on. The movement of the arc spot during the lifetime of each arc between the moments of the periodic change of the current polarity is uneven. The movement of the spot under the action of only gas-dynamic forces (without a solenoid) in the tangential direction is characterized by lower velocities than the corresponding component of the gas flow velocity. The superimposition of a magnetic field allows for a short-term acceleration of the arc movement in the near-electrode region.

The obtained experimental data on the dynamics of the arc motion under various operating modes of the plasma torch (arc current, power, gas supply modes and additional magnetic stabilization of the arc), together with data on the geometric dimensions of the area of interaction of the electric arc with the electrode surface, will allow us to determine the heat fluxes acting on the electrode and, using known methods for calculating near-electrode processes, will allow us to predict the erosion of the electrodes.

Calculations of the operating parameters of a high-voltage plasma torch on an improved two-dimensional model for different areas of the conducting surface of the electrode did not show any significant changes in the parameters. Changes in the interelectrode distance and the length of the plasma torch channel with the introduction of a variable dependence of the spot position on time allowed us to obtain minor changes in the parameters in the out jet but did not significantly affect their values in the channel. It seems appropriate to switch to a non-stationary three-dimensional model that allows more realistic resolution of spatial changes in parameters.

The obtained data are necessary for constructing advanced models for calculating AC plasma torches. At this stage, it seems that the data obtained is sufficient to be able to reconstruct the laws of motion of the arc spot under different operating modes. At the next stage of the work, the methods of generative design of analytical models will be used to construct differential equations that characterize the motion of the electric arc binding spot on the electrode surface. This will allow us to improve the model for calculating the operation of the plasma torch, which considers the interaction of the arc discharge with the plasma-forming gas.

IV. CONCLUSION

Data for various operating modes of a high-voltage AC plasma torch are obtained experimentally. The use of vortex and magnetic stabilization of the position of the spot of the electric arc allows one to change the speed of its movement along the inner surface of the cylindrical electrode. The range of changes in the average speed of the arc spot was from 4.5 to 11.1 m/s for 10 full revolutions. At the same time, the motion of the arc spot is uneven, and the range of speed

changes was from 0 to 80 m/s during one half-period of current change (10 ms). The obtained experimental data on the dynamics of the arc motion for various modes of operation of the plasma torch (arc current, power, gas supply modes and additional magnetic stabilization of the arc) with the use of known methods for calculating near-electrode processes will allow one to predict the erosion of the electrodes.

The law of movement of the electric arc spot is not reduced to a simple model, for example, moving at a constant speed along a simple trajectory. A sufficiently large amount of data is available to reconstruct the law of motion, and it is planned to do this in the next stage to perform a complete calculation of the AC plasma torch. The inclusion of the arc spot motion in the calculation will allow one to obtain information about the importance or, conversely, not the importance of taking into account the near-electrode processes occurring in high-voltage AC plasma torches for different technological purposes.

ACKNOWLEDGMENT

The authors thank the IEERAS and partially SPbPU for the data provided.

This research is financially supported by The Russian Science Foundation, Agreement No. 21-11-00296, https://rscf.ru/en/project/21-11-00296/.

REFERENCES

[1] Snoeckx, R., Bogaerts, A.: Plasma technology – a novel solution for CO 2 conversion? Chem. Soc. Rev. 46, 5805–5863 (2017). https://doi.org/10.1039/C6CS00066E.

[2] Samal, S.: Thermal plasma technology: The prospective future in material processing. J. Clean. Prod. 142, 3131–3150 (2017). https://doi.org/10.1016/j.jclepro.2016.10.154.

[3] Xu, Y., Zhang, X., Yang, C., Zhang, Y., Yin, Y.: Recent Development of CO2 Reforming of CH4 by "Arc" Plasma. Plasma Sci. Technol. 18, 1012–1019 (2016). https://doi.org/10.1088/1009-0630/18/10/08.

[4] Mapamba, L.S., Conradie, F.H., Fick, J.I.J.: Technology assessment of plasma arc reforming for greenhouse gas mitigation: A simulation study applied to a coal to liquids process. J. Clean. Prod. 112, 1097–1105 (2016). https://doi.org/10.1016/j.jclepro.2015.07.104.

[5] Rutberg, P.G., Kuznetsov, V.A., Popov, V.E., Popov, S.D., Surov, A. V., Subbotin, D.I., Bratsev, A.N.: Conversion of methane by CO2+H2O+CH4 plasma. Appl. Energy. 148, 159–168 (2015). https://doi.org/10.1016/j.apenergy.2015.02.087.

[6] Mostaghimi, J., Boulos, M.I.: Thermal Plasma Sources: How Well are They Adopted to Process Needs? Plasma Chem. Plasma Process. 35, 421–436 (2015). https://doi.org/10.1007/s11090-015-9616-y.

[7] Fulcheri, L., Fabry, F., Takali, S., Rohani, V.: Three-Phase AC Arc Plasma Systems: A Review. Plasma Chem. Plasma Process. 35, 565–585 (2015). https://doi.org/10.1007/s11090-015-9619-8.

[8] Surov, A.V., Popov, S.D., Popov, V.E., Subbotin, D.I., Serba, E.O., Spodobin, V.A., Nakonechny, G.V., Pavlov, A.V.: Multi-gas AC plasma torches for gasification of organic substances. Fuel. (2017). https://doi.org/10.1016/j.fuel.2017.02.104.

[9] Rutberg, P.G., Kuznetsov, V.A., Serba, E.O., Popov, S.D., Surov, A.V., Nakonechny, G.V., Nikonov, A.V.: Novel three-phase steam-air plasma torch for gasification of high-caloric waste. Appl. Energy. 108, (2013). https://doi.org/10.1016/j.apenergy.2013.03.052.

[10] Kim, K.S., Park, J.M., Choi, S., Kim, J., Hong, S.H.: Enthalpy probe measurements and three-dimensional modelling on air plasma jets generated by a non-transferred plasma torch with hollow electrodes. J. Phys. D. Appl. Phys. 41, (2008). https://doi.org/10.1088/0022-3727/41/6/065201.

[11] Cunha, M.D., Hartmann, W., Wenzel, N., Benilov, M.S., Almeida, P.G.C.: Simulating Propagation of Spots over Cathodes of High-Power Vacuum Circuit Breakers. Proc. - Int. Symp. Discharges Electr. Insul. Vacuum, ISDEIV. 2, 385–388 (2018). https://doi.org/10.1109/DEIV.2018.8537028.

[12] Delachaux, T., Fritz, O., Gentsch, D., Schade, E., Shmelev, D.L.: Numerical simulation of a moving high-current vacuum arc driven by

a transverse magnetic field (TMF). Proc. - Int. Symp. Discharges Electr. Insul. Vacuum, ISDEIV. 1, 273–276 (2006). https://doi.org/10.1109/DEIV.2006.357284.

[13] Marotta, A., Sharakhovsky, L.I.: Heat transfer and cold cathode erosion in electric arc heaters. IEEE Trans. Plasma Sci. 25, 905–912 (1997). https://doi.org/10.1109/27.649588.

[14] Kumar, A.S., Okazaki, K.: Electrode Surface Control of the Transition from Micro-Arcs to High-Current Arcs in Atmospheric-Pressure Plasmas. IEEE Trans. Plasma Sci. 23, 735–741 (1995). https://doi.org/10.1109/27.467996.

[15] Benilov, M.S.: A Self-Consistent Analytical Model of Arc Spots on Electrodes. IEEE Trans. Plasma Sci. 22, 73–77 (1994). https://doi.org/10.1109/27.281554.

[16] Rutberg, P.G., Safronov, A.A., Popov, S.D., Surov, A.V., Nakonechny, G.V.: Multiphase stationary plasma generators working on oxidizing media. Plasma Phys. Control. Fusion. 47, (2005). https://doi.org/10.1088/0741-3335/47/10/006.

[17] Rutberg, P.G., Nakonechny, G. V, Pavlov, A. V, Popov, S.D., Serba, E.O., Surov, A. V: AC plasma torch with a H2O/CO2/ CH4 mix as the working gas for methane reforming. J. Phys. D. Appl. Phys. 48, (2015).

[18] Surov, A.V., Popov, S.D., Serba, E.O., Nakonechny, G.V., Spodobin, V.A., Ovchinnikov, R.V., Kumkova, I.I., Shabalin, S.A.: The investigation of movement dynamics of an AC electric arc attachment along the working surface of a hollow cylindrical electrode under the action of gas-dynamic and electromagnetic forces. J. Phys. Conf. Ser. 406, (2012). https://doi.org/10.1088/1742-6596/406/1/012007.

[19] Bykov, N.Y., Obraztsov, N. V., Kobelev, A.A., Surov, A. V.: Modeling of an AC Plasma Torch Part I: Electrical Parameters and Flow Temperature. IEEE Trans. Plasma Sci. 49, 3, 1017-1022 (2021). https://doi.org/10.1109/TPS.2020.3040845.

[20] Bykov, N.Y., Obraztsov, N. V., Kobelev, A.A., Surov, A. V.: Modeling of an AC Plasma Torch—Part II: Gasdynamic Pattern and Effect of Flow Rate. IEEE Trans. Plasma Sci. 49, 3, 1023–1027 (2021). https://doi.org/10.1109/TPS.2021.3066035.

One-hot Programming LUT for FPGAs

Ivan Vasenin
Department of AT
PNRPU
Perm.Russia
Vasenin.Ioann@yandex.ru

Sergey F. Tyurin
Department of AT
PNRPU
Perm.Russia
tyurinsergfeo@yandex.ru

Abstract — **This article discusses the advantages of single-hot state automata, including ease of design, simple synchronization analysis, and high clock speeds. The article describes the designed and simulated novel one-hot coding input-one-hot coding output LUT. One of the important steps in state encoding is the choice between binary encoding and single hot encoding. In one-time encoding, a separate status bit is used for each state. It is called one-hot because only one bit is «hot» or true at any time. Each bit of the state is stored in a trigger, so more triggers are required for single encoding than for binary encoding. However, with a single encoding, the logic of the next state and output is often simpler, so fewer elements are required. The state decoder is simplified because the state bits themselves can be used directly to indicate whether the machine is in a certain state. They are particularly suitable for modern register-rich FPGA architectures.**

Keywords — one-hot, LUT, FPGA, component.

I. INTRODUCTION

FPGAs (Field-Programmable Gate Array) are widely used in various devices: consumer electronics, telecommunications equipment, various boards for use in data centers, various robotics, as well as in prototyping ASIC chips (application-specific integrated circuit).

One of the most important characteristics of FPGAs (programmable logic integrated circuits) is its logical capacity. Capacity determines how complex digital devices can be synthesized. The logical capacity shows how much can fit in the crystal. The abbreviation LUT stands for Look Up Table or simply Lookup Table, which can literally be translated as "reference table" or "search table". LUT is more than a table, LUT is a method of implementing a function in which direct calculation is replaced by a search through a table of ready–made solutions.

II. LITERATURE REVIEW

A. LUT

The basis of configurable logic in FPGA is an elementary computing unit. The LUT has a single-bit output signal, which is calculated based on the values of the input signal and the entries in the configurable table (or memory), as shown in the figure below.

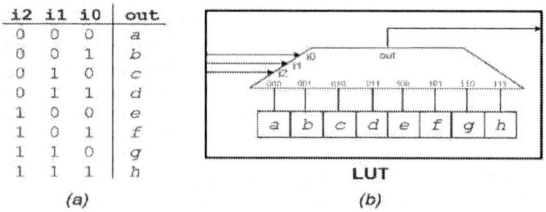

Fig. 1. (a) The relationship of the LUT truth table (b) The diagram of the logical behavior of the LUT.

The application of the LUT method to the FPGA allows you to implement any logical function in the form of SRAM (static random access memory). The address is the argument, and the contents of the cell are the value. In order to describe the logical function of three variables (DATAA, DATAB and DATAC), there is enough memory for 8 cells. The truth table is stored as a mask (LUT-mask) in the corresponding CRAM (Configuration RAM) cell. With the help of multiplexers, the desired value is selected. Multiplexers are controlled by input port signals to build a LUT (k-LUT) that implements any logical function from k variables, 2^k bits of SRAM and 2^{k-1} multiplexers are required. The device is shown in the figure below.

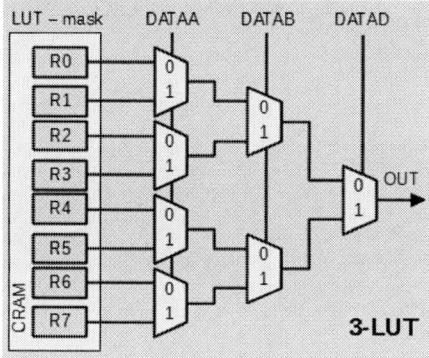

Fig. 2. LUT device

With this approach, it is possible to predict the signal transit time quite accurately, and it will not depend on the implemented logical function. This important feature makes possible the temporal analysis of the scheme..

B. The principle of LUT implementation.

There is some Boolean function y = (a & b) |~c. The schematic representation and the truth table are shown in Figure 3. The function has three arguments, so it takes $2^3 = 8$ values. Each of them corresponds to its own combination of input signals. These values are calculated by the FPGA firmware development program and recorded in special configuration memory cells.

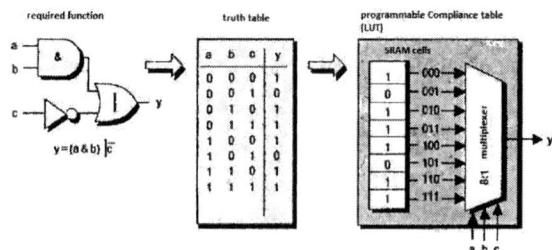

Fig. 3. The principle of LUT implementation

979-8-3503-8358-4/23 $31.00 © 2023 IEEE

The value of each of the cells is fed to its input of the output multiplexer LUT, and the input arguments of the Boolean function are used to select one or another value of the function.

LUT can implement any Boolean logic equation. The limit is only in the number of inputs. This characteristic is shown in the figure below. LUTs are usually chained or combined sequentially to implement larger logical equations.

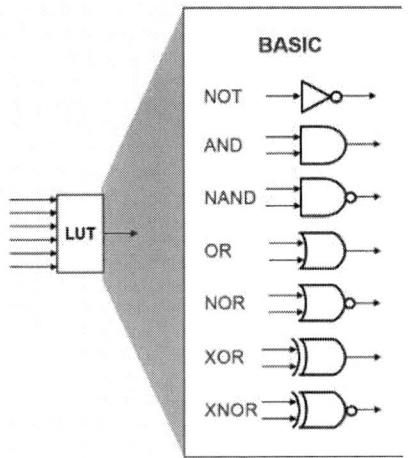

Fig. 4. Examples of several logical functions that LUT can potentially implement

C. One-hot encoding

In a single encoding, only one bit of the state vector is asserted for any given state. All other bits of the state are zero. Thus, if there are n states, then n state dips are required. The state decoder is simplified because the state bits themselves can be used directly to indicate whether the machine is in a certain state. No additional logic is required.

- Machines with one hot state usually work faster.
- The speed does not depend on the number of states and instead depends only on the number of transitions to a certain state.
- Characteristic designs of finite state machines minimize the number of triggers
- Machines with one hot state use one trigger for each state and therefore require much less decoding logic
- Schematics can be entered into computer memory, and code can be written directly from the state diagram without encoding the state table.

In one hot encoding, a separate status bit is used for each state. It is called one-time because only one bit is "hot" or TRUE at any given time.

Also, unitary encoding (one-hot encoding) is used in the State Machine Processing of the Quartus FPGA CAD by Intel and is performed by default when automatically synthesizing circuits, but there is a possibility of using another encoding.

III. PROBLEM STATEMENT AND SOLUTION METHOD

In my research, I put forward an idea, which is to try to use a unitary code to calculate logical functions. The purpose of applying such an idea is to increase the speed of logic elements, both in circuits and individually. When calculating

logical functions, the tuning takes place of one transistor, if there is a unitary code.

IV. EXPERIMENTS

To conduct simulation experiments, I used the scheme shown in Figure 5. The LUT-1 scheme is presented, when modeling, encoding occurs for 1 variable. The input can be a combination of 01 or 10. And in Figures 6 and 7, there are already 2 input variables, respectively, possible combinations of 0001, 0010, 0100, 1000.

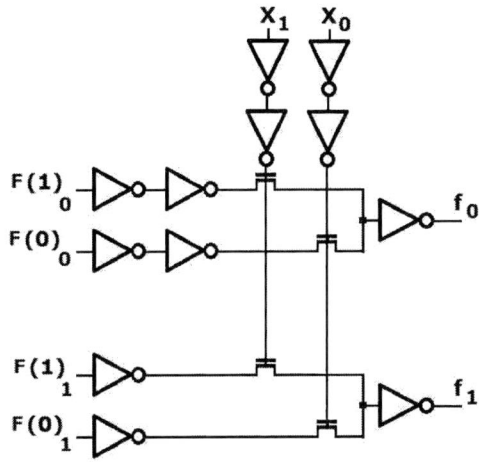

Fig. 5. LUT-1 with unitary encoding of the input variable and function

To test the proposed method, a circuit simulation of a LUT circuit was carried out using Multisim software. Modeling of the work with the use of unitary coding of the input variable and function is performed.

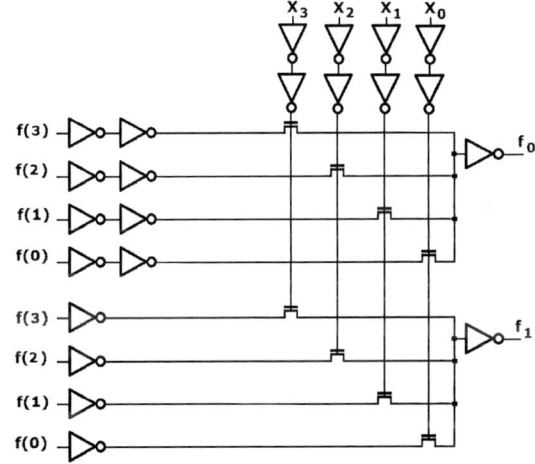

Fig. 6. LUT-2 with unitary encoding of input variables and functions

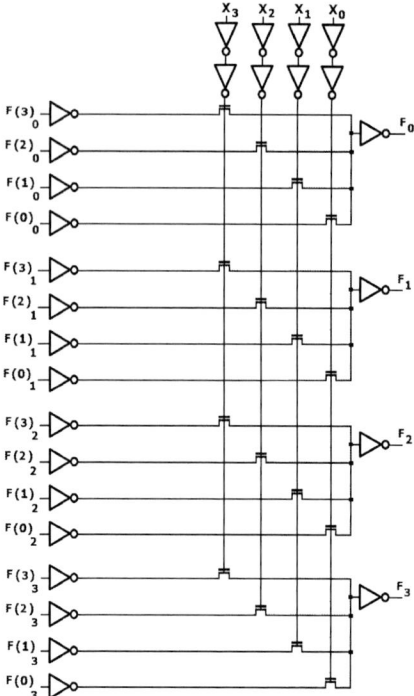

Fig. 7. LUT-2 with unitary encoding of input variables and functions

As a result of the simulation, the following results were obtained, presented in Table 1.

TABLE I. EXAMPLE OF ENCODING INPUT VARIABLES AND TWO FUNCTIONS

X3	X2	X1	X0	Z1 =OR	Z2=XOR	F_0	F_1	F_2	F_3
0	0	0	1	0	0	1	0	0	0
0	0	1	0	1	1	0	0	0	1
0	1	0	0	1	1	0	0	0	1
1	0	0	0	1	0	0	0	1	0

TABLE II. EXAMPLE OF ENCODING INPUT VARIABLES AND TWO FUNCTIONS (CONTINUATION OF TABLE 1)

X3	X2	X1	X0	F_0	F_1	F_2	F_3
0	0	0	1	1	0	0	0
0	0	1	0	0	0	0	1
0	1	0	0	0	0	0	1
1	0	0	0	0	0	1	0

TABLE III. EXAMPLE OF ENCODING INPUT VARIABLES AND TWO FUNCTIONS (CONTINUATION OF TABLE 1)

$F(0)_0$	$F(0)_1$	$F(0)_2$	$F(0)_3$	$F(1)_0$	$F(1)_1$	$F(1)_2$	$F(1)_3$
1	0	0	0	0	0	0	0
0	0	0	0	0	0	0	0
0	0	0	0	0	0	0	0
0	0	0	0	0	0	0	0

TABLE IV. EXAMPLE OF ENCODING INPUT VARIABLES AND TWO FUNCTIONS (CONTINUATION OF TABLE 1)

X3	X2	X1	X0	F_0	F_1	F_2	F_3
0	0	0	1	1	0	0	0
0	0	1	0	0	0	0	1
0	1	0	0	0	0	0	1
1	0	0	0	0	0	1	0

TABLE V. EXAMPLE OF ENCODING INPUT VARIABLES AND TWO FUNCTIONS (CONTINUATION OF TABLE 1)

$F(2)_0$	$F(2)_1$	$F(2)_2$	$F(2)_3$	$F(3)_0$	$F(3)_1$	$F(3)_2$	$F(3)_3$
0	0	0	0	0	0	0	0
0	0	0	0	0	0	0	1
0	0	0	0	0	0	1	0
0	0	0	1	0	0	0	0

V. RESEARCH RESULTS

Circuit modeling of the LUT circuit was performed using Multisim software. A combination of 01 or 10 was supplied to the input, combinations were set using keys so that the function combination was also equal to 01 or 10 at the output. The simulation was carried out with different variants of key closure. The inverter was also assembled in this program on the basis of transistors.

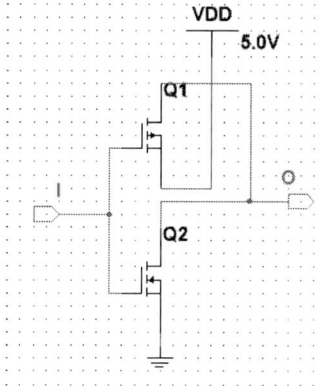

Fig. 8. Not gate

Figure 9 shows a scheme with the help of which modeling with unitary coding was performed.

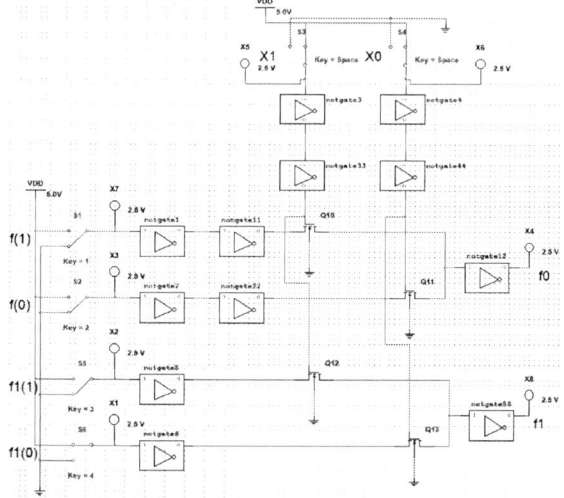

Fig. 9. LUT-1 created in Multisim software

In Figure 10, a combination of 10 is supplied and additionally the keys are not closed. The output is the correct combination of 10.

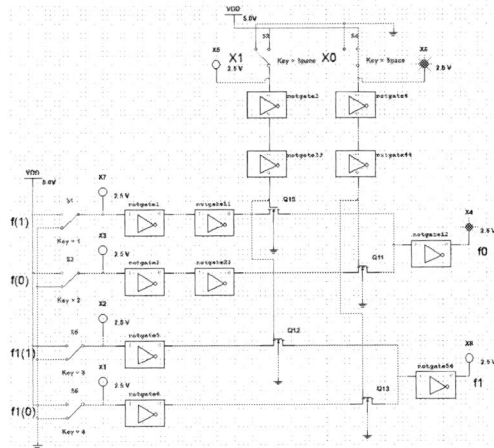

Fig. 10. LUT-1 and combination 10

In Figure 11, the combination 01 is supplied and the keys are additionally closed. The output is the correct combination 01.

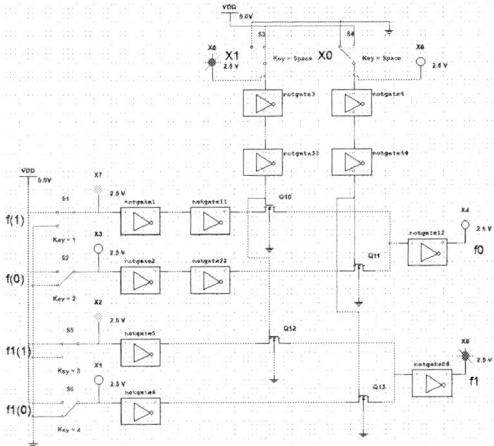

Fig. 11. LUT-1 and combination 01

In Figure 12, a combination of 10 is supplied and the keys are additionally closed. The output is the correct combination 01.

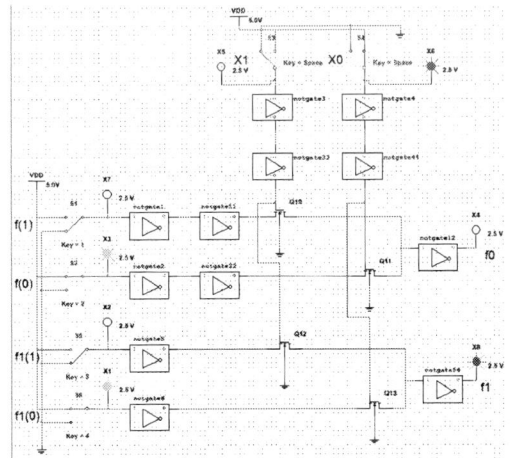

Fig. 12. LUT-1 and a combination of 10 at the input and 01 at the output

In Figure 13, the combination 01 is supplied and the keys are not additionally closed. The output is the correct combination of 10.

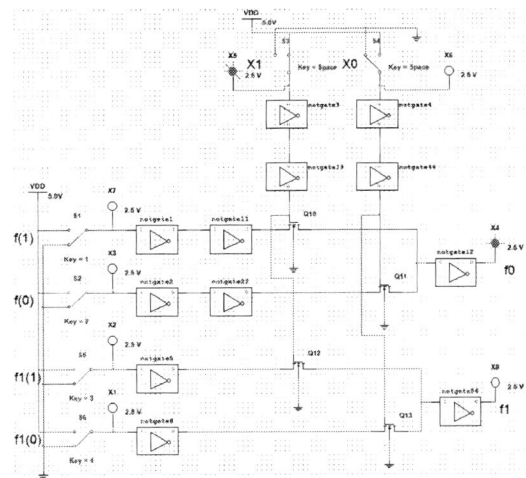

Fig. 13. LUT-1 and a combination of 01 at the input and 10 at the output

VI. CONCLUSION

However, when calculating functions of a large number of variables, the unitary code becomes too large and cumbersome. This may be an additional load on the system.

Therefore, the task of this study is to consider and apply the implementation of logical functions in unitary coding with the greatest positive effect.

Thanks to this study, it will be possible in the future to determine the complexity of the relevant schemes. And the opportunity to explore an approach for applying a combined representation. This will be the special case when the unitary code is partially used. This is the subject of future research.

ACKNOWLEDGMENT

This study was conducted with the support of the Department of Automation and Remote Control of Perm National Research Polytechnic University (Head of the Department Prof. Alexander A. Yuzhakov).

Special thanks to the Honoured Inventor of Russia, Professor of the Department of Automation and Telemechanics of Perm National Research Polytechnic University (PNRPU) Tyurin S.F.

REFERENCES

[1] FPGA architecture. Part 1. Logic element, Nov. 2022, [online] Available: http://www.labfor.ru/articles/fpga_arch_le

[2] FPGA Architecture Basics, Nov. 2022, [online] Available: https://www.rapidwright.io/docs/FPGA_Architecture.html

One-hot state machine design for FPGAs, Steve Golson, Nov. 2022, [online] Available: https://trilobyte.com/pdf/golson_pldcon93.pdf

[3] FPGA, Nov. 2022, [online] Available: https://xakep.ru/2018/11/15/fpga/

[4] Multisim, Nov. 2022, [online] Available: https://www.ni.com/ru-ru/support/downloads/software-products/download.multisim.html#452133

One-Hot Encoding, Nov. 2022, [online] Available: https://www.sciencedirect.com/topics/computer-science/one-hot-encoding

Thermal and Thermoelectric Properties of Bi-Te-Sb and Bi-Te-Se Doped by Carbon Additive

Irina A. Voloschuk
Institute of Advanced Materials and Technologies
National Research University of Electronic Technology
Moscow, Russia
raccoon13vol@gmail.com

Alexey V. Babich
Institute of Advanced Materials and Technologies
National Research University of Electronic Technology
Moscow, Russia
drent@yandex.ru

Maxim S. Rogachev
Institute of Advanced Materials and Technologies
National Research University of Electronic Technology
Moscow, Russia
rmaks1988@yahoo.com

Daria D. Glebova
Institute of Advanced Materials and Technologies
National Research University of Electronic Technology
Moscow, Russia
dariag99@mail.ru

Anastassiya S. Bozhedomova
Institute of Advanced Materials and Technologies
National Research University of Electronic Technology
Moscow, Russia
bozhedomova98@gmail.com

Tatiana A. Babich
Institute of Advanced Materials and Technologies
National Research University of Electronic Technology
Moscow, Russia
tibloko4545@gmail.com

Abstract— In this work, the thermoelectric and thermal properties of materials based on Bi-Te-Sb and Bi-Te-Se with various concentrations of carbon additive were studied. The mass percentage of alloying additives was from 0.1% to 2.0%. The electrical conductivity, Seebeck coefficient, power factor and thermal properties were measured. The highest values of electrical conductivity were possessed by legs of thermoelements with an additive concentration of 0.1% and 0.5% (for n- and p-type respectively). However, for the Seebeck coefficient, the maximum values were observed for a concentration of 1%, which also determines the maximum values of the power factor, so this composition is the most optimal. Thermal stability studies have shown that there is no phase transformations the investigated materials in the required temperature range (from 20 °C to 150 °C). Thus, it was shown that C45 is a promising additive for optimizing the properties of thermoelectric materials.

Keywords—thermoelectricity, materials testing, heat treatment, calorimetry

I. INTRODUCTION

In recent years, one can notice a strong interest in the development of flexible thermoelectric generators (TEG). Such devices have many possible applications. In particular, they can be used to power different wearable electronics (phones, watches etc) [1-9].

In such devices, low-intensity heat fluxes can be used for conversion into electricity. Another topical trend is the use of similar thermoelectric devices to power heat flow sensors, which can significantly improve the energy efficiency of buildings and structures.

However, despite the recent results, the currently obtained devices do not have the most optimal characteristics, which hinders their widespread use. In this regard, for the further development of this direction, it is necessary to develop methods and approaches that will make it possible to obtain thermoelectric generators with improved characteristics.

The work was supported by the Russian Science Foundation: 21-19-00312 (synthesis of materials) and 18-79-10231 (study of material properties).

Currently, flexible TEG technology uses a rather limited range of materials. They can be inorganic and organic However, for organic materials, degradation of properties over time can be observed, which will negatively affect the performance of the device, in addition, organic materials in most cases have worse characteristics than inorganic ones. [1-7].

Recently, some interesting achievements have been made in the field of creating new thermoelectric materials, and research is still ongoing, but at present, materials based on Bi-Te-Se and Bi-Te-Sb, proposed quite a long time ago, remain among the most effective in the low-temperature range, at which, flexible TEGs operate (from room temperature to about 100-150 °C). The first material is used as n-type, the second as p-type. To ensure the maximum efficiency of thermoelectric materials, it is necessary to optimize the composition of solid solutions. To ensure the optimal concentration of charge carriers in thermoelectric material, an optimal choice of dopants is required.

Preliminary results showed that carbon additives can affect the properties of such materials. However, such information in literature is quite little, in addition, there is lack of information about the effect of these additives on the thermal properties and stability of thermoelectric materials.

For this reason, the purpose of this research was to investigate the thermal stability and thermoelectric properties of Bi-Te-Sb and Bi-Te-Se doped by carbon additive.

II. MATERIALS AND METHODS

Four different compositions of suspensions based on powders of thermoelectric materials were investigated. The mass percentage of carbon additive was: 0.1 %, 0.5 %, 1 % and 2 %. $Bi_2Te_{2.8}Se_{0.2}$ (n-type) and $Bi_{0.5}Sb_{1.5}Te_3$ (p-type) were used as thermoelectric materials, an aqueous alkaline solution of sodium silicate was used as a binder. Thermoelectric materials of Bi-Te-Sb and Bi-Te-Se systems were synthesized by direct synthesis in quartz ampoules from the elements with semiconductor purity. An insulin syringe was used to dose the binder. The resulting compositions are presented in Table I.

TABLE I. SUSPENSION COMPOSITIONS

№	Additive, mass %	Thermoelectric material, mass %	Binder, mass %
1	0.1	84.9	15
2	0.5	84.5	15
3	1.0	84.0	15
4	2.0	83.0	15

To form the legs of n- and p-type thermoelements, the thermoelectric material, binder and additive (Super C45) were mixed in the required ratio (Table I), after which a metal mask with 6 windows was filled with the finished suspension. The geometric parameters of the window were: $10.0 \times 5.0 \times 0.3$ mm^3.

Next, the samples were dried at room temperature for 24 h, after which two-stage annealing was carried out: 4 hours at a temperature of 110 °C and 15 minutes at a temperature of 200 °C. Since the sample is deformed during drying (a slight bending of the legs appeared), after drying, the samples were subjected to grinding on both sides successively on sanding paper.

The formation of electrical contacts to the legs was carried out using the method of electrochemical deposition.

In order to evaluate the behavior of the studied materials and to make sure that there are no undesirable processes during heating, the thermal properties of each studied composition were investigated. For this, a Netzsch DSC 204 F1 Phoenix differential scanning calorimeter was used. Five measurements were carried out for each sample from room temperature up to 300 °C at a heating rate of 10 degrees per minute in a nitrogen atmosphere. Al pans were used. An empty crucible was used as a comparison sample.

The study of thermoelectric and electrophysical parameters was carried out on a hardware-software complex that allows creating the necessary temperature gradient at the cold and hot sides of thermoelements, and heating the investigated thermoelement legs in the range from -5 to 130 °C with a temperature gradient of 5 °C.

III. RESULTS AND DISCUSSION

The results of measurements using differential scanning calorimetry (DSC) can be seen in Figures 1 and 2 as an example. For other compositions, a similar behavior is observed.

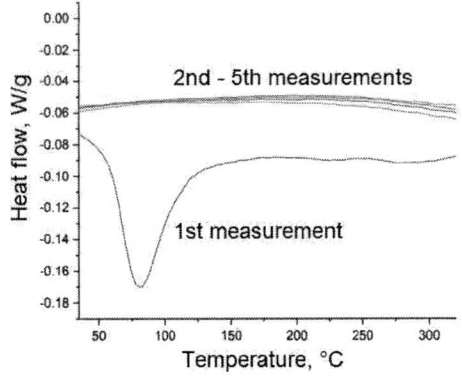

Fig. 1. DSC results for n-type material with 0.5 mass % of C45

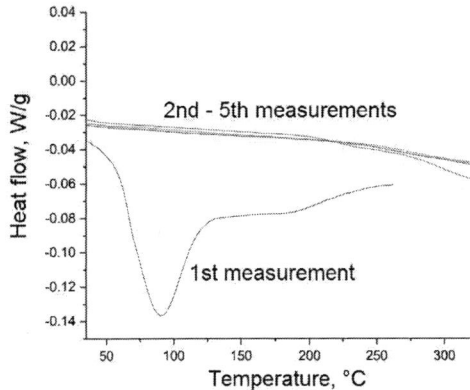

Fig. 2. DSC results for p-type material with 0.5 mass % of C45

It can be noted that for all compositions after the first DSC measurement, there is a noticeable endothermic peak in the temperature range of about 100°C, which may be due to the removal of moisture from the samples. At the same time, no thermal effects are observed in the DSC curves after subsequent measurements at all temperatures. And it should be noted that the DSC curves are changing little during the measurements. This may indicates a high thermal stability of the samples in the studied temperature range after heat treatment and removal of moisture.

Next, the Seebeck coefficient was studied. One of the obtained graphs for p-type material is shown in Figure 3.

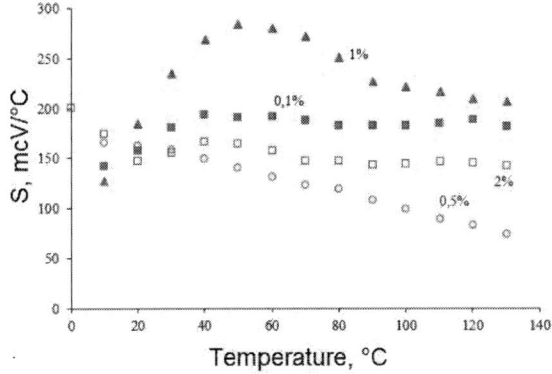

Fig. 3. Seebeck coefficient for p-type thermoelectric legs

The results of studying the effect of various concentrations of carbon additive on the Seebeck coefficient showed that suspensions with an additive concentration of 1% showed the best results for the legs of n- and p-type thermoelements.

The results of studying the the conductivity of thermoelement legs made from suspensions with carbon additive are presented in Figure 4.

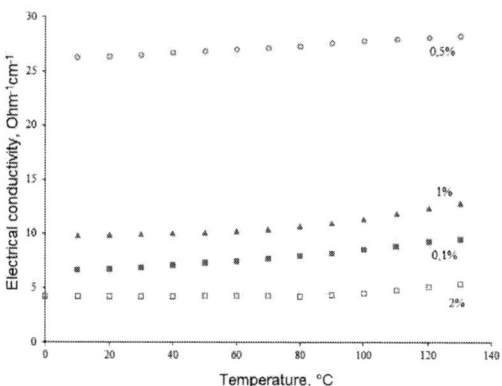

Fig. 4. Temperature dependences of electrical conductivity for p-type thermoelectric legs

The highest values of electrical conductivity were possessed by thermoelement legs with an additive concentration of 0.1% (for n-type) and 0.5% (for p-type). It should be noted that for all suspensions used, the dependences in the studied temperature range are close to linear.

Based on the results obtained, the power factor was calculated for the investigated thermoelement legs. Results are shown in the Figures 5 and 6.

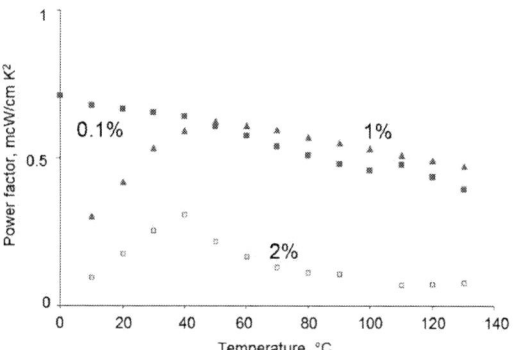

Fig. 5. Power factor for n-type thermoelectric legs

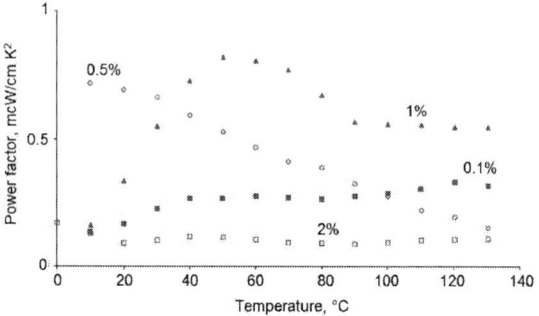

Fig. 6. Power factor for p-type thermoelectric legs

The power factor calculation results showed that suspensions with an additive concentration of 1% showed the best results: for n-type samples, the maximum power factor was 0.62 $\mu W/cm \cdot K^2$, for p-type samples, 0.82 $\mu W/cm \cdot K^2$.

Thus, according to the combination of characteristics, it can be noted that the thermoelectric legs (for both n- and p-

type) with an 1% of C45 additive has the best properties for using in thermoelectric generators.

IV. CONCLUSION

Thus, in this work, the thermal and thermoelectric properties of materials based on Bi-Te-Sb and Bi-Te-Se doped by carbon additive (C45) were studied.

Four different additive concentrations were tested (from 0 to 2.0 mass. %). The highest values of electrical conductivity were possessed by legs of thermoelements with an additive concentration of 0.1% (for n-type) and 0.5% (for p-type).

However, for the Seebeck coefficient, the maximum values were observed for a concentration of 1%, which also determines the maximum values of the power factor, so this composition is the most optimal. The study of thermal properties showed high stability of materials after heat treatment.

Thus, it can be noted that the studied materials are promising for the use in thermoelectric generators.

ACKNOWLEDGMENT

The work was supported by the Russian Science Foundation: 21-19-00312 (synthesis of materials) and 18-79-10231 (study of material properties).

REFERENCES

[1] Jaziri N., Boughamoura A., Muller J., Mezghani B., Tounsi F., Ismail M. A comprehensive review of thermoelectric generators: technologies and common applications. Energy Reports, 2020, V. 6, pp. 264-287.

[2] Sherchenkov A., Shtern Yu., Mironov R., Shtern M., Rogachev M. Current state of thermoelectric material science and the search for new effective materials. Nanotechnologies in Russia, 2015, V. 10, pp. 827–840.

[3] He W., Zhang G., Zhang X., Ji J., Li G., Zhao X. Recent development and application of thermoelectric generator and cooler. Applied Energy, 2015, V. 143, pp.: 1-25.

[4] He R., Schierning G., Nielsch K. Thermoelectric devices: a review of devices, architectures, and contact optimization. Advanced Materials Technologies, 2018, V. 3, pp. 1700256.

[5] Champier D. Thermoelectric generators: a review of applications. Energy Conversion and Management, 2017, V. 140, pp. 167-181.

[6] Siddique A.R.M., Mahmud S., Heyst B.V. A review of the state of the science on wearable thermoelectric power generators (TEGs) and their existing challenges. Renewable and Sustainable Energy Reviews, 2017, V. 73, pp. 730-744.

[7] Suarez F., Parekh D.P., Ladd C., Vasaee D., Dickey M.D., Ozturk M.C. Flexible thermoelectric generator using bulk legs and liquid metal interconnects for wearable electronics. Applied Energy, 2017, V. 202, pp. 736-745.

[8] Qing S., Rezania A., Rosendahl A., Enkeshafi A. Gou X. Characteristics and parametric analysis of a novel flexible ink-based thermoelectric generator for human body sensor. Energy Conversion and Management, 2018, V. 156, P. 655-665.

[9] Pelz U., Jaklin J., Rostek R., Kroner M., Woias P. Novel fabrication process for micro thermoelectric generators (μTEGs). Journal of Physics: Conference Series, 2015, V. 660, pp. 012084.

Modeling and Analysis of the Functioning of the Sensitive Element of a Capacitive Microaccelerometer Based on the Microstructure of Silicon Carbide(Sic) and Silicon(Si)

Ye Ko Ko Aung
Institute of nano and microelectronic
National Research University of
Electronic Technology
Zelenograd, Moscow, Russia
yekokoaung64675@gmail.com

Boris M. Simonov
Institute of nano and microelectronic
National Research University of
Electronic Technology
Zelenograd, Moscow, Russia
serborsel@mail.ru

Sergey P. Timoshenkov
Institute of nano and microelectronic
National Research University of
Electronic Technology
Zelenograd, Moscow, Russia
spt111@miee.ru

Phyo Win Tun
Institute of nano and microelectronic
National Research University of
Electronic Technology
Zelenograd, Moscow, Russia
kophyowinhtun0@gmail.com

Paing Soe Thu
Institute of nano and microelectronic
National Research University of
Electronic Technology
Zelenograd, Moscow, Russia
paingsthu7@gmail.com

Abstract— A comb-type capacitive accelerometer based on the microstructure of silicon carbide (SiC) and silicon (Si) is studied by the finite element method. The deformation of the movable mass during accelerations with different temperatures from room temperature to 100°C was studied. The maximum deformation changes of the movable mass (SiC) due to temperature changes is greater than the movable mass (Si). This means that Si accelerometer has more stabilized due to the temperature change than SiC. The maximum residual stress occurs in the moving mass (SiC) due to temperature changes. According to the results, the resonant frequency of the movable mass (SiC) is higher than the movable mass (Si). This means that it cannot only effectively avoid structural resonance that may cause structural damage, but also extend the operating frequency range of the accelerometer. The article shows the characteristics of the two materials and which one is better suited for different conditions.

Keywords—MEMS - comb-type capacitive accelerometer; deformation; maximum residual stress; resonant frequency

I. INTRODUCTION

One of the most important stages in the development of MEMS inertial sensors is their thermal modeling. Its necessity is due to the fact that micro devices operate in an environment with changing temperature, and they are temperature sensitive due to the properties of the materials used in their designs. Temperature has a particularly strong effect on MEMS micro devices due to the small size and distances between their individual moving elements. Changes in temperature change the technical parameters and characteristics of these devices and hence their usefulness. As an example, let's pay attention to the fact that with the electrostatic method of capacitance measurement (more precisely, capacitance difference), the measured values are small, ranging from units to 10-14 pF, which obviously requires measurement of accuracy, and even very small temperature fluctuations can worsen measurement process. On the other hand, it is known that the capacitance is a function of the dimensions of the electrodes, which change with temperature, therefore, in conditions of temperature instability, probing is difficult and potential errors must be compensated for in various ways. Since inertial MEMS

sensors are based in many cases on measurements of capacitance change, research in this area is well justified. When one or another inertial sensor operates under different temperature conditions, these changes will lead to thermal deformations, thermal stresses, and a change in the Young's modulus of materials used in micro devices. These changes will lead to deviations in their output (electrical) signals. The problem of the influence of mechanical deformations caused by temperature changes has been raised in many publications [1-2], however, in these works, only the mechanical response of the device is considered - the spring stiffness, damping coefficient, natural frequencies and quality factors.

In this article, we consider a differential capacitive accelerometer using a SiC and a Si microstructure by the FEM method. The principle of operation and the technological process of manufacturing the accelerometer are presented. In the modal analysis, we found that the accelerometer using the SiC inertial mass has a higher natural oscillation frequency compared to the Si-based accelerometer with the same structure. The deformations of the moving inertial mass during acceleration (10g-50g) in the temperature range from 20 to 100°C are analyzed. An accelerometer using SiC had a maximum stress due to temperature change compared to an accelerometer based on Si with the same structure. The modeling of the detection capacity and analysis of the differential capacitive accelerometer based on SiC and Si were carried out. Thus, the purpose of the article is to identify differences in the characteristics of two differential capacitive accelerometers based on SiC and Si materials and to determine which one is more thermally stable.

II. DESIGN OF A COMB-TYPE CAPACITIVE ACCELEROMETER

In Figure 1 shows a structural model of a comb-type one-axis capacitive microaccelerometer, consisting of a sensitive element mass with comb-shaped finger electrodes, fixed electrodes, and two pairs of folded beams. Sensitive element mass is located in the middle, which is fixed on the substrate with four folded beams. The fixed electrode plates and the movable electrode plates are arranged opposite side with

979-8-3503-8358-4/23 $31.00 © 2023 IEEE

each other and form differential acceleration detection electrodes.

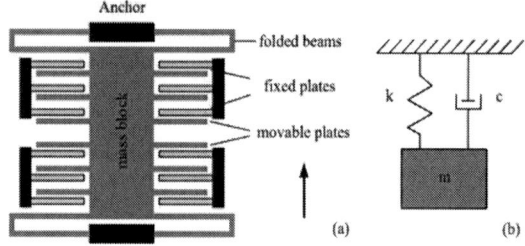

Fig. 1. Model of a one axis capacitive microaccelerometer: physical (a) and mechanical models of the accelerometer (b)

Under the influence of acceleration, the inertial mass is subjected to the action of an internal force and has a certain displacement in the direction of the acceleration. This deformation is converted into the capacitances changes and the acceleration value can be obtained by measuring the changes in capacitances. On fig. 2 and 3 schematically show the manufacturing process of the accelerometer [3-2].

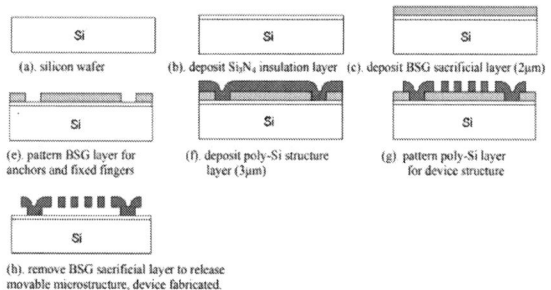

Fig. 2. Block diagram of the main operations of the technological process of manufacturing a MEMS comb accelerometer (from Si material)

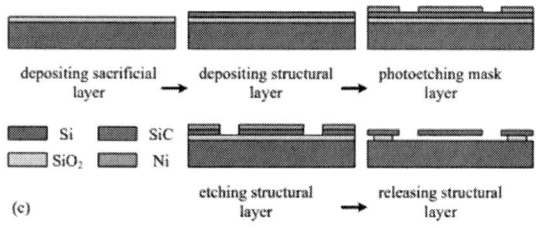

Fig. 3. Block diagram of the manufacture process of a MEMS comb accelerometer (from SiC materials)

III. THEORETICAL ANALYSIS OF THE DEVICE

When acceleration is applied to the accelerometer in a horizontal direction parallel to the mass, the inertial mass is deflected by the force of inertia. Its deviation occurs in the direction opposite to the action of the applied acceleration. The sensitivity of the device to displacement is defined as the displacement of the movable mass (and movable finger electrodes) per unit of free fall acceleration g (g=9.8 m/s2) along the direction of the sensitivity axis of the device.

Let us designate for each section of width of the beam and length of the beam as W_b and L_b, separately for each beam. We denote the width and length of the central inertial mass as W_m and L_m, and the thickness of the device as t. In total there are N_f groups of fingers, among which are N_s groups of sensitive fingers and N_d groups of adductors ($N_f=N_s+N_d$). For each movable finger, the width and length

of the finger are equal to W_f and L_f. In the absence of acceleration, the capacitive gap between each movable finger and its left (right) fixed finger is equal to d_0. The density ρ and Young's modulus E of the silicon material are shown below.

When there is no acceleration, the static (sensing) capacitance of the MEMS comb accelerometer is:

$$C_1 = C_2 = C_0 = \frac{\varepsilon.N_s.L_f.t}{d_0} \qquad (1)$$

Let acceleration to the left horizontally act, the moving mass experiences the force of inertia to the right by x, as shown in Figure 2. In the small deviation approximation ($x \ll d_0$), the left (right) capacities C_1 (C_2) has changed to [4]:

$$C_1 = \frac{\varepsilon.N_s.L_f.t}{d_0+x} = \frac{\varepsilon.N_s.L_f.t}{d_0.\left(1+\frac{x}{d_0}\right)} \approx \frac{\varepsilon.N_s.L_f.t}{d_0}\cdot\left(1-\frac{x}{d_0}\right)$$
$$(2)$$

$$C_2 = \frac{\varepsilon.N_s.L_f.t}{d_0-x} = \frac{\varepsilon.N_s.L_f.t}{d_0.\left(1-\frac{x}{d_0}\right)} \approx \frac{\varepsilon.N_s.L_f.t}{d_0}\cdot\left(1-\frac{x}{d_0}\right)$$
$$(3)$$

The total perceiving mass of the accelerometer is equal to M_s, and the inertia force F_{in} acting on the sensitive inertial mass during acceleration along the direction of the sensitivity axis is equal to:

$$F_{ин} = -M_s.a \qquad (4)$$

We denote the total stiffness of the beams as K_{total}, then the displacement x of the moving mass can be calculated as:

$$K_{total} = \frac{2EW_bt^3}{L_b^3} \qquad (5)$$

A. Structural design and optimization

The geometrical design parameters of the sensitive element (SE) of the capacitive MMA are given in Table 1, the characteristic properties of silicon carbide (SiC) and (Si) - in Table 2.

TABLE I. GEOMETRIC PARAMETERS OF THE DESIGN OF THE SENSITIVE ELEMENT (SE) CAPACITIVE MMA

Geometric parameters	Parameters of the investigated designs of sensitive element of the capacitive MMA
Width, length, thickness of the movable inertial mass (μm)	700×400×20
Width, length, thickness folded beam (μm)	400×400×20

TABLE II. CHARACTERISTIC PROPERTIES OF SILICON CARBIDE (SiC) AND SILICON (SI)

Property of the material	SiC	Si
Density (g/cm³)	3.14	2.33
Young's modulus (GPa)	404	131
Poisson's ratio	0,159	0,27
Thermal expansion coefficient(10⁻⁶C⁻¹)	4.4	2.6

979-8-3503-8358-4/23 $31.00 © 2023 IEEE   140

IV. MODELING AND SIMULATION OF THE SENSITIVE ELEMENT OF THE CAPACITIVE ACCELEROMETER

A. Study of the deformation of a moving mass under the influence of acceleration and temperature change

A study was made of the deformation of a moving mass under the action of acceleration from 10 to 50g at various temperature changes from +20 °C to 100 °C. As expected, the accelerometer sensor is affected by the ambient temperature. At a temperature of 22 °C, there are no noticeable movements along the sensitivity axis X in the absence of acceleration. With increasing temperature, the change in the dimensions of the constructs, due to the expansion along the X axis, causes additional stresses and deformations of the suspensions on both sides of the device. Under conditions of variable ambient temperature, changing the dimensions of the inertial mass and individual electrodes naturally causes measurement inaccuracies, since it causes a change in capacitance. The results of calculating the deformations of moving masses made of Si and SiC materials under the action of acceleration from 20 to 50g and when the temperature changes from +20 °C to 100 °C were obtained and analyzed, shown below in Fig. 3, a-e.

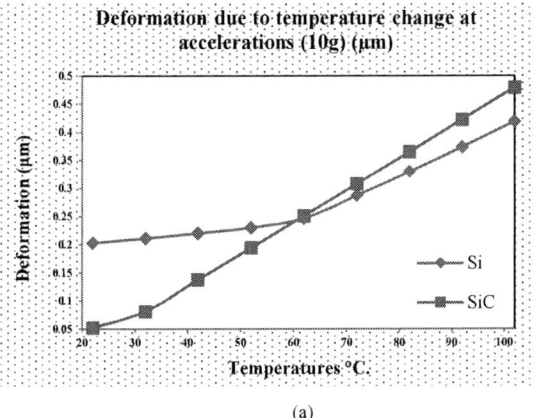

(a)

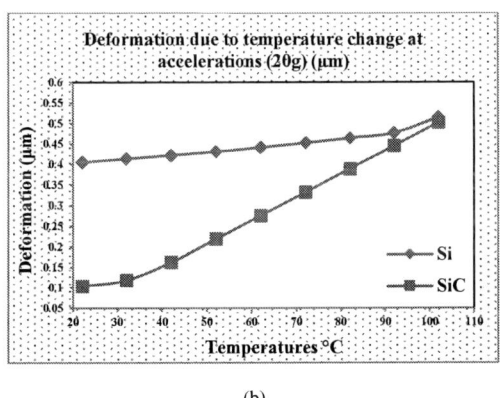

(b)

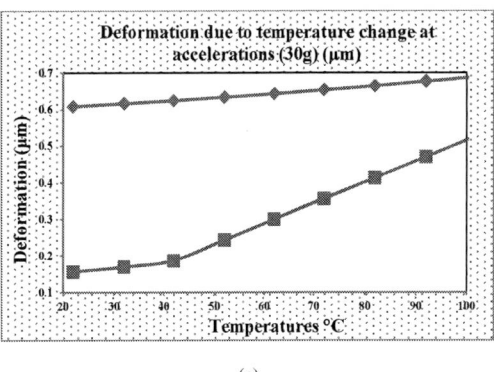

(c)

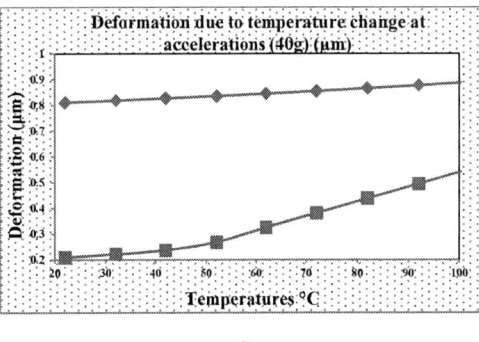

(d)

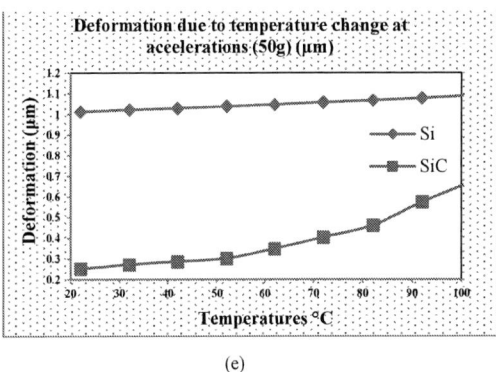

(e)

Fig. 4. Results of calculating the deformations of a moving inertial mass under the influence of acceleration (a) 10 g ,(b) 20g, (c) 30g, (d) 40g, (e)-50 g and temperature (from 20 to 100 °C)

The results of deformation of the moving mass during acceleration were calculated using the Ansys program. According to the results obtained, the deformation of the moving mass (Si) at an acceleration of 50g at 20°C is 0.20272 μm, and at 100°C it is 0.41879 μm. The deformation of the moving mass (Si) at an acceleration of 50g at 20°C is 0.052028 μm, and at 100°C it is 0.47869 μm. In all of the graphs shown, the deformation of the movable Si mass changes slightly due to temperature changes, and the deformation of the SiC movable mass changes greatly at temperatures below 40°C. This means that the different deformation from 20°C to 100°C is small, the accelerometer capacitance will be linear when the temperature changes. Therefore, we can say that the deformation of the inertial mass of Si changes more linearly than that of the inertial mass of SiC with temperature. On Fig. 5 and 6 shows the distribution of mechanical stresses arising in the sensitive element (SE) of a micromechanical accelerometer (MMA) It can be seen that when the temperature changes, the

979-8-3503-8358-4/23 $31.00 © 2023 IEEE

distribution of stresses in the SE remains unchanged, only their values increase.

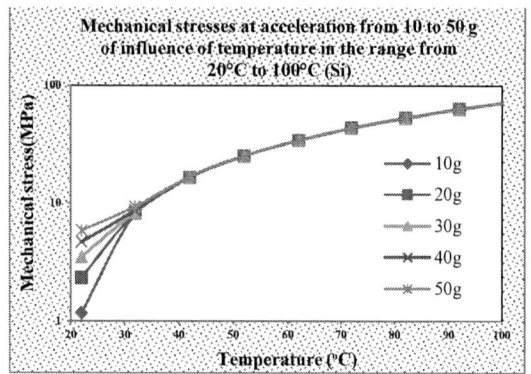

Fig. 5. The results of mechanical stresses in the inertial mass of Si under the influence of acceleration and temperature variation

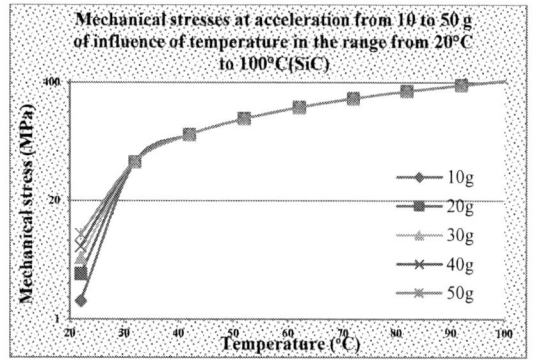

Fig. 6. The results of the calculation of mechanical stresses in inertial mass of SiC under the influence of acceleration and temperature variation

It can be seen from the obtained results that the maximum mechanical stress arising in the movable inertial mass of Si under the action of temperature and acceleration does not exceed 100 MPa. The maximum mechanical stress in the moving inertial mass of SiC is greater, at a temperature of 100°C it is 424.87 MPa. More mechanical stress can reduce the resistance of the device. The greater the value of mechanical stress due to temperature variation can cause the failure of the device [6].

B. Calculations of the natural frequency of the Sensitive element of the capacitive accelerometer

A modal analysis of the accelerometer with different widths of folded support beams was carried out, and the results showed that the width of the folded beams had a noticeable effect on the natural frequency and mode shape of the accelerometer. As shown in Figure 5a, the natural oscillation frequency for different oscillation modes of the silicon carbide accelerometer increases with the beam width. When the width of the beam is in the range of 5 to 25 μm, the oscillatory motion of the inertial mass along the sensitivity axis x is the first mode, and the oscillatory motion along the y axis is the second vibration mode. However, when the beam width is more than 25 μm, the order of vibration modes changes, i.e. movement along the y axis, oscillations of the first mode are set, and along the x axis of the second mode. When the accelerometer is energized, it will first vibrate along the Y-axis. However, the Y-axis is not the sensitive axis of the accelerometer, so fluctuations along this axis do not contribute to obtaining accurate measurements of external accelerations.

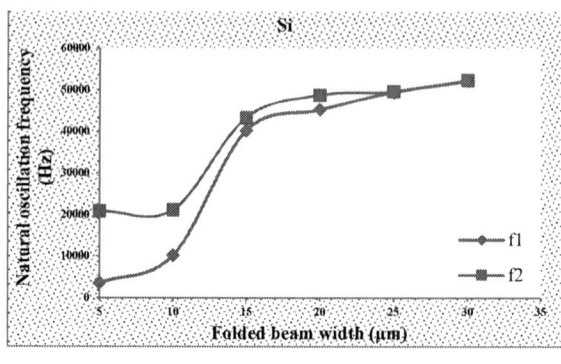

(a)

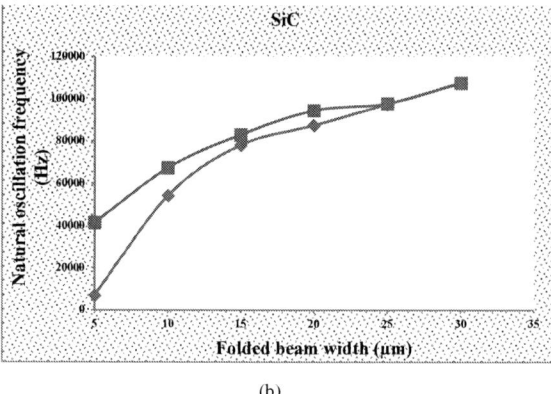

(b)

Fig. 7. Dependences of the eigenfrequencies of oscillations of the first and second modes on the width of the folded springs for the SE capacitive accelerometer based on the SiC microstructure (a) and the microstructure based on Si (b). (The blue line is the natural frequency of the oscillation mode along the x-axis, and the red line is the natural frequency of the oscillation mode along the y-axis.)

From fig. 7, and it is also seen that with an increase in the width of the folded spring, the natural frequency of the first mode gradually approaches the natural frequency of oscillations of the second mode. That is, the greater the width of the folded spring, the more easily cross-coupling occurs between vibration modes. So, when vibrating along the x-axis, increasing the width of the folded spring causes vibrations along the y-axis, and this is an undesirable effect. Therefore, in order to reduce cross-coupling between the x and y oscillations, the width of the folded beam should be less than 25 μm. According to the analysis carried out, the width of the folded beam must be 5 μm in order for the accelerometer to simultaneously achieve two advantages: a large frequency range and low cross-parasitic coupling.

Comparison of natural oscillation frequencies in the first accelerometers with inertial mass from SiC and from Si of the same structure, as shown (see Fig. 6), is performed. The results show that when the width of the folded beam is from 5 to 30 μm, the ratios of the natural frequencies of the first mode were equal to 1.28, and the ratios of the natural frequencies of the second mode were approximately equal to 1.29. Therefore, an accelerometer with an inertial mass of SiC has a higher natural oscillation frequency than a silicon accelerometer of the same structure. The high value of the natural oscillation frequency not only effectively prevents the structure from resonance, which can lead to its damage, but also expands the operating frequency range of the accelerometer [5].

C. Calculations of the capacitance changes of moveable electordes and fixed electrodes of the capacitive accelerometer

When the inertial mass is displaced from the equilibrium state, the movable electrodes are also displaced and their position changes relative to the static electrodes. In the presence of non-zero voltage (in relation to this type of capacitor) - the capacitance varies depending on the applied external acceleration. This capacitance is read by the measuring device and is ready for further processing in the ROIC (readout integrated circuit The dependence of the capacitance of the capacitors associated with the moving inertial mass at an acceleration of 10g at different temperatures is shown in Fig.8.

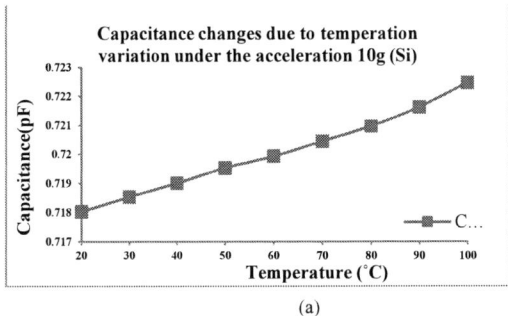

(a)

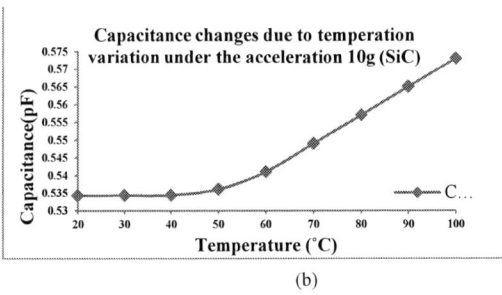

(b)

Fig. 8. The dependence of the measured capacitance of the capacitors associated with the moving inertial mass on temperature: for the inertial mass of Si (a) and of SiC (b)

According to the results obtained, the change in the capacitance of the moving electrodes at an acceleration of 10 g in the case of a Si target at a temperature of 20°C is 0.718 pF and 0.723 pF at 100°C, and the change in the capacitance of the moving electrodes for a SiC target at a temperature of 20°C is 0.535 pF and at 100°C - 0.575pF. The difference in capacitance changes for the Si cell is 0.005 pF, and for the SiC cell it is 0.04 pF. It can be concluded that the capacitance values under the influence of temperature in the SE of the capacitive accelerometer from Si almost do not change, and in the SE of the capacitive accelerometer from SiC there is a noticeable change in the measured capacitance due to temperature changes.

V. CONCLUSIONS

In this article, the principle of operation and technological schemes for manufacturing the sensitive elements of a capacitive accelerometer with an inertial mass of silicon and silicon carbide were considered. Thanks to the modal analysis of the microaccelerometer, the structural parameters of the elastic microbeam were optimized, which not only made it possible to ensure the operation of the microaccelerometer (MMA) according to the first oscillation mode, but also expanded the difference in natural frequencies between the first and second oscillation modes in order to avoid cross parasitic coupling.

It was found that the ratio of natural frequencies of vibrations of the first and second modes of an accelerometer with an inertial mass of SiC compared to an accelerometer of silicon with the same structure is approximately equal to 1.3. The SiC inertial mass accelerometer has a higher natural oscillation frequency, for this reason, it can not only effectively avoid structural resonance that may cause damage to it, but also extend the operating frequency range of the accelerometer.

It has been established that the deformation of the SE with the inertial mass of Si depends more linearly on temperature than that of the SE with the inertial mass of SiC. The results show that the maximum mechanical stress in the SE made of Si at temperatures up to 100°C does not exceed 100 MPa, and in the SE with inertial mass of SiC it is 424.87 MPa. More mechanical stress due to temperature fluctuations can reduce the resistance of the device, i.e. Si MMA is more resistant to temperature changes than SiC inertial mass MMA. The values of the measured capacitances under the influence of temperature in the capacitive accelerometer based on Si almost do not change, and in the capacitive accelerometer based on Si and SiC, noticeable changes in the measured capacitances occur.

REFERENCES

[1] Hong Lianjin, Fang Erzheng, *"Model Of Frequency Characteristics Of Capacitive Accelerometers,"* IEEE International Conference on Measuring Technology and Mechatronics Automation (March 13-14, 2010, Changsha, China), 2010, pp. 668-670.

[2] Xiang Tian, Wei Sheng, Zhanshe Guo, Weiwei Xing, Runze Tang *"Modeling and Analysis of a SiC Microstructure-Based Capacitive Micro-Accelerometer,"* Materials 2021,vol-14, (August 27, 2021), 2021, pp. 01-18.

[3] YunboShi, YongqiZhao, HengzhenFeng, HuiliangCao. *"Design, fabrication and calibration of a high-G MEMS accelerometer,"* ScienceDirect, vol-1966, (August 15, 2018), pp.733-742.

[4] Jacek Nazdrowicz, Andrzej Napieralski, *"An Analysis of Temperature Variation Effect on Response and Performance of Capacitive Microaccelerometer Inertial Sensors,"* IEEE 36th SEMI-THERM Symposium (August 18,2020,), 2020, pp.91-96.

[5] Jacek Nazdrowicz, Andrzej Napieralski." *Analysis of Temperature Variation Influence on Capacitance Inertial Sensors Parameters.,"* 19th IEEE ITHERM Conference , (September 27,2020),2020, pp.1364-1373.

[6] Q.J. Tang, X.J. Wang, Q.P. Yang, and C.Z. Liu, "Static temperature analysis and compensation of MEMS gyroscopes", International Journal of Metrology and Quality Engineering, pp. 209–214, 2013.

Fractal Character of the Conductor Destruction in the Zone of Local Increase of the Pulsed Current Density

Dmitry V. Zhukov
Institute of Energy
Peter the Great St.Petersburg
Polytechnic University (SPbPU)
St.Petersburg, Russia
s_zhukov51@bk.ru

Sergey I. Krivosheev
Institute of Energy
Peter the Great St.Petersburg
Polytechnic University (SPbPU)
St.Petersburg, Russia
ksi.mgd@gmail.com

Sergey G. Magazinov
Institute of Energy
Peter the Great St.Petersburg
Polytechnic University (SPbPU)
St.Petersburg, Russia
magazinov_sg@mail.ru

Dmitry V. Kiesewetter
Institute of Energy
Peter the Great St.Petersburg
Polytechnic University (SPbPU)
St. Petersburg, Russia
dmitrykiesewetter@gmail.com

Victor I. Malyugin
Institute of Energy
Peter the Great St.Petersburg
Polytechnic University (SPbPU)
St. Petersburg, Russia
vim@spbstu.ru

Abstract— **Under the conditions of the pulsed current carrying with an amplitude of up to 200 kA along the flat conductor with the defects of various initial lengths, the main factors influencing the propagation of the defects are determined. As a result of comparison of experiments and simulation results, the influence of thermal stresses on the conductor destruction was revealed. An analysis of the defects after repeated carrying of the pulsed current demonstrates the fractal character of the destruction zones formed after exposure. The possibility of determining some fracture parameters using the fiber Bragg gratings was evaluated by numerical simulation**

Keywords— Local inhomogeneity of current density, pulsed heating, thermal stress, fractal character of fracture

I. INTRODUCTION

High-power pulsed systems are widely used in various, not only technological tasks, but also in experimental techniques. These systems provide the realization of high densities of electromagnetic energy in a wide range of pulse duration: 10^{-10} - 10^{-5} s. High energy densities are provided by the flow of pulsed currents with an amplitude of 0.1-2MA, and the current-carrying elements of these systems may be affected by various factors, such as Lorentz forces, Joule heating, affecting the efficiency and resource of these elements. Thus, during generation of strong pulsed magnetic fields, the magnetic energy density can approach the conductor sublimation energy, which requires taking into account the nonlinear current (field) diffusion, which distribution significantly affects the distribution of Lorentz forces (magnetic pressure)[1] acting on current-carrying elements of the magnetic system. In open magnetic systems (solenoid type) in fields of 100-300 T this process leads to acceleration of magnetic system destruction, [2,3], and in quasi-force-free magnetic systems [4,5] significantly increases the requirements for inter-turn isolation.

A number of works note the possibility of using pulse systems to form controlled pressure pulses as applied to the problems of investigating the properties of various materials under high-speed loads [6,7,8]. In these studies, the formation of pressure pulses with an amplitude of up to 1-3 Gpa is realized by flowing pulsed currents through thin flat bifilar conductors, providing control over the parameters of pressure pulses while keeping the current carrying bars intact during flowing pulsed currents [9]. For reusable systems, the fracture process starts when the limiting plastic deformations are reached [10] with the formation of initial defects, leading to a local increase in the current density near the defect [11]. Under subsequent influences, the development of these defects is observed, leading to the destruction of the current-carrying layer. In magnetic systems, this leads to the so-called sawtooth effect and destruction of the magnetic system [12]. As noted in [13,14]. The sawtooth effect in [15] is used to produce directional thin cuts in metal plates and foils. But there are also examples where, due to the flow of pulse currents through the conductor, the crack development was slowed down [16] or as in the study [17] there was a partial healing of the crack. All of the above phenomena are caused by a local increase in current density at the top of the defect.

II. EXPERIMENT

Experimental study of the pulse current interaction with a crack-type defect was carried out at the GIT-50/12 (PCG-50/12) High School of High Voltage Power Engineering. A flat conductor made of bronze 300 mm long, 0.25 mm thick and 25 mm wide was used as an object of research. On the edges of the conductor, 0.2 mm wide notches of different lengths perpendicular to the current flow were made. The limit value of the action integral J_{sm}, at which the conductor material transitions to the liquid phase, is $4{,}75 \times 10^{16}$ [$A^2 sm^{-4}$]. The sample shown in Fig. 1 after the tests demonstrates the geometry of the conductor and the location of the investigated defects

Fig. 1. Specimen geometry after the experiment.

To identify the peculiarities of interaction of pulse current with defects two modes of pulse current generator discharge were realized: oscillatory discharge mode and mode with generation of unipolar current pulse. Typical oscillograms of the experimental current are shown in Fig. 2.

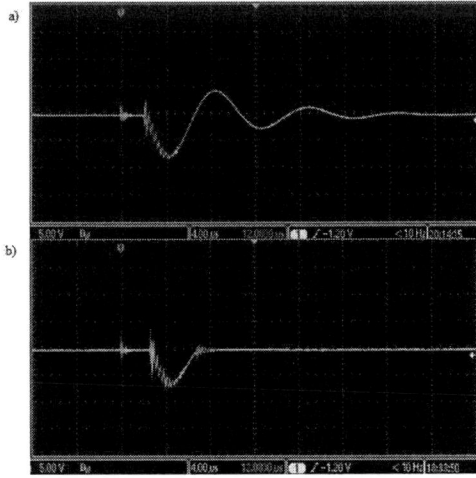

Fig. 2. Oscillograms of current in oscillatory (a) and unipolar (b) modes.

The current in the oscillatory discharge mode has the form:

$$I(t)=I_0*\sin(2\pi t/T)*\exp(-t/\tau),\qquad(1)$$

where $I_0=200\times10^3$A; $T=8,8\times10^{-6}$s; $\tau=7\times10^{-6}$ s. Unipolar current pulse formation was performed by installing oxyzinc nonlinear resistors in the discharge circuit of the PCG.

The samples used in the experiment had identical geometry, the cuts were made by laser cutting with an initial length of 0.2 to 2 mm equally on each sample.

Five series of oscillatory and nine series of unipolar current loading were performed. After each stage, imaging was performed to monitor the developmental features of each defect.

Figs. 3 and 4 show the evolution of the defect with an initial length of 2 mm under sequential exposure to pulsed current in the oscillatory and unipolar discharge modes, respectively. Analysis of images of the defect apex after pulse current exposure, obtained using microscopy after each experiment, showed an increase in the defect size, localized in the defect apex. The presence of droplets of molten material carried out of this zone indicates a high temperature and magnetic pressure sufficient for formation of hydrodynamic flow of molten metal.

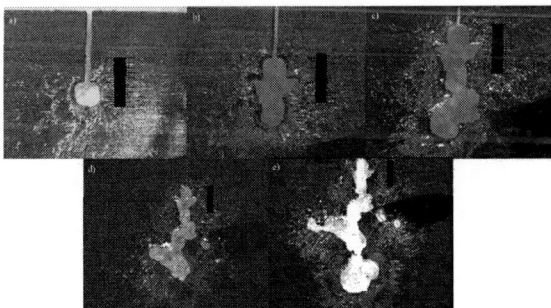

Fig. 3. Development of defects after a)one, b)two, c)three, d)four, and e)five pulses of oscillatory current with an applied sample millimeter.

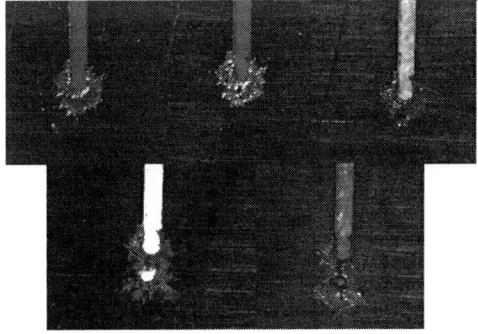

Fig. 4. Development of defects after 5 consecutive loadings with a unipolar pulse.

In addition, the appearance of additional cracks starting from the boundary of the craters formed as a result of current exposure is observed. These cracks do not show traces of melting or other interaction with the flowing current, which makes it likely that they appear due to sufficient thermal stresses for material rupture, resulting from a sharp, localized in the high current density zone, joule heating.

The complex shape of the defects, especially characteristic of repetitive pulses, makes it difficult to analyze the experimentally obtained data. However, from the point of view of influence on durability and service life of the current-carrying element (sample) the important factor is the defect advance in the direction across the current. It should be noted that the elongation of the defect for the oscillatory mode significantly exceeds the elongation under unipolar action. For the realized exposure modes, these dependences of defect elongation on the number of exposures demonstrate sensitivity to the initial defect length, i.e. to the initial degree of inhomogeneity of current distribution in the defect vertex, see Fig.5a. At the same time, the sensitivity to the initial size of the defect of its relative elongation is very weak, and for the oscillatory mode is absent within the accuracy of measurements (Fig.5b).

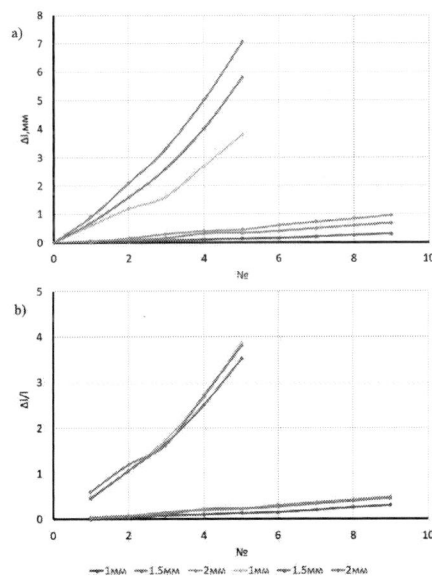

Fig. 5. Plots of (a) absolute and (b) relative elongation of the defect along the incision under the action of an oscillatory (yellow, blue and green) and unipolar (blue, orange and gray) pulse.

979-8-3503-8358-4/23 $31.00 © 2023 IEEE

III. MODELING

For a thin copper conductor with a crack defect, a numerical calculation of the electric and magnetic fields in the three-dimensional formulation was performed in the Comsol Mutiphysics program.

To simulate the electric field, the "AC/DC-electric currents (ec)" model was used. to simulate the magnetic field, the module "AC / DC-magnetic feld (mf)" was used; the connection between the electric and magnetic fields was set through external current densities, similarly to what was done in [18]. The thermal field was calculated in the module "Heat transfer in solid (ht)", similar to what was done in [18]. The mechanical stresses are calculated in the module "Solid Mechanics (solid)" with the bilinear dependence of the material deformation, similar to that in [14].

A unipolar current pulse with an amplitude of 150 kA and a pulse duration of 5.6 µs was set as an external action.

The calculated geometry is a fragment of a flat conductor with one defect, with the following parameters: length of the calculated conductor - l=30 mm, width - c=25 mm, thickness of the conductor h=0.25 mm; length of the defect - 2 mm, width - 0.2 mm.

The calculated grid in the program Comsol Mutiphysics is shown in Fig.6. According to the conductor thickness, the element size is 0.03 mm, see Fig.7, which is 7 times less than the skin layer thickness of 0.2 mm. The maximum size of the element near the top of the defect on the conductor surface was set to 0.01 mm.

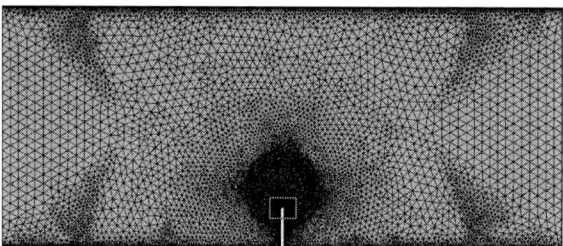

Fig. 6. Calculation grid.

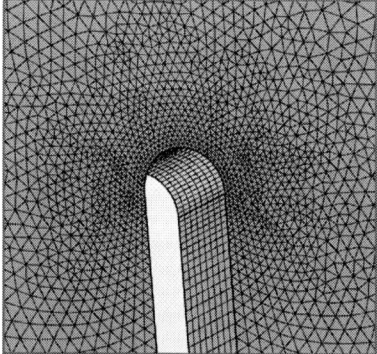

Fig. 7. . Calculation grid near the defect.

The cross-section of the simulated conductor is S=6.25 mm², and the current density amplitude in the area without the defect is 24 kA/mm².

As a result of the simulation, relative strain distribution patterns were obtained for X in Fig. 8 and for Z in Fig. 9.

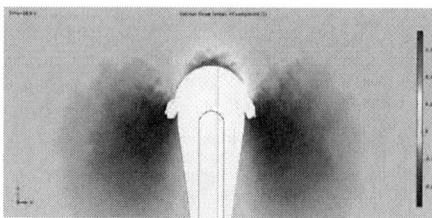

Fig. 8. Distribution of relative strains along the X-axis with the exception of the zone where the temperature and mechanical stresses exceed the melting temperature and the tensile strength, respectively, at the time of 5µs.

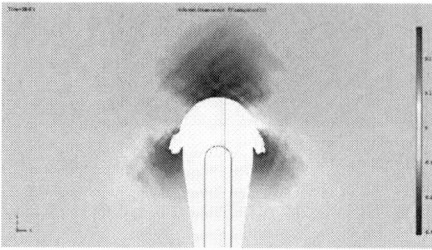

Fig. 9. Distribution of relative strains along the Z axis with the exception of the zone in which the temperature and mechanical stresses exceed the melting temperature and the tensile strength, respectively, at a time of 5µs.

The given distributions of relative strains along the x and z directions, coupled with the allocation of zones in which the temperature and mechanical stresses exceed the melting temperature and the ultimate strength of the material, demonstrate the presence of large relative strains in the directions characteristic of the cracks developing during the experimental exposure. The presence of such a distribution can lead to a violation of the flatness of the sample, which is also observed in the experiment in the form of the appearance of some waviness of the surface. Investigation of such deformations in dynamics by traditional methods is extremely difficult. However, the development of optical techniques using Bragg gratings [19], which allow to register the deformation at the level of 5-10 microns, makes the prognosis for obtaining such information very optimistic.

IV. FRACTAL ANALYSIS

One type of description of patterns at the macroscopic level is fractal analysis. This is demonstrated in the works [20-28].

According to the results of comparing the calculated and experimental data, we can conclude that the process of tire fracture is a complex function, which is a difficult task to describe. It is safe to say that it cannot be considered only from the side of current integral and neglect other factors, such as thermal stresses, mechanical stresses, magnetic field and electric explosion.

In order to determine the dimensionality of a defect in metric space, you can calculate its fractional dimensionality, and the values may take a non-integer numerical value.

According to the characteristic geometry of the defects, we can assume that they are fractal in nature. From the snapshots from the experimental data, we can see that the fractures are similar, that is, self-similar. With each pulse, the shape becomes more complex. Let us determine and estimate the fractal dimensionality of the defects.

The algorithm works according to the "box-counting dimension" principle, i.e. counts the Minkowski dimension:

$$D = \frac{\log N(\varepsilon)}{\log \frac{1}{\varepsilon}} \leftrightarrow D \log \frac{1}{\varepsilon} = \log N(\varepsilon) \leftrightarrow D \log \frac{1}{\varepsilon} - \log N(e) = 0$$

where, $N(\varepsilon)$ is the minimum number of n-dimensional squares with long side ε needed to cover the set.

To calculate the fractal dimensionality, we cut the shape of the defect after 4 series of oscillatory loading using tools and translate it into black-and-white format. As a result, to calculate the fractal dimension the defect has the form shown in Fig.10

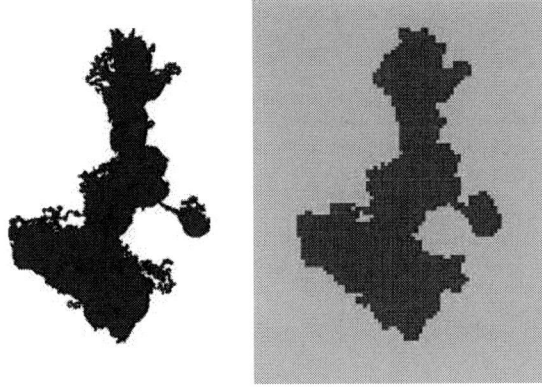

Fig. 10. Characteristic view of the defect after 4 pulses of oscillating load for the box-counting algorithm.

We start to cover the obtained image step by step with squares of different size, changing the side length upwards, i.e. increasing the number of iterations, the number of squares falling into the required set will decrease and the regression line graph will be displayed, the fractal dimension in the obtained case is D=1.794. (Fig. 11).

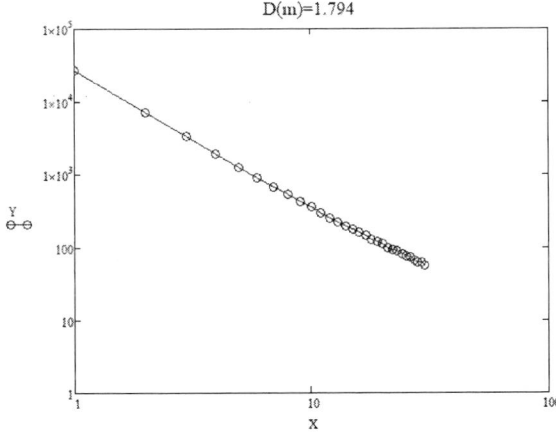

Fig. 11. Defect regression line.

Where the coordinates on the X-axis is the number of iterations associated with the size of the side from smaller to larger, and on the Y-axis the number of squares.

V. Conclusion

The weak sensitivity of the relative elongation of the defect from its initial length when exposed to pulse current in the unipolar and oscillatory mode of exposure is revealed.

The appearance of the observed defects after exposure has a significantly more complex shape than the described calculation model. Analysis of the aggregate images of the formations in the crack apex revealed a fractal character, demonstrating the insensitivity of the fractal dimension, equal to 1.75 ± 0.05, to the initial length of the notch.

The simulation results showed the possibility of developing the defect by sprouting additional cracks due to the resulting thermal stresses, the duration of which significantly exceeds the exposure time.

The effects observed in the experiment, caused by high local energy input, determine the need to expand the set of acting factors taken into account in modeling, for example, the magnetohydrodynamic flow of the molten material.

Acknowledgments

This research was done by Peter the Great St. Petersburg Polytechnic University and supported under the strategic academic leadership program 'Priority 2030' of the Rus-sian Federation (Agreement 075-15-2021-1333 dated 30.09.2021);

The results of the work were obtained using computational resources of the supercomputer center in Peter the Great Saint-Petersburg Polytechnic University Supercomputing Center (www.spbstu.ru).

References

[1] Shneerson G.A., Dolotenko M.I., and Krivosheev S.I., "Strong and Superstrong Pulsed Magnetic Fields Generation," Strong Superstrong Pulsed Magn. Fields Gener., 2014, doi: 10.1515/9783110252576.

[2] Krivosheev S.I., Titkov V.V., and Shneerson G.A., "Two-dimensional field diffusion and magnetohydrodynamic flow in an electric explosion of a miniature single-turn solenoid in a megagauss magnetic field," Tech. Phys., vol. 42, no. 4, pp. 352–366, 1997, doi: 10.1134/1.1258833.

[3] Krivosheev S.I., Magazinov S.G., and Alekseev D.I., "On the impact of the elastic-plastic flow upon the process of destruction of the solenoid in a super strong pulsed magnetic field," J. Phys. Conf. Ser., vol. 946, no. 1, 2018, doi: 10.1088/1742-6596/946/1/012040.

[4] Shneerson G.A., "Configurations of axisymmetric quasi-force-free magnetic systems. i," Tech. Phys., vol. 53, no. 10, pp. 1267–1277, 2008, doi: 10.1134/S1063784208100034.

[5] Shneerson G.A., Vecherov I.A., Degtev D.A., Koltunov O.S., Krivosheev S.I., and Shishigin S.L., "Configurations of axisymmetric quasi-force-free magnetic systems. II," Tech. Phys., vol. 53, no. 10, pp. 1278–1288, 2008, doi: 10.1134/S1063784208100046.

[6] Ravi-Chandar K. and Knauss W.G., "An experimental investigation into dynamic fracture: IV. On the interaction of stress waves with propagating cracks," Int. J. Fract., vol. 26, no. 3, pp. 189–200, 1984, doi: 10.1007/BF01140627.

[7] Morozov V.A., Petrov Y.V., and Sukhov V.D., "Experimental Evaluation of Structural and Temporal Characteristics of Material Fracture Based on Magnetic Pulse Loading of Ring Samples," Tech. Phys., vol. 64, no. 5, pp. 642–646, 2019, doi: 10.1134/S1063784219050165.

[8] Song H., Wang Z.J., He X.D., and Duan J., "Self-healing of damage inside metals triggered by electropulsing stimuli," Sci. Rep., vol. 7, no. 1, pp. 1–11, 2017, doi: 10.1038/s41598-017-06635-9.

[9] Krivosheev S.I., "Pulsed magnetic technique of material testing under impulsive loading," Tech. Phys., vol. 50, no. 3, pp. 334–340, 2005, doi: 10.1134/1.1884733.

[10] Magazinov S.G., Krivosheev S.I., Adamyan Y.E., Alekseev D.I., Titkov V.V., and Chernenkaya L.V., "Adaptation of the magnetic pulse method for conductive materials testing," Mater. Phys. Mech., vol. 40, no. 1, pp. 117–123, 2018, doi: 10.18720/MPM.4012018_14.

[11] Emel'yanov O.A., "Local fracture of thin metallic films during electromagnetic loading," Tech. Phys., vol. 53, no. 7, pp. 866–874, 2008, doi: 10.1134/S1063784208070098.

[12] Spirin A.V. et al., "Effect of structural steel ion plasma nitriding on material durability in pulsed high magnetic fields," J. Phys. Conf. Ser., vol. 830, no. 1, 2017, doi: 10.1088/1742-6596/830/1/012080.

[13] [Krivosheev S.I., Adamian Y.E., Alekseev D.I., Magazinov S.G., Chernenkaya L.V., and Titkov V.V., "The impact of local current density increase on conductor destruction," J. Phys. Conf. Ser., vol. 1147, no. 1, 2019, doi: 10.1088/1742-6596/1147/1/012033.

[14] Adamyan Y.E., Alekseev D.I., Chernenkaya L.V., Krivosheev S.I., Magazinov S.G., and Titkov V.V., "Interaction the high-density pulse current with material in the zone of local conduction disturbance at the edge of a thin wall magnetic system," 2018 16th Int. Conf. Megagauss Magn. F. Gener. Relat. Top. MEGAGAUSS 2018 - Proc., vol. 6, pp. 1–4, 2018, doi: 10.1109/MEGAGAUSS.2018.8722688.

[15] Sitzman A.J., Stefani F., Bourell D.L., and Treviño E., "Use of the magnetic saw effect for manufacturing," IEEE Trans. Plasma Sci., vol. 42, no. 5, pp. 1173–1178, 2014, doi: 10.1109/TPS.2013.2286178.

[16] Finkel V.M., Golovin Y. I., Sletkov A.A.. On the possibility of inhibition of fast cracks by current pulses. Tekhnicheskaya fizika[Technical Physics], 1976, no.227, pp. 848–851 (in Russia)

[17] Hosoi A., Kishi T., and Ju Y., "Healing of fatigue crack treated with surface-activated pre-coating method by controlling high-density electric current," 13th Int. Conf. Fract. 2013, ICF 2013, vol. 2, pp. 1686–1695, 2013.

[18] Ostropiko E.S., Magazinov S.G., and Krivosheev S.I., "Uniaxial Magnetic Pulse Tension of TiNi Alloy with Experimental Strain Rate Evaluation," Exp. Mech., vol. 62, no. 6, pp. 1027–1036, 2022, doi: 10.1007/s11340-022-00864-4.

[19] Kiesewetter D.V., et al., "Application of Fiber Bragg Gratings as a Sensor of Pulsed Mechanical Action," Sensors, vol. 22, no. 19, p. 7289, 2022, doi: 10.3390/s22197289.

[20] Bulat A.F., Dyrda V.I., Fraktaly v geomekhanike [Fractals in geomechanics], Naukova dumka, Kiev, Ukraine., 2005.

[21] Bulat A.F., and Dyrda V.I., "Fractal nature of destruction of elastomers under prolonged cyclic loading", Geo-Technical Mechanics, 2003. no. 45, pp. 137-144.

[22] Shcholokova M.O., "Fractal generalization of the energy criterion for quasi-violent destruction of solids", Abstract of Ph.D. dissertation, National Academy of Sciences of Ukraine, Zaporizhia, Ukraine, 2007.

[23] Shchelokova M.A., "Fractal generalization of the Griffith equation", Innovative Materials and Technologies in Metallurgy and Mechanical Engineering, 2004. no. 2, p. 86-89.

[24] Shchelokova M.A. (b), "The application of fractal geometry to the description of the mechanism of destruction", Problems of computational mechanics and strength of structures, 2004. no. 8, p. 137-144.

[25] Shchelokova M.A., "Investigation of fractal singularities of the vertex of a fracture-like structural defect", Problems of computational mechanics and strength of structures, 2003. no. 7, p. 134-141.

[26] Mosolov A.B., and Dinariev O.Yu., "Self-similarity and fractal geometry of destruction", Strength of materials, 1988. no. 1, p. 3- 7.

[27] Miclashevich I.A., Mikromekhanika razrusheniya v obobshchennykh prostranstvakh [Micromechanics of fracture in generalized spaces], Logvinov, Mink, Belarus, 2003.

[28] Turbin A.F., and Pratsevity N.V., Fraktalnyye mnozhestva, funktsii, raspredeleniya [Fractal sets, functions, distributions], Naukova dumka, Kyiv, Ukraine, 1992.

Autors Index

Aivazyan Vagarshak M. 63
Arkhipenko Viktoriya A. 7
Artemyev Ilya A. 91
Babich Alexey V. 136
Babich Tatiana A. 136
Bagan Gontrand S. S. 3
Bagautdinova Liliya N. 43
Belko Victor O. 27, 31, 35
Belyakov Igor A. 71
Bezverkhniy Vladislav P. 7, 11
Bogdan Alexey 14
Bolshakov Aleksey D. 40
Borisova Margarita E. 99
Bozhedomova Anastassiya S. 136
Bykov Nikolay 127
Chemerev Ilya 18
Chernyshov Dmitriy A. 27
Chukanova Olga B. 23
Egorkin Vladimir I. 23
Fedotov Nikita A. 27, 31
Feklistov Efrem G. 52
Feng Shengxi 35
Fleyteng Vladimir A. 82
Gadzhiev Samir R. 82
Gagarina Alena Yu. 40, 60
Gaisin Almaz F. 43
Gaisin Azat F. 43
Gaysin Fivzat M. 43
Glebova Daria D. 136
Goldade Victor A. 99
Govor Vladislav M. 47
Guketlov Aslan M. 7
Guminov Nikolay V. 115
Hojamov Ahmet A. 103
Ivanov Ivan O. 52, 56
Kalimov Alexander G. 3, 47
Kamalov Almaz M. 99
Karamov Artem R. 14
Karelin Alexandr M. 63
Khalugarova Kamilya 60
Khlybov Alexander I. 115
Khmelnitskiy Ivan K. 63
Kiesewetter Dmitry V. 67, 112, 144
Kobzar Evgenii N. 47
Kochergin Mikhail D. 71
Kondratev Valeriy M. 60
Korlyakov Andrey V. 63
Kostelov Andrey M. 27, 31
Kotlyarov Evgeny Yu. 75, 115
Kovalenko Mikhail A. 99
Kovalev Andrey V. 82
Krivosheev Sergey I. 144
Kurakina Natalia 86
Kuzmin Vladislav S. 82
Kuznetsov Alexey 40, 60
Le Sun 67
Libin Lev N. 82
Litvinov Danila 67
Luchinin Victor V. 63
Magazinov Sergey G. 144
Malyugin Victor I. 144

Mannanov Emil R. 27, 31
Mastyukov Karim Sh. 43
Migalin Mikhail M. 82
Mikhailov Viktor 75
Moshnikov Vyacheslav A. 7, 40
Murashov Iurii 86
Naborshikov Anton A. 91
Nakonechny Ghennady 127
Nalimova Svetlana S. 7
Obraztsov Nikita 86, 127
Oputin Nikita E. 95
Paing Soe Thu 139
Pavlov Andrey A. 99
Pechnikov Alexey V. 103
Phyo Win Tun 109
Posyagin Anton I. 91
Poyarkov Pavel A. 112
Putrya Mikhail 75
Reznik Alexandr S. 31, 67, 112
Rodionov Denis V. 115
Rogachev Maxim S. 136
Sapego Evgeny N. 14
Shikova Tatyana M. 27, 31
Shomakhov Zamir V. 7
Simonov Boris M. 109, 139
Smirnova Ekaterina G. 112
Sorvina Maria A. 119
Sovetov Stanislav I. 122
Spivak Julia M. 40
Surov Alexander 127
Testov Dmitriy O. 63
Timoshenkov Pavel V. 115
Timoshenkov Sergey P. 109, 139
Timoshin Sergey 75
Trubin Denis 67
Tsyganstsev Vyacheslav A. 91
Tumarkin Andrei V. 14
Tyurin Sergey F. 95, 122
Vasenin Ivan 132
Vazhnov Sergey 3
Vertyanov Denis V. 71
Voloschuk Irina A. 136
Ye Ko Ko Aung 139
Zakirov Dzhaudat U. 43
Zemlyakov Valery E. 23
Zhiligotov Ruslan 86
Zhukov Dmitry V. 144
Zhumagali Raiymbek N. 71
Zhuravlev Maxim N. 23
Zhuravleva Natalia M. 67, 112
Zotov Sergey V. 99
Zubov Igor 75

IEEE
445 Hoes Lane
Piscataway, NJ 08854-4141

ISBN 979-8-3503-8358-4